T0342061

INTRODUCTION TO COMPUTATION AND MODELING FOR DIFFERENTIAL EQUATIONS

INTRODUCTION TO COMPUTATION AND MODELING FOR DIFFERENTIAL EQUATIONS

Second Edition

LENNART EDSBERG
Department of Numerical Analysis and Computing Science
KTH - Royal Institute of Technology
Stockholm, Sweden

Copyright © 2016 by John Wiley & Sons, Inc. All rights reserved

Published by John Wiley & Sons, Inc., Hoboken, New Jersey
Published simultaneously in Canada

No part of this publication may be reproduced, stored in a retrieval system, or transmitted in any form or by any means, electronic, mechanical, photocopying, recording, scanning, or otherwise, except as permitted under Section 107 or 108 of the 1976 United States Copyright Act, without either the prior written permission of the Publisher, or authorization through payment of the appropriate per-copy fee to the Copyright Clearance Center, Inc., 222 Rosewood Drive, Danvers, MA 01923, (978) 750-8400, fax (978) 750-4470, or on the web at www.copyright.com. Requests to the Publisher for permission should be addressed to the Permissions Department, John Wiley & Sons, Inc., 111 River Street, Hoboken, NJ 07030, (201) 748-6011, fax (201) 748-6008, or online at http://www.wiley.com/go/permissions.

Limit of Liability/Disclaimer of Warranty: While the publisher and author have used their best efforts in preparing this book, they make no representations or warranties with respect to the accuracy or completeness of the contents of this book and specifically disclaim any implied warranties of merchantability or fitness for a particular purpose. No warranty may be created or extended by sales representatives or written sales materials. The advice and strategies contained herein may not be suitable for your situation. You should consult with a professional where appropriate. Neither the publisher nor author shall be liable for any loss of profit or any other commercial damages, including but not limited to special, incidental, consequential, or other damages.

For general information on our other products and services or for technical support, please contact our Customer Care Department within the United States at (800) 762-2974, outside the United States at (317) 572-3993 or fax (317) 572-4002.

Wiley also publishes its books in a variety of electronic formats. Some content that appears in print may not be available in electronic formats. For more information about Wiley products, visit our web site at www.wiley.com.

Library of Congress Cataloging-in-Publication Data:

Edsberg, Lennart, 1946–
Introduction to computation and modeling for differential equations / Lennart Edsberg. – Second edition.
pages cm
Includes bibliographical references and index.
ISBN 978-1-119-01844-5 (cloth)
1. Differential equations–Data processing. 2. Mathematical models. I. Title.
QA371.5.D37E37 2016
515′.350285–dc23
2015018724

Typeset in 10/12pt TimesLTStd by SPi Global, Chennai, India

10 9 8 7 6 5 4 3 2 1

1 2016

CONTENTS

PREFACE

In the Second Edition of this book, corrections and a large number of additions and modifications have been made. Chapter 3 contains a new section with special methods for special problems. The emphasis is on the numerical preservation of invariants of the solution, such as positivity and quadratic expressions of, e.g., energy. Chapters 4 and 7 have new sections presenting the weak formulations of ordinary differential equations and partial differential equations as starting point for Galerkin's method followed by the finite element method. Chapter 8 has a new section on the finite volume method that presents the wave equation with clear connections to Chapter 3 where the leap-frog method with staggered grid is discussed. Chapter 10 has been newly added in this edition where well-tested student projects are presented for use in, e.g., a course based on this book. The projects have been used at KTH, Stockholm, for many years, and the projects have been developed and modified for several decades by my colleague Gerd Eriksson who has given me the permission to publish them in this new edition. I would like to extend my sincere thanks to her and also to my colleague Jesper Oppelstrup who has contributed with many valuable comments and suggestions concerning the new stuff of this book for the past many years.

LENNART EDSBERG
KTH
Stockholm
August 2015

1

INTRODUCTION

It is probably no exaggeration to say that differential equations are the most common and important mathematical model in science and engineering. Whenever we want to model a system where the state variables vary with time and/or space, differential equations are the natural tool for describing its behavior. The construction of a differential equation model demands a thorough understanding of what takes place in the process we want to describe.

However, setting up a differential equation model is not enough, we must also solve the equations. The process of finding useful solutions of a differential equation is much a symbiosis of modeling, mathematics and choosing a method, analytical or numerical. Therefore, when you are requested to solve a differential equation problem from some application, it is useful to know facts about its modeling background, its mathematical properties, and its numerical treatment. The last part involves choosing appropriate numerical methods, adequate Software, and appealing ways of visualizing the result.

The interaction among modeling, mathematics, numerical methods, and programming is nowadays referred to as *scientific computing* and its purpose is to perform *simulations* of processes in science and engineering.

1.1 WHAT IS A DIFFERENTIAL EQUATION?

A differential equation is a relation between a function and its derivatives. If the function u depends on only one variable t, i.e., $u = u(t)$, the differential equation is called

Introduction to Computation and Modeling for Differential Equations, Second Edition. Lennart Edsberg.
© 2016 John Wiley & Sons, Inc. Published 2016 by John Wiley & Sons, Inc.

ordinary. If u depends on at least two variables t and x, i.e., $u = u(x, t)$, the differential equation is called *partial*.

1.2 EXAMPLES OF AN ORDINARY AND A PARTIAL DIFFERENTIAL EQUATION

An example of an elementary ordinary differential equation (ODE) is

$$\frac{du}{dt} = au \tag{1.1}$$

where a is a *parameter*, in this case a real constant. It is frequently used to model, e.g., the growth of a population ($a > 0$) or the decay of a radioactive substance ($a < 0$). The ODE (1.1) is a special case of differential equations called *linear with constant coefficients* (see Chapter 2).

The differential equation (1.1) can be solved analytically, i.e., the solution can be written explicitly as an *algebraic formula*. Any function of the form

$$u(t) = Ce^{at} \tag{1.2}$$

where C is an arbitrary constant satisfies (1.1) and is a solution. The expression (1.2) is called the *general* solution. If C is known to have a certain value, however, we get a unique solution, which, when plotted in the (t, u)-plane, gives a trajectory (solution curve). This solution is called a *particular* solution.

The constant C can be determined, e.g., by selecting a point (t_0, u_0) in the (t, u)-plane through which the solution curve shall pass. Such a point is called an *initial point* and the demand that the solution shall go through this point is called the *initial condition*. A differential equation together with an initial condition is called an *initial value problem* (IVP) (Figure 1.1).

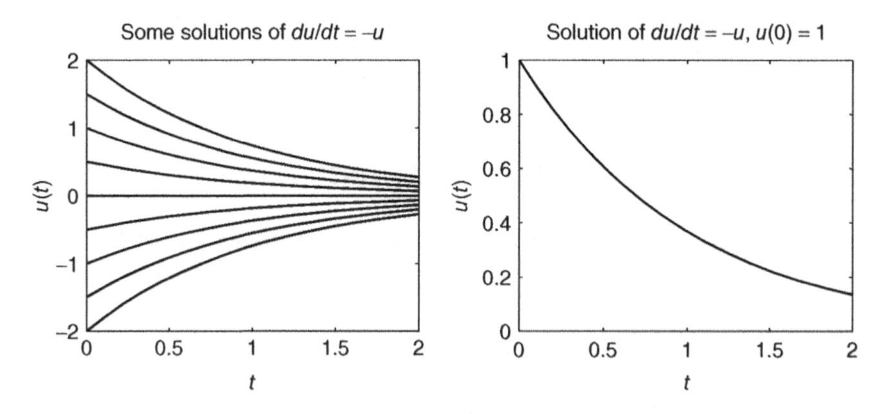

Figure 1.1 General and particular solution

Observe that the differential equation alone does not define a unique solution, we also need an initial condition or other conditions. A plot of all trajectories, i.e., all solutions of the ODE (1.1) in the (t, u)-plane will result in a graph that is totally black as there are infinitely many solution curves filling up the whole plane.

In general, it is not possible to find analytical solutions of a differential equation. The "simple" differential equation

$$\frac{du}{dt} = t^2 + u^2 \tag{1.3}$$

cannot be solved analytically. If we want to plot some of its trajectories, we have to use numerical methods.

An example of an elementary partial differential equation (PDE) is

$$\frac{\partial u}{\partial t} + a\frac{\partial u}{\partial x} = 0 \tag{1.4}$$

where a is a *parameter*, in this case a real constant. The solution of (1.4) is a function of two variables $u = u(x, t)$. This differential equation is called the *1D* (one space dimension, x) *advection equation*. Physically it describes the evolution of a scalar quantity, e.g., temperature $u(x, t)$ carried along the x-axis by a flow with constant velocity a. It is also known as the linear convection equation and is an example of a *hyperbolic* PDE (see Chapter 5).

The *general* solution of this differential equation is (see Exercise 1.2.4)

$$u(x, t) = F(x - at) \tag{1.5}$$

where F is any arbitrary differentiable function of one variable. This is indeed a large family of solutions! The three functions

$$u(x, t) = x - at, \quad u(x, t) = e^{-(x-at)^2}, \quad u(x, t) = sin(x - at)$$

are just three solutions out of the infinitely many solutions of this PDE.

To obtain a unique solution for $t > 0$ we need an *initial condition*. If the differential equation is valid for all x, i.e., $-\infty < x < \infty$ and $u(x, t)$ is known for $t = 0$, i.e., $u(x, 0) = u_0(x)$ where $u_0(x)$ is a given function, the initial value function, we get the *particular* solution (Figure 1.2)

$$u(x, t) = u_0(x - at) \tag{1.6}$$

Physically, (1.6) corresponds to the propagation of the initial function $u_0(x)$ along the x-axis with velocity $|a|$. The propagation is to the right if $a > 0$ and to the left if $a < 0$.

The graphical representation can alternatively be done in 3D (Figure 1.3).

When a PDE is formulated on a semi-infinite or finite x-interval, *boundary conditions* are needed in addition to initial conditions to specify a unique solution.

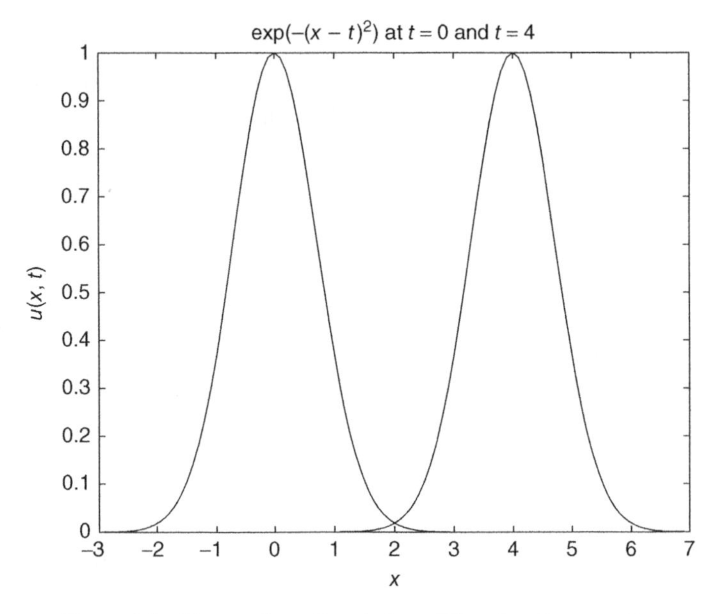

Figure 1.2 Propagation of a solution of the advection equation

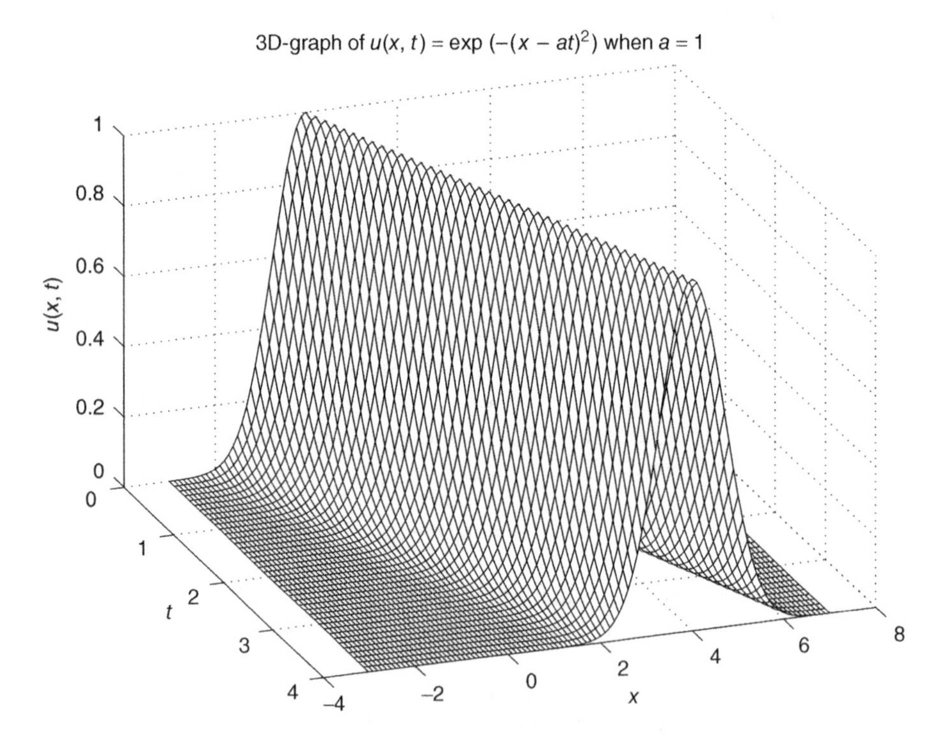

Figure 1.3 3D graph of a propagating solution

Most PDEs can only be solved with numerical methods. Only for very special classes of PDE problems it is possible to find an analytic solution, often in the form of an infinite series.

Exercise 1.2.1. If a is a complex constant $a = \mu + i\omega$ what is the real and imaginary part of e^{at}?

Exercise 1.2.2. What conditions are necessary to impose on μ and ω if $Re(e^{at})$ for $t > 0$ is to be

a) exponentially decreasing,
b) exponentially increasing,
c) oscillating with constant amplitude,
d) oscillating with increasing amplitude,
e) oscillating with decreasing amplitude?

Exercise 1.2.3. If a is a complex constant what condition on a is needed if e^{at} is to be bounded for $t \geq 0$?

Exercise 1.2.4. Show that the general solution of $u_t + au_x = 0$ is $u(x, t) = F(x - at)$ by introducing the transformation

$$\xi = x + at, \quad \eta = x - at$$

Transform the original problem to a PDE in the variables ξ and η, and solve this PDE. Sketch the two coordinate systems in the same graph.

Exercise 1.2.5. Show that a solution of (1.4) starting at $t = 0$, $x = x_0$ is constant along the straight line $x - at = x_0$. This means that the initial value $u(x_0, 0) = u_0(x_0)$ is transported unchanged along this line, which is called a *characteristic* of the hyperbolic PDE (1.4).

1.3 NUMERICAL ANALYSIS, A NECESSITY FOR SCIENTIFIC COMPUTING

In scientific computing, the numerical methods used to solve mathematical models should be *robust*, i.e., they should be reliable and give accurate values for a large range of parameter values. Sometimes, however, a method may fail and give unexpected results. Then, it is important to know how to investigate why an erroneous result has occurred and how it can be remedied.

Two basic concepts in numerical analysis are *stability* and *accuracy*. When choosing a method for solving a differential equation problem, it is necessary to have some knowledge about how to analyze the result of the method with respect to these concepts. This necessity has been well expressed by the late Prof. Germund Dahlquist, famous for his fundamental research in the theory of numerical treatment of differential equations: "There is nothing as practical as a little good theory."

As an example of what may be unexpected results, choose the well-known *vibration equation*, occurring in, e.g., mechanical vibrations, electrical vibrations, and sound vibrations. The form of this equation with initial conditions is

$$m\frac{d^2u}{dt^2} + c\frac{du}{dt} + ku = f(t), \qquad u(0) = u_0, \frac{du}{dt}(0) = v_0 \tag{1.7}$$

In mechanical vibrations, m is the mass of the vibrating particle, c the damping coefficient, k the spring constant, $f(t)$ an external force acting on the particle, u_0 the initial position, and v_0 the initial velocity of the particle. The five quantities m, c, k, u_0, v_0 are referred to as the *parameters* of the problem.

Solving (1.7) numerically for a set of values of the parameters is an example of *simulation* of a mechanical process and it is desirable to choose a robust method, i.e., a method for which the results are reliable for a large range of values of the parameters. The following two examples based on the vibration equation show that unexpected results depending on instability and/or bad accuracy may occur.

Example 1.1. Assume that $f(t) = 0$ (free vibrations) and the following values of the parameters: $m = 1, c = 0.4, k = 4.5, u_0 = 1, v_0 = 0$. Without too much knowledge about mechanics, we would expect the solution to be oscillatory and damped, i.e., the amplitude of the vibrations is decreasing. If we use the simple *Euler method with constant stepsize* $h = 0.1$ (see Chapter 3), we obtain the following numerical solution, visualized together with the exact solution (Figure 1.4).

The graph shows a numerical solution that is oscillatory but unstable with increasing amplitude. Why? The answer is given in Chapter 3. For the moment just accept that insight in *stability* concepts and experience in handling unexpected results are needed for successful simulations.

Example 1.2. When the parameters in equation (1.7) are changed to $m = 1$, $c = 10$, $k = 10^3$, $u_0 = 0$, $v_0 = 0$, and $f(t) = 10^{-4}\sin(40t)$ (forced vibrations) we obtain the following numerical result with a method from a commercial software product for solving differential equations (Figure 1.5).

The graph shows that the numerical result is not correct. Why? In this example there is an *accuracy* problem. The default accuracy used in the method is not sufficient; the corresponding *numerical parameter* must be *tuned* appropriately. Accuracy for ODEs is discussed in Chapter 3.

Numerical solution of PDEs can also give rise to unexpected results. As an example consider the PDE (1.4), which has the property of propagating the initial function along the x-axis. One important application of this equation occurs in gas dynamics where simulation of *shock waves* is essential. A simple 1D model of a

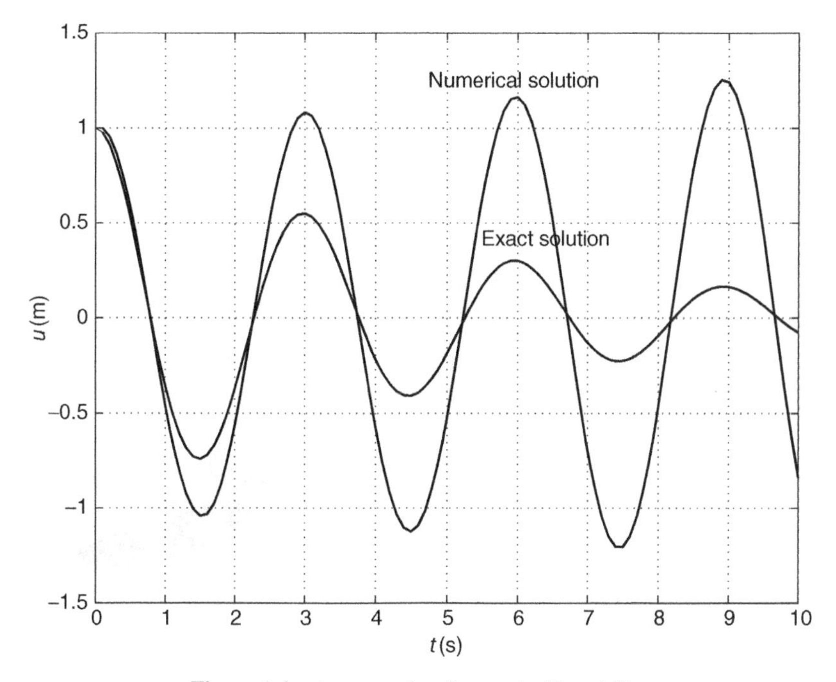

Figure 1.4 An example of numerical instability

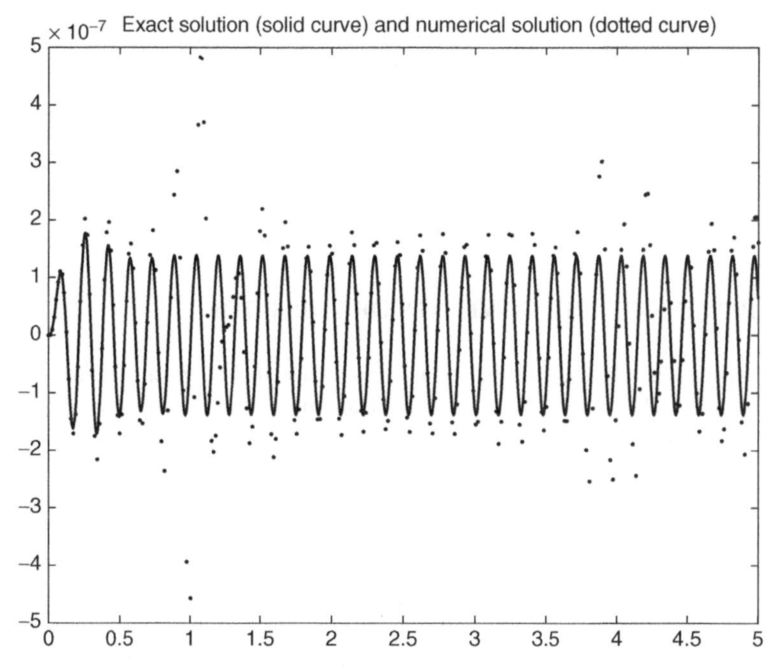

Figure 1.5 An example of insufficient accuracy

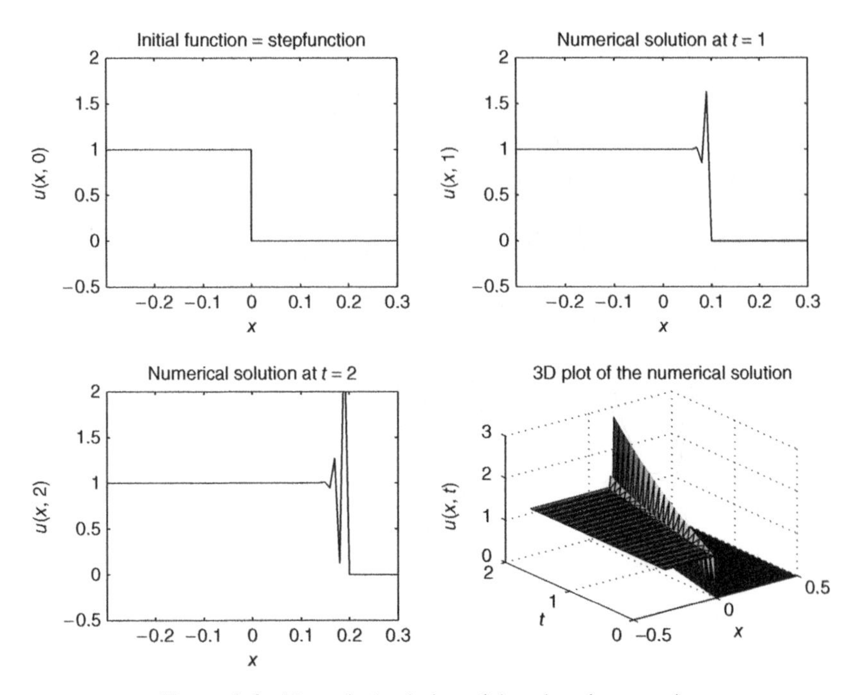

Figure 1.6 Numerical solution of the advection equation

shockwave is a *stepfunction*. Assume the initial function $u_0(x)$ is a stepfunction (Figure 1.6). In the exact solution of (1.4) with a stepfunction as initial condition, the solution propagates along the x-axis *without changing shape*.

However, using a numerical method, where simple difference approximations are used in both the t- and the x-direction, wiggles are generated as the solution propagates (see the graphs in Figure 1.6). The shape of the initial function is distorted. Why? Answers will be given in Chapter 8.

1.4 OUTLINE OF THE CONTENTS OF THIS BOOK

After this introductory chapter, the text is organized so that *ODEs* are treated first, followed by *PDEs*. The aim of this book is to be an introduction to *scientific computing*. Therefore, not only *numerical methods* are presented but also

1. how to set up a *mathematical model* in the form of an ODE or a PDE;
2. an outline of the *mathematical properties* of differential equation problems and explicit analytical solutions (when they exist); and
3. examples of how results are presented with proper *visualization*.

The ODE part starts in Chapter 2 presenting some mathematical properties of ODEs, first the basic and important problem class of ODE systems which are *linear with constant coefficients* applied to important models from classical mechanics, electrical networks, and chemical kinetics. This is followed by numerical treatment of ODE problems in general, following the classical subdivision into *IVPs* in Chapter 3, and *boundary value problems*, BVPs, in Chapter 4. For IVPs, the *finite difference method* (FDM) is described starting with the elementary Euler method. Important concepts brought up for ODEs are *accuracy* and *stability* which is followed up also for PDEs in later chapters. For BVPs, both the FDM and the *finite element method* (FEM) are described.

Important application areas where ODEs are used as mathematical model are presented, selected examples are described in the chapters and exercises, sometimes suitable for computer labs, are inserted into the text.

PDEs are introduced in Chapter 5, which deals with some mathematical properties of their solutions. There is also a presentation of several of the important PDEs of science and engineering, such as the equations of Navier–Stokes, Maxwell and Schrödinger.

The three chapters to follow are devoted to the numerical treatment of PDEs following the classical subdivision into *parabolic, elliptic*, and *hyperbolic* problems. Concepts from the ODE chapters such as accuracy and stability are treated for time-dependent, parabolic and hyperbolic PDEs. For stationary problems (elliptic PDEs), sparse linear systems of algebraic equations are essential and hence discussed.

Selected models introduced in Chapters 2, 5, and 9 are used as illustrations of the different methods introduced. Models are taken from mechanics, fluid dynamics, electromagnetics, reaction engineering, biochemistry, control theory, quantum mechanics, solid mechanics, etc. and are suitable for computer labs.

In Chapter 9, an outline of *mathematical modeling* is brought up with the intention of giving a feeling of the principles used when a differential equation (ODE or PDE) is set up from *conservation laws* and *constitutive relations*. It is also shown by examples how a general differential equation model can be simplified by suitable assumptions. This chapter can be studied in parallel with Chapters 3, 4, 6, 7, and 8 if the reader wants to see how the models are constructed.

In a number of Appendices (A.1–A.6), different parts of mathematics and numerical mathematics that are essential for numerical treatment of differential equations are presented as summaries.

Appendix B gives an overview of existing software for scientific computing with emphasis on the use of MATLAB® for programming and COMSOL Multiphysics® for modeling and parameter studies. Many of the exercises in this chapter and in Chapters 2–8 are solved with MATLAB programs in this appendix.

Appendix C contains a number of computer exercises to support the chapters containing numerical solution of ODEs and PDEs.

In Chapter 10, a number of projects are suggested. These projects involve problems where knowledge from several chapters and appendices are needed to compute a solution.

BIBLIOGRAPHY

1. G. Dahlquist and Å Björck, "Numerical Methods", Dover, 2003
2. L. Råde, B. Westergren, "Mathematics Handbook for Science and Engineering", Studentlitteratur 1998

2

ORDINARY DIFFERENTIAL EQUATIONS

This chapter is not intended to be a thorough treatment of mathematical properties of ordinary differential equations (ODEs). It is rather an overview of some important definitions and concepts such as problem classification and properties of linear ODE systems with constant coefficients and stability. These matters are important for numerical analysis of methods solving ODE problems. For a more extensive treatment of ODEs, it is recommended to consult some mathematical textbook, e.g., one of those referenced at the end of this chapter.

2.1 PROBLEM CLASSIFICATION

A general way of formulating a *first-order scalar* ODE is

$$\frac{du}{dt} = f(t, u) \tag{2.1}$$

where t is defined in some interval, bounded or unbounded, and $u(t)$ is a solution if $u(t)$ satisfies (2.1) for all t in the interval. The variable t is called the *independent* variable and u is called the *dependent* variable. The derivative du/dt can also be denoted $\dot{u}$ or u_t.

As we have already pointed out, the *general* solution $u = u(t, C)$ of (2.1) contains an arbitrary constant C. Hence, the general solution is a family of infinitely many solutions.

Introduction to Computation and Modeling for Differential Equations, Second Edition. Lennart Edsberg.
© 2016 John Wiley & Sons, Inc. Published 2016 by John Wiley & Sons, Inc.

If we prescribe an initial condition (IC) $u(t_0) = u_0$, the constant C is determined and we get a unique solution, a *particular* solution.

A scalar ODE of *second order* can be written as

$$\frac{d^2u}{dt^2} = f\left(t, u, \frac{du}{dt}\right) \tag{2.2}$$

The general solution of (2.2) contains *two* arbitrary constants C_1 and C_2, i.e., $u = u(t, C_1, C_2)$. Therefore, we need *two* conditions to determine C_1 and C_2 to get a unique solution. These conditions can be given in various ways. One way is to specify u at two different values of t, i.e., at $t = a$ and $t = b$:

$$u(a) = \alpha \qquad u(b) = \beta \tag{2.3}$$

This way of specifying *boundary conditions* (BCs) to the second-order ODE (2.2) is called a *boundary value* formulation, and (2.2) and (2.3) define a *boundary value problem* (BVP). Other possible types of BCs will be shown in Chapter 4.

Another way to specify conditions is by ICs. Equation (2.2) can be written as a system of two first-order ODEs by introducing the auxiliary variable $v = du/dt$:

$$\frac{du}{dt} = v \tag{2.4a}$$

$$\frac{dv}{dt} = f(t, u, v) \tag{2.4b}$$

Hence, we get two differential equations of type (2.1) and a unique solution can be obtained by specifying the initial values for the two ODEs

$$u(a) = u_0 \qquad v(a) = v_0 \tag{2.5}$$

In the original formulation (2.2), the ICs (2.5) correspond to $u(a) = u_0$ and $u'(a) = v_0$.

Specifying ICs in this way together with an ODE system (2.4) defines an *initial value problem* (IVP) (Figure 2.1).

For a *system* of ODEs

$$\begin{aligned}
\frac{du_1}{dt} &= f_1(t, u_1, u_2, \ldots, u_n) \\
\frac{du_2}{dt} &= f_2(t, u_1, u_2, \ldots, u_n) \\
&\;\;\vdots \\
\frac{du_n}{dt} &= f_n(t, u_1, u_2, \ldots, u_n)
\end{aligned} \tag{2.6}$$

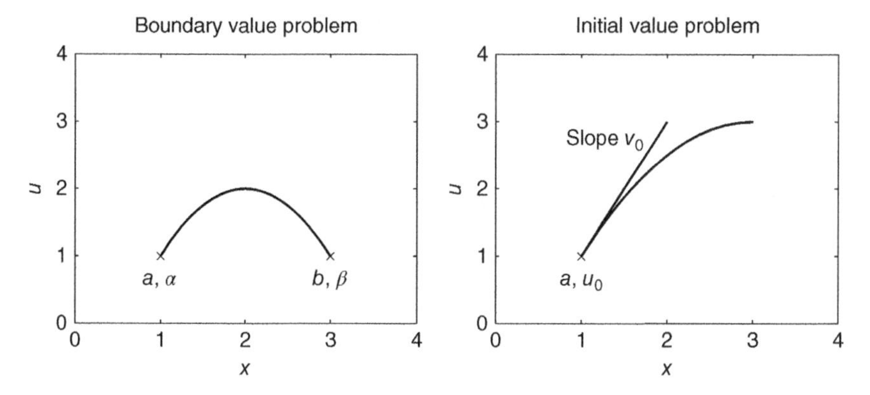

Figure 2.1 A BVP compared to an IVP

it is convenient to use column vector formulation

$$\frac{d\mathbf{u}}{dt} = \mathbf{f}(t, \mathbf{u}) \tag{2.7}$$

where the dependent variables and the right hand sides are collected in the vectors **u** and **f**

$$\mathbf{u} = (u_1, u_2, \dots, u_n)^T, \qquad \mathbf{f} = (f_1, f_2, \dots, f_n)^T$$

Observe that there are as many dependent variables as differential equations.

The *general* solution $\mathbf{u} = \mathbf{u}(t, \mathbf{C})$, where $\mathbf{C}$ is an arbitrary constant vector in R^n, is a family of solution curves in $[t_0, t_{end}] \times R^n$, where $[t_0, t_{end}]$ is the time interval where we want the solution. If $\mathbf{C}$ is determined, e.g., by the IC $\mathbf{u}(t_0) = \mathbf{u}_0$, we get a unique solution, a *particular* solution. Numerical methods are for most problems necessary to produce graphs or tables of solutions.

The standard general formulation of an IVP is

$$\frac{d\mathbf{u}}{dt} = \mathbf{f}(t, \mathbf{u}), \quad \mathbf{u}(t_0) = \mathbf{u}_0, \quad t \in [t_0, t_{end}] \tag{2.8}$$

and a fairly general formulation for a BVP is

$$\frac{d\mathbf{u}}{dt} = \mathbf{f}(t, \mathbf{u}), \quad \mathbf{g}_1(\mathbf{u}(a)) = 0, \quad \mathbf{g}_2(\mathbf{u}(b)) = 0, \quad t \in [a, b] \tag{2.9}$$

where $\mathbf{g}_1$ and $\mathbf{g}_2$ together make up n BC equations.

A scalar ODE of *order n*

$$\frac{d^n u}{dt^n} = f\left(t, u, \frac{du}{dt}, \frac{d^2u}{dt^2}, \dots, \frac{d^{n-1}u}{dt^{n-1}}\right) \tag{2.10}$$

can be written as a system of ODEs by introducing auxiliary variables $u_1 = u$, $u_2 = du/dt$, $u_3 = d^2u/dt^2$, $\ldots$, $u_n = d^{n-1}u/dt^{n-1}$:

$$\begin{aligned}
\frac{du_1}{dt} &= u_2 \\
\frac{du_2}{dt} &= u_3 \\
&\vdots \\
\frac{du_n}{dt} &= f(t, u_1, u_2, \ldots, u_n)
\end{aligned} \tag{2.11}$$

A particular solution is obtained either by specifying an initial vector $\mathbf{u}_0$ or by specifying n BCs for u. We will not present the general case here but give different ways of specifying BCs together with problems from applications in Chapters 3 and 4.

For the IVP (2.8) to have a unique solution, there is a condition on the right hand side $\mathbf{f}(t, \mathbf{u})$ of the ODE to be differentiable with respect to t and $\mathbf{u}$. This means that f_i is differentiable with respect to t and u_j $(i, j = 1, 2, \ldots, n)$ for all $t \in [t_0, t_{end}]$ and for all points $\mathbf{u}$ in a certain domain that contains $\mathbf{u}_0$ as an interior point. This condition is a little stronger than *Lipschitz continuity* (see a mathematical textbook on ODEs).

For a BVP, the question of a unique solution is more complicated. However, unicity for BVPs is not a big issue as many such problems occurring in applications can have several solutions which all make sense (see Example 2.3 in Section 2.5).

Exercise 2.1.1. Given the linear differential equation $\dot{u} = -tu$.

a) Find the general solution of the ODE. Observe that the arbitrary constant C enters linearly in the solution.
b) Find the particular solution satisfying the IC $u(0) = 1$.

Exercise 2.1.2. Given the nonlinear differential equation $\dot{u} = -u^2$.

a) Give the general solution of the ODE. Observe that the arbitrary constant C enters nonlinearly in the solution.
b) Find the particular solution satisfying $u(0) = 1$.
c) Find the particular solution satisfying $u(0) = 0$.

Exercise 2.1.3. Given the linear second-order ODE $u'' + 2u' - 3u = 0$

a) Find the general solution of the ODE. Observe that the arbitrary constants enter linearly.
b) Find the particular solution satisfying the following IC $u(0) = 1, u'(0) = -1$.

c) Find the particular solution satisfying the following BC $u(0) = 0, u(1) = 1$.
d) Find the particular solution satisfying the following BC $u(0) = 1, u(\infty) = 0$.

Exercise 2.1.4. A particle is thrown vertically upwards. During its motion, it is influenced by two forces: gravity and air resistance. The following second- order ODE models the motion along the y-axis:

$$m\ddot{y} = -mg - c\dot{y}|\dot{y}|, \quad y(0) = y_0, \quad \dot{y}(0) = v_0$$

Rewrite this problem as a system of two ODEs.

Exercise 2.1.5. The following system of two second-order ODEs models the vibrations of a mechanical system with two degrees of freedom.

$$\begin{pmatrix} m_1 & 0 \\ 0 & m_2 \end{pmatrix} \frac{d^2\mathbf{x}}{dt^2} + \begin{pmatrix} d_{v1} + d_{v2} & -d_{v2} \\ -d_{v2} & d_{v2} \end{pmatrix} \frac{d\mathbf{x}}{dt} + \begin{pmatrix} \kappa_1 + \kappa_2 & -\kappa_2 \\ -\kappa_2 & \kappa_2 \end{pmatrix} \mathbf{x} = \begin{pmatrix} 0 \\ \hat{F}_2 \sin(\omega t) \end{pmatrix}$$

At the initial point $t = 0$, the ICs are

$$\mathbf{x}(0) = 0, \qquad \frac{d\mathbf{x}}{dt}(0) = 0$$

Formulate these differential equations as a system of first-order ODEs on vector form.

Exercise 2.1.6. The following BVP models the concentration of a chemical substance in a long cylindrical catalyst pellet

$$\frac{1}{r}\frac{d}{dr}\left(r\frac{dc}{dr}\right) = kc^2, \quad \frac{dc}{dr}(0) = 0, \quad c(R) = c_0$$

The parameters k, R, and c_0 are positive parameters. The point $r = 0$ is a singular point as the left hand side term is not defined there. Use l'Hôpital's rule to find the form of the ODE at that point.

Exercise 2.1.7. The following mixture of an ODE system and a system of algebraic equations is called a *DAE system* (Differential Algebraic Equations)

$$\frac{d\mathbf{x}}{dt} = \mathbf{f}(\mathbf{x}, \mathbf{y}), \quad \mathbf{x}(0) = \mathbf{x}_0$$
$$0 = \mathbf{g}(\mathbf{x}, \mathbf{y})$$

Assume that $\mathbf{x}, \mathbf{f} \in R^{n_1}$ and $\mathbf{y}, \mathbf{g} \in R^{n_2}$, where $n_1 + n_2 = n$, the number of unknowns of the system.

Formulate this DAE system as an ODE system with corresponding initial values. Hint: Differentiate the algebraic system with respect to t. Assume that the matrix $\partial \mathbf{g}/\partial \mathbf{y}$ is nonsingular. This DAE system is called an *index-1 system*, as a standard IVP can be obtained by differentiating part of the system only once.

2.2 LINEAR SYSTEMS OF ODES WITH CONSTANT COEFFICIENTS

There is one type of ODE system that can be treated by analytical tools namely linear systems with constant coefficients. This problem class is very important in engineering and scientific applications, such as mechanical vibrations, electric circuits, and first-order chemical kinetics. Linear ODE systems with constant coefficients (LCC systems) take the following form:

$$\frac{d\mathbf{u}}{dt} = A\mathbf{u} + \mathbf{g}(t), \quad \mathbf{u}(0) = \mathbf{u}_0 \tag{2.12}$$

where A is a constant $n \times n$ matrix and $\mathbf{g}(t)$ is a vector of n functions of t called the *driving function*.

If $\mathbf{g}(t) \equiv 0$, the system is *homogeneous*, i.e.,

$$\frac{d\mathbf{u}}{dt} = A\mathbf{u}, \quad \mathbf{u}(0) = \mathbf{u}_0 \tag{2.13}$$

Otherwise, it is *inhomogeneous*.

In the homogeneous case, $n = 1$, i.e., the scalar equation

$$\frac{du}{dt} = au, \quad u(0) = u_0$$

we know that the solution is (see Section 1.2)

$$u(t) = e^{at}u_0$$

where u_0 is the initial value. A generalization of this algebraic solution form to a system leads us to the following solution formula for (2.13):

$$\mathbf{u}(t) = e^{At}\mathbf{u}_0 \tag{2.14}$$

However, what is e^{At}, e raised to the power of the matrix At?

Let's *assume* that a solution to (2.13) can be written

$$\mathbf{u}(t) = \mathbf{c}e^{\lambda t} \tag{2.15}$$

where $\mathbf{c}$ and λ are to be determined. Inserting this expression into (2.13) gives the following relation to be fulfilled for all t:

$$\mathbf{c}\lambda e^{\lambda t} = A\mathbf{c}e^{\lambda t}$$

This leads to the *eigenvalue problem*:

$$A\mathbf{c} = \lambda\mathbf{c} \qquad or \qquad (A - \lambda I)\mathbf{c} = 0 \tag{2.16}$$

We are interested in solutions where $\mathbf{c} \neq 0$ ($\mathbf{c} = 0$ gives us the trivial solution $\mathbf{u} = 0$). This is possible only if

$$det(A - \lambda I) = 0 \tag{2.17}$$

This equation is called the *characteristic equation* and its roots are known as the *eigenvalues* of the matrix A. If A is an $n \times n$ matrix, (2.17) is a polynomial equation of degree n, and consequently has n roots $\lambda_1, \lambda_2, \dots. \lambda_n$. The roots can be real or complex, single or multiple.

When an eigenvalue λ_i is known, we compute the corresponding *eigenvector* $\mathbf{c}_i$ by solving the homogeneous linear equation system

$$(A - \lambda_i I)\mathbf{c}_i = 0 \tag{2.18}$$

The solution $\mathbf{c}_i$ is not unique; $\mathbf{c}_i$ is determined only up to a scalar factor. In addition, $\mathbf{c}_i$ can be complex even if A is real. If A is real and symmetric, however, both λ_i and $\mathbf{c}_i$ are real.

From now on, we assume that the n eigenvectors of the matrix A are linearly independent. Such a matrix is called a *diagonalizable* or *nondefective* matrix. (If A has fewer than n linearly independent eigenvectors, it is called *defective*). For a nondefective matrix, the relation (2.18) can be written in the form

$$A\mathbf{c}_i = \lambda_i \mathbf{c}_i \rightarrow AS = S\Lambda \rightarrow A = S\Lambda S^{-1} \quad \text{or} \quad S^{-1}AS = \Lambda$$

where S is a matrix composed of the eigenvectors as columns and Λ is a diagonal matrix of the eigenvalues:

$$S = (\mathbf{c}_1, \mathbf{c}_2, \dots, \mathbf{c}_n), \qquad \Lambda = diag(\lambda_1, \lambda_2, \dots, \lambda_n)$$

Going back to (2.15), we now have n solutions to (2.13)

$$\mathbf{u}_i(t) = \mathbf{c}_i e^{\lambda_i t}, i = 1, 2, \dots, n \tag{2.19}$$

As (2.13) is a linear homogeneous problem, the solutions (2.19) can be superposed, i.e., linearly combined to a *general solution*:

$$\mathbf{u}(t) = \sum_{i=1}^{n} \alpha_i \mathbf{c}_i e^{\lambda_i t} \tag{2.20}$$

where α_i are arbitrary constants. Formula (2.20) can also be written in matrix form:

$$\mathbf{u}(t) = Se^{\Lambda t}\alpha \tag{2.21}$$

$e^{\Lambda t}$ is a diagonal matrix with $e^{\lambda_i t}$ as diagonal elements:

$$e^{\Lambda t} = diag(e^{\lambda_1 t}, e^{\lambda_2 t}, \dots, e^{\lambda_n t})$$

and α is an arbitrary column vector with components $\alpha_1, \alpha_2, \dots \alpha_n$. When the initial vector $\mathbf{u}(0) = \mathbf{u}_0$ is known, we get by inserting $t = 0$ into (2.21):

$$\mathbf{u}(0) = S\alpha = \mathbf{u}_0 \quad \rightarrow \quad \alpha = S^{-1}\mathbf{u}_0$$

and (2.21) can be written

$$\mathbf{u}(t) = Se^{\Lambda t}S^{-1}\mathbf{u}_0 \tag{2.22}$$

Hence, comparing with (2.14), we define e^{At}, the *exponential matrix*, as

$$e^{At} = Se^{\Lambda t}S^{-1} \tag{2.23}$$

The expression (2.23) can be used only if S has an inverse (S is nonsingular). It can be shown that if all eigenvalues of A are single (all λ_i are different) then A is nondefective. This is true also when A is symmetric.

However, there exist matrices that are defective. In that case, the formula (2.22) does not make sense as S has no inverse, but we can always use the Taylor series expansion as definition of e^{At}:

$$e^{At} = I + At + \frac{t^2}{2!}A^2 + \frac{t^3}{3!}A^3 + \cdots + \frac{t^n}{n!}A^n + \dots \tag{2.24}$$

It can be shown that if S has an inverse, the two definitions (2.23) and (2.24) are equivalent.

If the ODE system is LCC and inhomogeneous, i.e., the right hand side contains a driving function $\mathbf{g}(t)$ as in equation (2.12)

$$\frac{d\mathbf{u}}{dt} = A\mathbf{u} + \mathbf{g}(t), \quad \mathbf{u}(0) = \mathbf{u}_0$$

we obtain the analytic solution in the form of an integral, the Duhamel formula:

$$\mathbf{u}(t) = e^{At}\mathbf{u}_0 + \int_0^t e^{A(t-\tau)}\mathbf{g}(\tau)d\tau \tag{2.25}$$

In case the driving vector function $\mathbf{g}(t)$ is a constant vector $\mathbf{g}$, the integral can be solved and the solution of the ODE system is (if A has an inverse)

$$\mathbf{u}(t) = e^{At}\mathbf{u}_0 + A^{-1}(e^{At} - I)\mathbf{g} \tag{2.26}$$

Exercise 2.2.1. Compute the eigenvalues and eigenvectors of the matrix

$$A = \begin{pmatrix} -2 & 1 \\ -4 & 3 \end{pmatrix}$$

Exercise 2.2.2. Compute e^{At} for the matrix given in Exercise 2.1.1.

Exercise 2.2.3. Solve the differential equation $\dot{\mathbf{u}} = A\mathbf{u}, \quad \mathbf{u}_0 = (1,1)^T$, where A is the matrix in Exercise 2.2.1, using the result obtained in Exercise 2.2.2.

Exercise 2.2.4. Compute eigenvalues and eigenvectors of the matrix

$$A = \begin{pmatrix} 0 & 1 \\ -5 & 2 \end{pmatrix}$$

Exercise 2.2.5. Solve the differential equation $\dot{\mathbf{u}} = A\mathbf{u}, \quad \mathbf{u}_0 = (1,1)^T$, where A is the matrix in Exercise 2.2.4.

Exercise 2.2.6. Verify that the matrices

$$A = \begin{pmatrix} -1 & 0 & 0 \\ 1 & -1 & 0 \\ 0 & 1 & -1 \end{pmatrix}, \quad B = \begin{pmatrix} 0 & 1 & 1 \\ 0 & 0 & 1 \\ 0 & 0 & 0 \end{pmatrix}$$

are defective. Verify also that B is *nilpotent*, i.e., $B^k = 0, k = 3, 4, \ldots$.

Exercise 2.2.7. Compute e^{At} and e^{Bt} for the matrices given in Exercise 2.2.6.

Exercise 2.2.8. Verify the solution formula (2.25) for the inhomogeneous LCC problem (2.12).

Exercise 2.2.9. Modify the formulas (2.14) and (2.25) to the case where the initial value is $\mathbf{u}(t_0) = \mathbf{u}_0$, i.e., $\mathbf{u}$ is known in the initial point $t = t_0$.

Exercise 2.2.10. In what sense is the system (2.13) and its solutions linear?

2.3 SOME STABILITY CONCEPTS FOR ODEs

The goal of stability analysis is to investigate if small perturbations in a system will cause large changes in the behavior of the system, in which case the system is *unstable*. In this chapter, we present two definitions of stability for ODE systems. The first definition refers to what is meant by a stable *solution trajectory* and the second concerns the stability of the special solutions called *critical points* (also called equilibrium points). In this textbook, let us call these two stability concepts *analytical stability*.

Another stability concept deals with the stability of numerical solutions. It is important to distinguish between analytical stability and numerical stability. Numerical stability is introduced in Chapter 3.

2.3.1 Stability for a Solution Trajectory of an ODE System

As already mentioned, analytic solution of a nonlinear ODE system (2.7) is not possible in general. However, for small perturbations, the stability analysis can be based on a linear approximation leading to an LCC system. First we need a definition of stability. In the definition below we restrict ourselves to *autonomous* systems, i.e., ODE systems of the form

$$\frac{d\mathbf{u}}{dt} = \mathbf{f}(\mathbf{u}) \tag{2.27}$$

Definition 1: A solution $\mathbf{u}(t)$ of (2.27) is *stable* if every solution that starts sufficiently close to $\mathbf{u}(t)$ at $t = t_0$ remains close to $\mathbf{u}(t)$ for all $t > t_0$. The solution $\mathbf{u}(t)$ is *unstable* if there exists at least one solution starting close to $\mathbf{u}(t)$ at $t = t_0$, which does not remain close to $\mathbf{u}(t)$ for all $t > t_0$. ◊

The stability question can be completely answered for LCC systems

$$\frac{d\mathbf{u}}{dt} = A\mathbf{u} \tag{2.28}$$

Consider first the scalar ODE

$$\frac{du}{dt} = \lambda u \tag{2.29}$$

Assume that $u(t)$ is a solution of (2.29). At the point (t_0, u_0) on the solution trajectory, a small perturbation δu_0 takes us to a new solution trajectory $u(t) + \delta u(t)$ starting at $(t_0, u_0 + \delta u_0)$. This trajectory is followed for $t > t_0$. As the perturbed solution satisfies the ODE

$$\frac{d(u + \delta u)}{dt} = \lambda(u + \delta u) \tag{2.30}$$

(2.30) is reduced to an ODE for the perturbation only ($t_0 = 0$):

$$\frac{d\delta u}{dt} = \lambda \delta u, \quad \delta u(t_0) = \delta u_0 \tag{2.31}$$

As the constant λ can be complex, i.e., $\lambda = \mu + i\omega$, where $i = \sqrt{-1}$, the solution of the scalar ODE (2.31) can be written as (assume that $t_0 = 0$).

$$\delta u(t) = e^{\lambda t}\delta u_0 = e^{\mu t}e^{i\omega t}\delta u_0 = e^{\mu t}(\cos(\omega t) + i\sin(\omega t))\delta u_0 \tag{2.32}$$

The only part in the expression (2.32) that can increase infinitely as t increases is $e^{\mu t}$. Hence, we see that the solution is stable iff $\mu = Re(\lambda) \leq 0$. To be more specific, if $Re(\lambda) < 0$, the perturbation $\delta u(t) \to 0$ as $t \to \infty$. This is called *asymptotic stability*. If $Re(\lambda) = 0$, the perturbation $\delta u(t)$ will stay close to $u(t)$ as $t \to \infty$ but $\delta u(t)$ will not vanish, see Figure 2.2 where a perturbation is introduced at the point marked o.

A similar analysis of (2.28) gives the following LCC system for the perturbation $\delta\mathbf{u}$, ($t_0 = 0$)

$$\frac{d\delta\mathbf{u}}{dt} = A\delta\mathbf{u}, \quad \delta\mathbf{u}(t_0) = \delta\mathbf{u}_0 \tag{2.33}$$

From the explicit solution formula (2.20), where $t_0 = 0$, it is clear that stability depends on the eigenvalues $\lambda_i, i = 1, 2, \ldots, n$ of the matrix A. Hence, the LCC system (2.33) is asymptotically stable if $Re(\lambda_i) < 0$ for $i = 1, 2, \ldots n$.

If $Re(\lambda_i) = 0$, the system is stable if λ_i is *simple*. If $Re(\lambda_i) = 0$ and λ_i is a *multiple* eigenvalue, the stability problem is more intricate. The solution can be stable or unstable and each case must be investigated separately. We will not treat this case here (see Exercise 2.3.7 for an example).

If we plot the eigenvalues of A in the complex plane, asymptotic stability means that the eigenvalues are situated in the *left half* of the complex plane.

Example 2.1. The eigenvalues of the matrices

$$A = \begin{pmatrix} -1 & 0 & 0 \\ 3 & 1 & -2 \\ 2 & 2 & -2 \end{pmatrix} \qquad B = \begin{pmatrix} -1 & 0 & 0 \\ 3 & 1 & -2 \\ 2 & 2 & 1 \end{pmatrix}$$

have the following location in the complex plane (Figure 2.3).

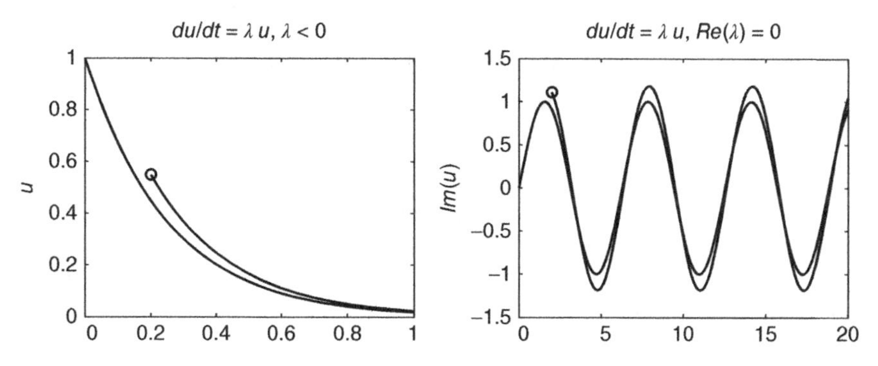

Figure 2.2 Asymptotic stability and stability

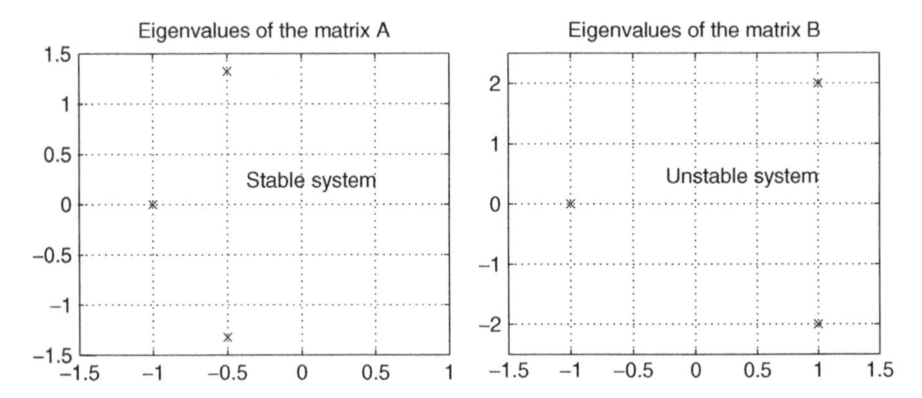

Figure 2.3 Eigenvalues in the complex plane

For the nonlinear autonomous system (2.27), the stability of a solution can be examined if the system is approximated by an LCC system. Assume that $\mathbf{u}(t)$ is a solution of (2.27). At time $t = t_0$, where $\mathbf{u} = \mathbf{u}(t_0)$, a small perturbation $\delta\mathbf{u}_0$ brings us to a new solution trajectory, $\mathbf{u}(t) + \delta\mathbf{u}(t)$, which is followed for $t > t_0$. The perturbed solution satisfies the ODE system

$$\frac{d(\mathbf{u} + \delta\mathbf{u})}{dt} = f(\mathbf{u} + \delta\mathbf{u}) \tag{2.34}$$

Taylor expansion of the right hand side up to the first-order term gives

$$\frac{d\mathbf{u}}{dt} + \frac{d\delta\mathbf{u}}{dt} = \mathbf{f}(\mathbf{u}) + \frac{\partial \mathbf{f}}{\partial \mathbf{u}}\delta\mathbf{u} + O(\|\ \delta\mathbf{u}\|^2) \tag{2.35}$$

In (2.35), the *jacobian* is introduced

$$J(\mathbf{u}) = \frac{\partial \mathbf{f}}{\partial \mathbf{u}} = \begin{pmatrix} \frac{\partial f_1}{\partial u_1} & \frac{\partial f_1}{\partial u_2} & \cdots & \frac{\partial f_1}{\partial u_n} \\ \frac{\partial f_2}{\partial u_1} & \frac{\partial f_2}{\partial u_2} & \cdots & \frac{\partial f_2}{\partial u_n} \\ .. & .. & \cdots & .. \\ .. & .. & \cdots & .. \\ \frac{\partial f_n}{\partial u_1} & \frac{\partial f_n}{\partial u_2} & \cdots & \frac{\partial f_n}{\partial u_n} \end{pmatrix} \tag{2.36}$$

As $\mathbf{u}$ is a solution of (2.27) we obtain, after neglecting the higher order term, the (linearized) *variational equation*

$$\frac{d\delta\mathbf{u}}{dt} = J(\mathbf{u})\delta\mathbf{u}, \quad \delta\mathbf{u}(t_0) = \delta\mathbf{u}_0 \tag{2.37}$$

If the jacobian is frozen at $t = t_0$ to a constant matrix J_0, i.e.,

$$J_0 = \frac{\partial \mathbf{f}}{\partial \mathbf{u}}(\mathbf{u}(t_0)) \tag{2.38}$$

we can use the stability results for LCC systems above to answer questions about the stability of solution trajectories in a close neighborhood of a given solution trajectory.

2.3.2 Stability for Critical Points of ODE Systems

For an autonomous nonlinear ODE system (2.27), it is of special interest to investigate the stability of *critical points*, i.e., points satisfying the algebraic equation

$$\mathbf{f}(\mathbf{u}) = 0 \tag{2.39}$$

Critical points are also called the stationary or steady-state solutions, as such solutions do not change with time t.

Let $\mathbf{u}^*$ be a critical point of (2.39). We want to investigate if this point is stable. The stability concept has to be modified a little for a critical point, as it is a constant vector.

Definition 2: Assume the critical point $\mathbf{u}^*$ is perturbed to a solution $\mathbf{u}^* + \delta\mathbf{u}(t)$ in the neighborhood of $\mathbf{u}^*$. If all perturbed solutions converge to the critical point $\mathbf{u}^*$, this point is said to be asymptotically stable. If at least one perturbed solution diverges away from the critical point, it is unstable (Figure 2.4). ◇

If the perturbation $\delta\mathbf{u}$ is small, we can approximate $\mathbf{f}(\mathbf{u})$ close to the point $\mathbf{u}^*$ using Taylor expansion up to the first-order term:

$$\mathbf{f}(\mathbf{u}^* + \delta\mathbf{u}) = \mathbf{f}(\mathbf{u}^*) + J(\mathbf{u}^*)\delta\mathbf{u} + O(\| \delta\mathbf{u}\|^2) \tag{2.40}$$

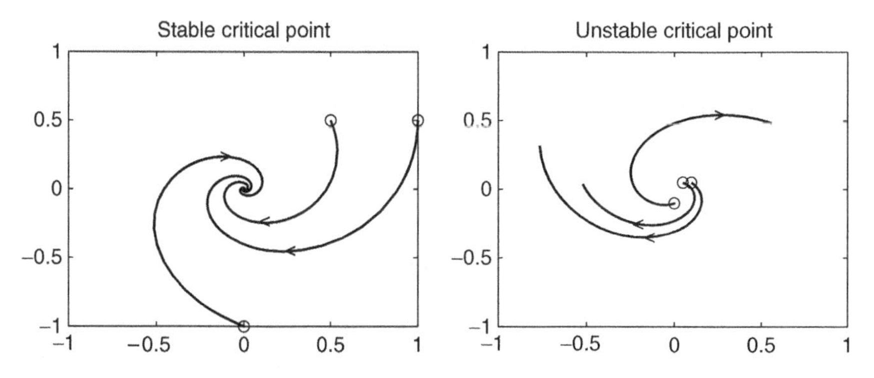

Figure 2.4 Stability of critical points

The perturbed solution satisfies the differential equation, i.e.,

$$\frac{d\mathbf{u}^*}{dt} + \frac{d\delta\mathbf{u}}{dt} = \mathbf{f}(\mathbf{u}^*) + J(\mathbf{u}^*)\delta\mathbf{u} \tag{2.41}$$

where we have neglected the higher order terms in (2.40). As $\mathbf{u}^*$ is a constant and $\mathbf{f}(\mathbf{u}^*) = 0$, we get an ODE system, the variational equation, for the perturbation term $\delta\mathbf{u}$:

$$\frac{d\delta\mathbf{u}}{dt} = J(\mathbf{u}^*)\delta\mathbf{u} \tag{2.42}$$

This is an LCC system of type (2.28). The perturbed solution will converge to the critical point $\mathbf{u}^*$ if the eigenvalues of $J(\mathbf{u}^*)$ are situated in the left half plane of the complex plane, so the behavior of solutions close to a critical point can be described in the same way as the solutions of an LCC problem.

Hence, to investigate the stability of a critical point, proceed as follows:

1. Compute a critical point $\mathbf{u}^*$ by solving $\mathbf{f}(\mathbf{u}) = 0$ with algebraic tools or with Newton's method (see Appendix A.1).
2. Form the jacobian $J(\mathbf{u})$ and compute $J^* = J(\mathbf{u}^*)$.
3. Compute the eigenvalues λ_i of J^* and check if $Re(\lambda_i) < 0$.

Example 2.2. A famous ODE system called the *Lotka–Volterra* differential equations or the *predator–prey* model is

$$\begin{aligned}\frac{dx}{dt} &= ax - bxy \\ \frac{dy}{dt} &= -cy + dxy\end{aligned}$$

where a, b, c, and d are positive parameters.

The critical points of this system are $(0, 0)$ and $(c/d, a/b)$. The jacobian of the system is

$$J = \begin{pmatrix} a - by & -bx \\ dy & -c + dx \end{pmatrix}$$

and the values of the jacobian at the critical points are

$$J(0,0) = \begin{pmatrix} a & 0 \\ 0 & -c \end{pmatrix}, \qquad J(c/d, a/b) = \begin{pmatrix} 0 & -bc/d \\ da/b & 0 \end{pmatrix}$$

The eigenvalues of $J(0, 0)$ are a and $-c$, i.e., one eigenvalues is positive and the critical point $(0, 0)$ is unstable.

The eigenvalues of $J(c/d, a/b)$ are $\pm i\sqrt{ac}$. They have a zero real part but are simple, which implies that the linearized system at the critical point $(c/d, a/b)$ is stable.

Exercise 2.3.1. Given the following ODE system

$$\dot{\mathbf{u}} = \mathbf{A}\mathbf{u}, \qquad \mathbf{A} = \begin{pmatrix} -2 & 1 \\ -4 & 3 \end{pmatrix}$$

Find the critical points and characterize the stability properties of the critical points.

Exercise 2.3.2. What is the critical point $\mathbf{x}_{cr}$ of the ODE system (A is nonsingular)?

$$\dot{\mathbf{x}} = \mathbf{A}\mathbf{x} - \mathbf{b}$$

Introduce the variable $\mathbf{y} = \mathbf{x} - \mathbf{x}_{cr}$. Derive the ODE for $\mathbf{y}$. This shows that stability does not depend on $\mathbf{b}$.

Exercise 2.3.3. Find the critical points and the solution of the ODE system

$$\frac{d\mathbf{w}}{dt} = k_t \begin{pmatrix} -1 & 1 \\ 1 & -1 \end{pmatrix} \mathbf{w} + k_i \begin{pmatrix} 1 \\ 0 \end{pmatrix}, \quad \mathbf{w}(0) = \begin{pmatrix} 0 \\ 0 \end{pmatrix}$$

The parameters k_t and k_i are both positive. What about the stability of the critical points?

Exercise 2.3.4. Given the following ODE

$$\frac{du}{dt} = 1 - u^2$$

a) Find the critical points of the ODE.
b) Decide for each critical point if it is stable or not.

Exercise 2.3.5. Given the second-order ODE, known as it van der Pol's equation

$$\frac{d^2x}{dt^2} + \epsilon(x^2 - 1)\frac{dx}{dt} + x = 0$$

where the parameter $\epsilon > 0$.

a) Find the critical point of the ODE system.
b) Is the critical point stable or unstable? Hint: Rewrite as a system of first order.

Exercise 2.3.6. Consider the homogenous vibration equation (see Section 1.3)

$$m\frac{d^2y}{dt^2} + d_v\frac{dy}{dt} + ky = 0$$

The parameters m, d_v and k are positive.

a) Find the critical point.
b) Sketch how the eigenvalues at the critical point 'move' in the complex plane as the parameter k changes in the interval $(0, \infty)$. Is the critical point always stable?

Exercise 2.3.7. Given the ODE system $\dot{\mathbf{u}} = A\mathbf{u}, \quad A = \begin{pmatrix} 0 & 1 \\ 0 & 0 \end{pmatrix}$.

Show, by solving the system, that the critical points are unstable!

Hence, when two eigenvalues $= 0$ (0 is a double root), the solution *can* be unstable.

2.4 SOME ODE MODELS IN SCIENCE AND ENGINEERING

Mathematical models based on ODE systems occur frequently in science and engineering. The independent variable usually denotes time t or space x.

A time-dependent ODE system is often referred to as a dynamic system. As such a model does not take into account any space variations of the dependent variables, it is sometimes called a lumped model. A space-dependent ODE system involves only one space variable but does not take time into account. Such models often occur by simplifying a stationary PDE model in several space variables into a one-dimensional (1D) space- dependent problem.

In mathematical models of processes in science and engineering, it is important to define the *units* of the variables and parameters, e.g., density ρ (kg/m^3), flow Q (m^3/s), and gravitational acceleration g (m/s^2). In this textbook, SI units will be used in most presentations of Laws of Nature, examples, and exercises.

2.4.1 Newton's Second Law

A system of moving particles with masses $m_1, m_2, \ldots, m_n$ (kg) move according to:

$$m_i\ddot{\mathbf{r}}_i = \mathbf{F}_i(t, \mathbf{r}_1, \mathbf{r}_2, \ldots, \mathbf{r}_n, \dot{\mathbf{r}}_1, \dot{\mathbf{r}}_2, \ldots, \dot{\mathbf{r}}_n), \quad i = 1, 2, \ldots, n \tag{2.43}$$

Here $\mathbf{r}_i = (x_i, y_i, z_i)^T$ (m) denotes the position of particle i, $\dot{\mathbf{r}}_i$ (m/s) its velocity, and $\ddot{\mathbf{r}}_i$ its acceleration (m/s^2). $\mathbf{F}_i$ (N) is the total force acting on particle i. The system (2.43) consists of $3n$ second-order ODEs.

This equation was formulated by the English mathematician and scientist Isaac Newton in his work Principia (Mathematical Principles of Science) from 1687.

Engineering applications of (2.43) occur in, e.g., mechanical vibrations. One important model consists of a system of masses connected by dampers and springs:

$$M\ddot{\mathbf{x}} + D_v\dot{\mathbf{x}} + K\mathbf{x} = \mathbf{F}(t) \tag{2.44}$$

Here M denotes the *mass* matrix, D_v the *damping* matrix, and K the *spring or stiffness* matrix. $\mathbf{x}$ is a vector of displacements of the masses along the x direction. $\mathbf{F}$ is a vector of external forces often of the form $\mathbf{F}(t) = \hat{\mathbf{F}}sin(\omega t)$, where ω is the angular velocity, see Exercise 2.1.5 for an example. Observe that the ODE system (2.44) is an LCC system.

2.4.2 Hamilton's Equations

A conservative mechanical system is defined as a system where the total energy is conserved. In such a system, the motions can, as an alternative to Newton's equations, be described by a scalar *hamiltonian* function H invented by the Irish mathematician W.R.Hamilton in 1835.

H is the sum of the kinetic energy T (J) and the potential energy U (J) of the system. For a system of n particles in 3D, the hamiltonian has the form:

$$H = T + U = H(\mathbf{q}_1, \mathbf{q}_2, \dots, \mathbf{q}_n, \mathbf{p}_1, \mathbf{p}_2, \dots \mathbf{p}_n), \tag{2.45}$$

where the $\mathbf{q}_i$ denote the generalized coordinates and $\mathbf{p}_i$ are the generalized momenta. Hamilton's equations of motion take the form

$$\frac{d\mathbf{q}_i}{dt} = \frac{\partial H}{\partial \mathbf{p}_i}, \quad \frac{d\mathbf{p}_i}{dt} = -\frac{\partial H}{\partial \mathbf{q}_i}, \quad i = 1, 2, \dots, n \tag{2.46}$$

The ODE system (2.45) consists of $6n$ first-order ODEs.

2.4.3 Electrical Networks

An electrical network consists of passive elements such as resistors, capacitors and inductors and active elements such as voltage and current sources. The model of a network is obtained using Kichhoff's laws and current–voltage relations owing to Ohm's law for all elements. If all elements are assumed to have constant impedance values, a linear *DAE* system of the following form can be set up, with an algorithm called *modified nodal analysis* (2.47):

$$\begin{pmatrix} A_C C A_C^T & 0 & 0 \\ 0 & L & 0 \\ 0 & 0 & 0 \end{pmatrix} \frac{d}{dt} \begin{pmatrix} \mathbf{e} \\ \mathbf{i}_L \\ \mathbf{i}_V \end{pmatrix} + \begin{pmatrix} A_R G A_R^T & A_L & A_V \\ -A_L^T & 0 & 0 \\ A_V^T & 0 & 0 \end{pmatrix} \begin{pmatrix} \mathbf{e} \\ \mathbf{i}_L \\ \mathbf{i}_V \end{pmatrix} = \begin{pmatrix} -A_I\mathbf{I} \\ \mathbf{0} \\ \mathbf{E} \end{pmatrix} \tag{2.47}$$

where C (F), L (H), and G (Ω^{-1}) are diagonal matrices with the capacitances, inductances, and conductances (inverses of resistances). A_C, A_R, A_L, A_V, and A_I are (reduced) incidence matrices, i.e., sparse matrices (see Appendix A.4) consisting of the integers $-1, 0, 1$. The vector $\mathbf{I}$ (A) consists of the current source values and $\mathbf{E}$ (V) the voltage source values.

The DAE system (2.47) consists of differential equations for $\mathbf{e}$ (V) (the node potentials) and $\mathbf{i}_L$ (A) (vector of branch currents through the inductors) and additional algebraic equations for $\mathbf{i}_V$ (A) (vector of branch currents through voltage sources).

Linear DAE systems in general have the following form (special case of the DAE system formulated in Exercise 2.1.7):

$$A\dot{\mathbf{u}} = B\mathbf{u} + \mathbf{b}(t) \tag{2.48}$$

where A and B are $n \times n$ matrices. If A is *singular*, the system has "index" ≥ 1: at least one differentiation is required to produce an ODE system.

2.4.4 Chemical Kinetics

Consider a mixture of N chemical substances $A_1, A_2, \dots, A_N$ reacting with each other in M reactions, where the substance concentrations change with time

$$\sum_{j=1}^{N} \alpha_{ij} A_j \rightarrow \sum_{j=1}^{N} \beta_{ij} A_j, \quad i = 1, 2, \dots, M$$

where α_{ij} and β_{ij} are integer numbers called the *stoichiometric coefficients*. The time-dependent substance concentrations $c_1, c_2, \dots, c_N$ (mol/m^3) of $A_1, A_2, \dots, A_N$ are obtained from the ODE system

$$\frac{d\mathbf{c}}{dt} = S\mathbf{r}(\mathbf{c}) \tag{2.49}$$

In (2.49), S is the stoichiometric matrix of integers, a sparse $N \times M$ matrix, and $\mathbf{r}(\mathbf{c})$ [mol/(m^3 · s)] an $M \times 1$ vector of reaction rates. In a model by the Norwegian chemists Guldberg and Waage from 1879 called the *mass action law*, r_i has the form

$$r_i = k_i \prod_{j=1}^{N} c_j^{\alpha_{ij}}, \quad i = 1, 2, \dots, M \tag{2.50}$$

where k_i (the unit depends on N and α_{ij} is called the *rate constant*, which is temperature dependent according to Arrhenius' equation $k = Ae^{-E/RT}$, where A is called *pre-exponential factor*, E (J/mol) is the activation energy, $R = 8.314472$ [J/(K · mol)] the gas constant, and T (K) the temperature. Svante Arrhenius was a Swedish chemist and Nobel laureate active about 100 years ago. The ODE system (2.49) is generally nonlinear. With slight modifications, it can be used to model reactors of great importance in applications.

2.4.5 Control Theory

A common model used in control theory is the *state space* model

$$\frac{d\mathbf{x}}{dt} = A\mathbf{x} + B\mathbf{u} \tag{2.51a}$$

$$\mathbf{y} = C\mathbf{x} + D\mathbf{u} \tag{2.51b}$$

where $\mathbf{x}$ are internal state variables of the system studied, $\mathbf{u}$ are control variables, and $\mathbf{y}$ are observed variables. A, B, C, and D are constant matrices. Of special interest is the study of *stability properties* of control systems. Another important problem in control theory is the *identification* problem, i.e., from measurements of the observed variables determine some of the parameters of the system.

In control theory, the models are often described by *block diagrams*, which can be translated into ODE systems of type (2.51).

Models in control theory can also be nonlinear ODE systems of type

$$\frac{d\mathbf{x}}{dt} = \mathbf{f}(\mathbf{x}, \mathbf{u}, \mathbf{p}) \tag{2.52a}$$

$$\mathbf{y} = \mathbf{g}(\mathbf{x}, \mathbf{u}, \mathbf{p}) \tag{2.52b}$$

where $\mathbf{p}$ is a parameter vector.

2.4.6 Compartment Models

Compartment models are frequently used in applications where simplified models of flows of gases or fluids are studied. Examples are found in biomedicine, pharmacokinetics, and air pollution. The model consists of a system of first-order ODEs, describing the exchange of substances between the compartments of the model. Assume the model has n compartments, which can be regarded as "boxes". If x_i is the amount of some substance in compartment i, the model is formulated as an IVP, based on the balance principle for the amounts of the substances called the *continuity equation* (see Chapter 9):

$$\frac{dx_i}{dt} = Q_i^I + \sum_{j=1}^{n} Q_{ij}^I - \sum_{j=1}^{n} Q_{ij}^O - Q_i^O, \qquad x_i(0) = x_{i0} \tag{2.53}$$

where the inflow consists of Q_i^I, which is the flow from the environment to compartment i and Q_{ij}^I the flow to compartment i from compartment j. The outflow consists of Q_i^O, which is the flow from compartment i to the environment and Q_{ij}^O being the flow from compartment i to compartment j. The initial value x_{i0} is the initial amount of the substance in compartment i.

If the flows are modeled as $Q_{ij}^I = k_{ij}^I x_j$, $Q_{ij}^O = k_{ij}^O x_i$, and Q_i^I, Q_i^O are constant or time dependent, the resulting ODE system will be an LCC system.

2.5 SOME EXAMPLES FROM APPLICATIONS

In this part of the chapter, some ODE applications are shown, exemplifying the models presented in Section 2.4. The examples will be referred to in Chapters 3 and 4, where numerical methods for IVPs and BVPs are described. To each example, some exercises are given. They are mainly intended for repetition of analytic treatment of ODEs and also to demonstrate analytic preprocessing necessary before numerical treatment of the problems should be performed.

Example 2.3. Particle dynamics

The following differential equation is a model of the motion in the xy-plane of a particle released with a certain initial velocity from a point and then moving under the influence of gravity and a force generated by the air resistance, which is proportional to the square of the velocity (see Exercise 2.1.4 for the corresponding 1D problem).

$$m\frac{d^2\mathbf{r}}{dt^2} = -mg\mathbf{e}_y - c\left\|\frac{d\mathbf{r}}{dt}\right\|_2 \frac{d\mathbf{r}}{dt}, \quad \mathbf{r}(0) = y_0\mathbf{e}_y, \quad \frac{d\mathbf{r}}{dt}(0) = \mathbf{v}_0 \tag{2.54}$$

where m (kg) is the mass of the particle, r (m) is the position of the particle, g (m/s^2) gravitational acceleration, c (N $\cdot$ s^2/ m^2) air resistance coefficient, y_0 (m) initial height of the particle, and $\mathbf{v}_0$ (m/s) the initial velocity vector, which has an elevation angle $\alpha > 0$ with the horizontal plane (see Figure 2.5).

The formulation (2.54) of the problem is an IVP. The computational task is to integrate step by step until the particle hits the ground ($y = 0$). However, this problem can also be formulated as a BVP if the task is to determine the elevation angle that causes the particle to hit a given point x_0 on the ground for a given value of the initial velocity. The ICs are replaced by BCs:

$$\mathbf{r}(0) = y_0\mathbf{e}_y, \quad \mathbf{r}(T) = x_0\mathbf{e}_x \tag{2.55}$$

where T (s) is the time when the particle hits the ground in the point $(x_0, 0)$. By adjusting the elevation angle α, the BC at $x = x_0$ can be satisfied. It turns

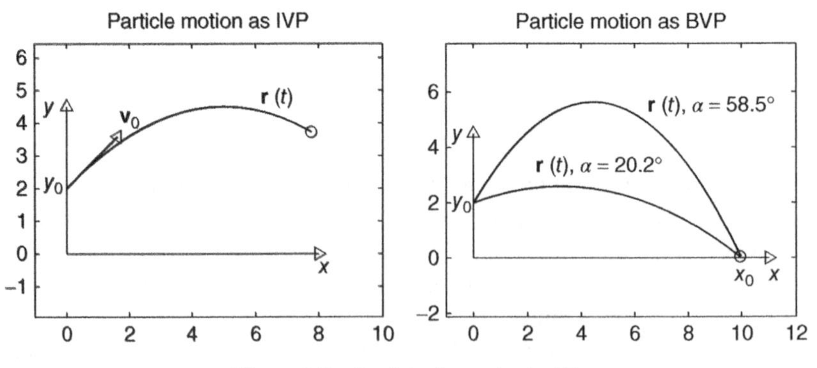

Figure 2.5 Particle dynamics in 2D

out that this BVP has two solutions, for certain values of the parameters, see Figure 2.5.

Exercise 2.5.1. Solve analytically the IVP and the BVP in the case $c = 0$.

Exercise 2.5.2. Rewrite the IVP as an ODE system of first order, i.e., in the form (2.8).

Example 2.4. **Planetary motion**

Consider the motion of a single planet about a heavy sun. The assumption heavy sun means that the sun influences the motion of the planet, but the relatively small mass of the planet has an influence on the sun that is negligible. Let $\mathbf{r}(t) = (x(t), y(t))^T$ (m) denote the coordinates of the planet with the sun in the origin. The only force acting on the planet is the gravitational force from the sun having the magnitude $F_r = \gamma Mm / \| \mathbf{r} \|_2^2$ (N), where $\gamma = 6.6742 \cdot 10^{-11}$ ($\mathrm{N} \cdot \mathrm{m}^2/\mathrm{kg}^2$) is the gravitational constant, M (kg) the mass of the sun, and m the mass of the planet.

Newton's equation (2.43) takes the form (for illustration, see Figure 2.6):

$$m\frac{d^2\mathbf{r}}{dt^2} = -\frac{\gamma Mm}{\| \mathbf{r} \|_2^2}\mathbf{e}_r \tag{2.56}$$

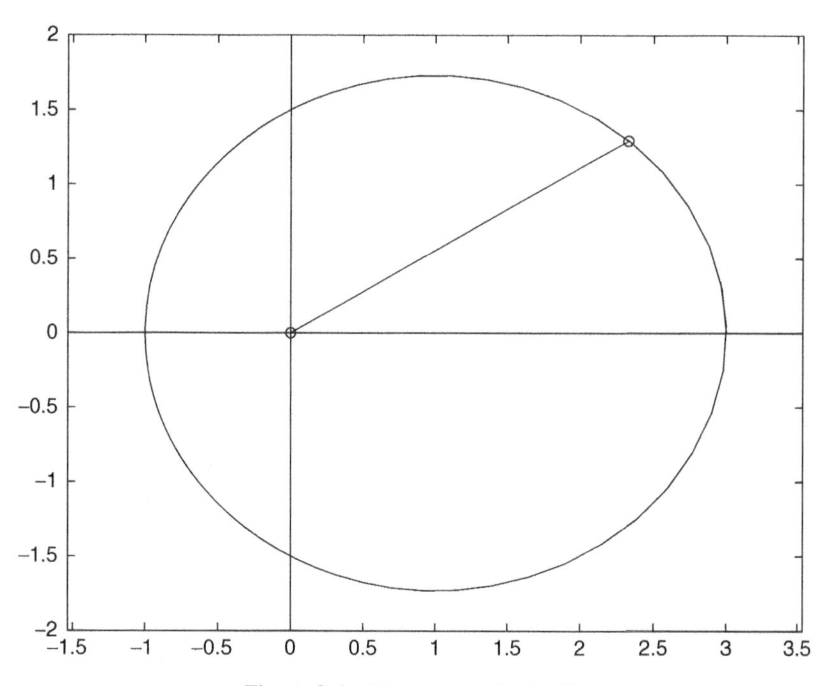

Figure 2.6 Planetary motion in 2D

***Exercise* 2.5.3.** Write Newton's equation (2.56) in component form, i.e., one second-order ODE for $x(t)$ and another for $y(t)$. Also write (2.56) as a system of first-order ODEs.

***Exercise* 2.5.4.** For planetary motion, the hamiltonian can be written

$$H(\mathbf{q}, \mathbf{p}) = T + U = \frac{1}{2m}(p_1^2 + p_2^2) - \frac{\gamma M m}{\sqrt{q_1^2 + q_2^2}} \tag{2.57}$$

where $\mathbf{q}$ and $\mathbf{p}$ are the (generalized) coordinates and momenta

$$\mathbf{q} = \mathbf{r}, \quad \mathbf{p} = m\frac{d\mathbf{r}}{dt}$$

a) Show that H, i.e., the total energy (sum of kinetic and potential energy) is conserved. Also show that the angular momentum $\mathbf{L} = \mathbf{q} \times \mathbf{p}$ is conserved. Hint: Form dH/dt and $d\mathbf{L}/dt$.
b) Show that Hamilton's equations applied to (2.57) imply Newton's equation.

***Example* 2.5. A simple RLC circuit**

Consider the following electrical network (Figure 2.7), consisting of two resistors, one capacitance, one inductance, one voltage source, and one current source.

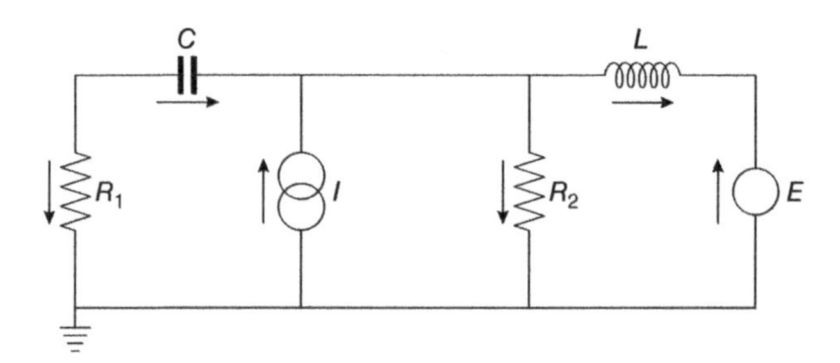

Figure 2.7 A simple electrical network

The DAE system corresponding to this electrical network is (2.58)

$$\begin{pmatrix} C & -C & 0 & 0 & 0 \\ -C & C & 0 & 0 & 0 \\ 0 & 0 & 0 & 0 & 0 \\ 0 & 0 & 0 & L & 0 \\ 0 & 0 & 0 & 0 & 0 \end{pmatrix} \frac{d}{dt} \begin{pmatrix} e_1 \\ e_2 \\ e_3 \\ i_L \\ i_V \end{pmatrix} + \begin{pmatrix} 1/R_1 & 0 & 0 & 0 & 0 \\ 0 & 1/R_2 & 0 & 1 & 0 \\ 0 & 0 & 0 & -1 & 1 \\ 0 & -1 & 1 & 0 & 0 \\ 0 & 0 & 1 & 0 & 0 \end{pmatrix} \begin{pmatrix} e_1 \\ e_2 \\ e_3 \\ i_L \\ i_V \end{pmatrix} = \begin{pmatrix} 0 \\ -I \\ 0 \\ 0 \\ E \end{pmatrix} \tag{2.58}$$

In a linear DAE system, there are linear relations between the dependent variables and the initial values must be consistent with these relations.

Exercise 2.5.5. Verify that the following linear algebraic relations are valid for a solution of the DAE system above:

$$\begin{cases} R_1^{-1}e_1 + R_2^{-1}e_2 + i_L = -I \\ i_L = i_V \\ e_3 = E \end{cases}$$

and that the following IC is consistent with these algebraic relations: $u_0 = (0, 0, E, -I, -I)^T$

Exercise 2.5.6. Find the steady-state solution of (2.58).

Example 2.6. **Biochemical reaction**

The following reaction system, consisting of two reversible reactions, occurs frequently in biochemistry

$$E + S \Longleftrightarrow C \Longleftrightarrow E + P$$

In the reaction formula, E stands for enzyme, S for substrate, C for complex, and P for product. The respective concentrations are denoted by c_1, c_2, c_3, and c_4. The ODE system modeling the kinetic behavior of the reaction system is

$$\frac{d\mathbf{c}}{dt} = \begin{pmatrix} -1 & 1 & 1 & -1 \\ -1 & 1 & 0 & 0 \\ 1 & -1 & -1 & 1 \\ 0 & 0 & 1 & -1 \end{pmatrix} \begin{pmatrix} k_1c_1c_2 \\ k_2c_3 \\ k_3c_3 \\ k_4c_1c_4 \end{pmatrix}, \quad \mathbf{c}_0 = \begin{pmatrix} E_0 \\ S_0 \\ 0 \\ 0 \end{pmatrix} \tag{2.59}$$

where k_1, k_2, k_3, and k_4 are the rate constants of the reactions and $E_0 > 0$ and $S_0 > 0$ are the initial concentrations of E and S.

Exercise 2.5.7. Verify that $c_1(t) + c_3(t) = E_0$ and $c_2(t) + c_3(t) + c_4(t) = S_0$.

When the reactions have come to chemical equilibrium, the concentrations satisfy the following nonlinear algebraic system of equations

$$\mathbf{f}(\mathbf{c}) = \begin{pmatrix} -1 & 1 & 1 & -1 \\ -1 & 1 & 0 & 0 \\ 1 & -1 & -1 & 1 \\ 0 & 0 & 1 & -1 \end{pmatrix} \begin{pmatrix} k_1c_1c_2 \\ k_2c_3 \\ k_3c_3 \\ k_4c_1c_4 \end{pmatrix} = 0 \tag{2.60}$$

One solution of this system is obviously $\mathbf{c} = (0, 0, 0, 0)^T$, which, however, cannot be the equilibrium concentrations as the relations in Exercise 2.5.7 are not fulfilled. Using Newton's method (see Appendix 1) to solve the system of equations (2.60) will not work, as the matrix is singular, which implies that the jacobian will also be singular.

***Exercise* 2.5.8.** Preprocess the equilibrium problem so that Newton's method can be used.

Example 2.7. Stability of a regulator

Consider a servo mechanism where the output signal $y(t)$ depends on an input signal $u(t)$. In an application, $u(t)$ can be the torsion angle of the steering wheel of a boat and $y(t)$ the torsion angle of the boat's rudder. The purpose of servo is to strengthen the torsion moment from a small value to a large value.

The following control system models a feedback regulator consisting of an electromechanical system described with a block diagram (Figure 2.8):

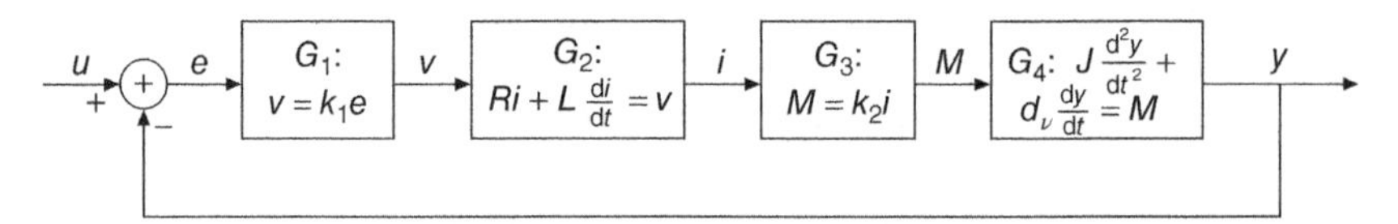

Figure 2.8 Block diagram of a servo mechanism

In mathematical terms, the model is formulated in the following way:

$\oplus:$ $e(t) = u(t) - y(t)$, the difference (error) between $u(t)$ and $y(t)$
$G_1:$ $v(t) = k_1 e(t)$, amplification of $e(t)$ into a voltage value
$G_2:$ $Ri(t) + L\frac{di}{dt} = v(t)$, induced current in a motor driving an axis.
$G_3:$ $M(t) = k_2 i(t)$, torsion moment of the axis.
$G_4:$ $J\frac{d^2y}{dt^2} + d_v\frac{dy}{dt} = M(t)$, the moment drives a load (the rudder) having moment of inertia J and damping coefficient d_v.

***Exercise* 2.5.9.** Write the equations above in state space form (2.51) with $\mathbf{x}(t) = (i(t), z(t), y(t))^T$, where $z(t) = dy/dt$, $\mathbf{u}(t) = (u(t), 0, 0)^T$, and $\mathbf{y}(t) = y(t)$. Hint: Eliminate $e(t)$, $v(t)$, and $M(t)$. Note that the resulting ODE system is of LCC type.

***Exercise* 2.5.10.** Verify that the equations above can also be written as a third-order ODE (of LCC type):

$$\frac{LJ}{R}\frac{d^3y}{dt^3} + \left(\frac{Ld_v}{R} + J\right)\frac{d^2y}{dt^2} + d_v\frac{dy}{dt} + Ky = Ku(t), \quad K = \frac{k_1k_2}{R} \tag{2.61}$$

The properties of this regulator model can be simulated in a parameter study. Particularly important is the parameter K, which is proportional to the amplification factors of the model.

To answer the question, if the system is stable for all values of K, it is enough to study the homogeneous version of (2.61), i.e., the case where $u = 0$

$$\frac{LJ}{R}\frac{d^3y}{dt^3} + \left(\frac{Ld_v}{R} + J\right)\frac{d^2y}{dt^2} + d_v\frac{dy}{dt} + Ky = 0 \tag{2.62}$$

If the parameter $L = 0$, (2.62) turns into a second-order ODE of the same type as in Exercise 2.3.6 and therefore stable for all values of K. However, what happens if the parameter $L \neq 0$?

Exercise 2.5.11. Assume the following values of the parameters: $J = 0.4$, $d_v = 1$, $R = 100$, $L = 10$. Write a program that makes a graph (root locus) of the curves generated by the eigenvalues of (2.62) as K varies in the interval $[0, 20]$. For which values of K is the system unstable?

Example 2.8. **Drug kinetics**

The following two-compartment model is an example from pharmacokinetics. Assume a drug, which has been taken orally, is present in the intestine during a certain time interval. The drug is absorbed with the constant flow rate q (mmol/s) into the first compartment, the blood plasma. In the blood plasma, the concentration of the drug is $c_1(t)$ (mmol/l). The second compartment is the organ where the drug is active. Between the first and second compartments, there is an drug exchange with rate $k_1c_1(t)$ (mmol/l/s) leading to the drug concentration $c_2(t)$ in the second compartment. In the organ, the drug is consumed with the rate $k_bc_2(t)$ (mmol/s) and the surplus is sent back to the blood with the rate $k_2c_2(t)$ (mmol/s). From the blood, finally, there is an elimination of the drug through the kidneys with the rate $k_ec_1(t)$ (mmol/s). Figure 2.9 illustrates the blood–organ compartment model of drug distribution:

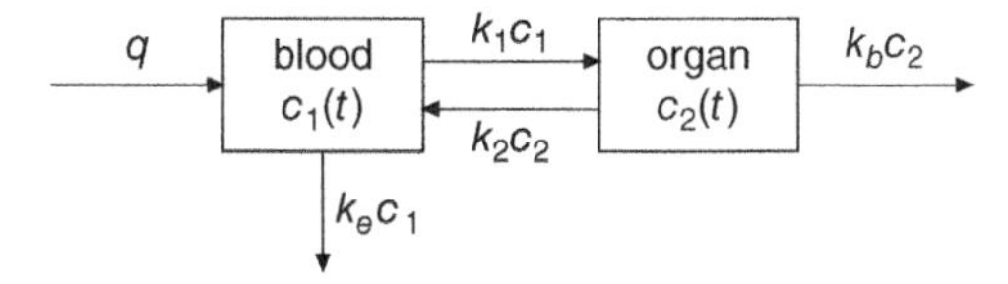

Figure 2.9 A simple compartment model

The ODE system (2.53) takes the following form in this case:

$$\frac{dc_1}{dt} = -k_1c_1 + k_2c_2 - k_ec_1 + q, \quad c_1(0) = 0$$

$$\frac{dc_2}{dt} = k_1c_1 - k_2c_2 - k_bc_2, \quad c_2(0) = 0$$

The steady-state solution, i.e., the solution obtained when the time derivatives are zero, is obtained from the algebraic linear system of equations:

$$\begin{pmatrix} k_1 + k_e & -k_2 \\ -k_1 & k_2 + k_b \end{pmatrix} \begin{pmatrix} c_1 \\ c_2 \end{pmatrix} = \begin{pmatrix} q \\ 0 \end{pmatrix}$$

***Exercise* 2.5.12.** Formulate the analytical solution of the ODE system using Duhamel's formula (2.25).

BIBLIOGRAPHY

1. D. Zill and M. Cullen, "Differential Equations with Boundary Value Problems", 5th edition, Brooks and Cole, 2005
2. W. Boyce and R. DiPrima, "Elementary Differential Equations and Boundary Values Problems", 8th edition, Wiley, 2005

   The following textbook is older but elementary and well-structured
3. M. Braun, "Differential Equations and Their Applications", Springer, 1975

   An old classical textbook, extensive and theoretical is
4. E. Coddington and N. Levinson, "Theory of Ordinary Differential Equations", McGraw-Hill, 1955

References to special application areas of ODEs

Particle dynamics: H.Goldstein et al, "Classical Mechanics", Addison Wesley, 2004

Electrical networks: J.D.Irwin, "Basic Engineering Circuit Analysis", Prentice Hall, 2005
Chemical kinetics: P.Erdi, J.Toth, "Mathematical models of chemical reactions", Manchester University Press, 1989

Control theory: L.Ljung, T.Glad, "Modeling of Dynamical Systems, Prentice Hall, 1994
Compartment models: K.Godfrey, "Compartmental Models and Their Application", Academic Press, 1983

3

NUMERICAL METHODS FOR INITIAL VALUE PROBLEMS

As we have seen in Chapters 1 and 2, ordinary differential equation(ODE) problems occur in numerous applications and therefore important for modeling processes whose evolution depends on one variable, usually time t or one spatial variable x. In this chapter, we treat the initial value problem (IVP).

The general formulation of an IVP is

$$\frac{d\mathbf{u}}{dt} = \mathbf{f}(t, \mathbf{u}), \quad \mathbf{u}(t_0) = \mathbf{u}_0, \quad t_0 < t \leq t_{end} \tag{3.1}$$

Often the solution depends on parameters, denoted by the vector $\mathbf{p}$, that occur in the right hand side function:

$$\frac{d\mathbf{u}}{dt} = \mathbf{f}(t, \mathbf{u}, \mathbf{p}), \quad \mathbf{u}(t_0) = \mathbf{u}_0, \quad t_0 < t \leq t_{end} \tag{3.2}$$

The initial values $\mathbf{u}_0$ can also be regarded as parameters in case the dependence of the initial values is studied: $\mathbf{u} = \mathbf{u}(t, \mathbf{p}, \mathbf{u}_0)$

Example 3.1. **(Newton's law for a particle)**
In Exercise 2.1.4, the following IVP was introduced

$$m\ddot{y} = -mg - c\dot{y}|\dot{y}|, \quad y(0) = y_0, \quad \dot{y}(0) = v_0 \tag{3.3}$$

This ODE models a particle thrown vertically from the position y_0 and with initial velocity v_0. The particle is influenced by gravity and an air resistance force being

Introduction to Computation and Modeling for Differential Equations, Second Edition. Lennart Edsberg.
© 2016 John Wiley & Sons, Inc. Published 2016 by John Wiley & Sons, Inc.

proportional to the square of the velocity. When the solution is studied as a function of, e.g., c, we make a *parameter study* of the problem. The solution of this problem depends on t and the parameters m, c, g, y_0, and v_0, i.e., $y = y(t, m, c, g, y_0, v_0)$.

This example indicates that there are several questions that can be posed concerning the solution of an IVP:

- How does the solution behave on a time interval $[t_0, t_{end}]$ for a given $\mathbf{p}$?
- How does the solution depend on parameters $\mathbf{p}$ in the problem?
- Which are the critical points and what are the stability properties of the critical points (stability analysis)?
- How sensitive is the solution with respect to the parameters (sensitivity analysis)?
- When a parameter varies over an interval, does the solution change character for some value of the parameter in this interval (bifurcation analysis)?
- When measurements $(t_k, \mathbf{u}_k^*), k = 1, 2, \ldots, M$, of the process are given from experiments, how do we estimate unknown parameters in the model (parameter estimation)?

To be able to answer the questions, it is necessary to have an accurate, efficient, and robust numerical solver for the IVP (3.1).

3.1 GRAPHICAL REPRESENTATION OF SOLUTIONS

To get a better understanding of the behavior of solutions to ODE problems, it is necessary to *visualize* the solutions graphically. This can be done in different ways:

- plot *trajectories*, i.e., $\mathbf{u}(t)$ is plotted as function of t in a coordinate system. A trajectory is appropriate when we want to study the time evolution of a process. Depending on the size of variables, it should be judged what scales should be used for the different coordinate axis, linear or logarithmic.
- plot *phase portraits* showing two components of the solution, $u_i(t)$ and $u_j(t)$ with t as a parameter along the curve $(u_i(t), u_j(t))$ plotted in a coordinate system. A phase portrait is appropriate when one wants to study how solutions behave as function of initial values, especially in the neighborhood of critical points. In 3D graphics, a phase portrait with three components can be plotted.

There are other types of graphical representation, e.g., the *root locus plot*, often used in control theory to visualize the stability properties of an ODE system, see Example 2.7.

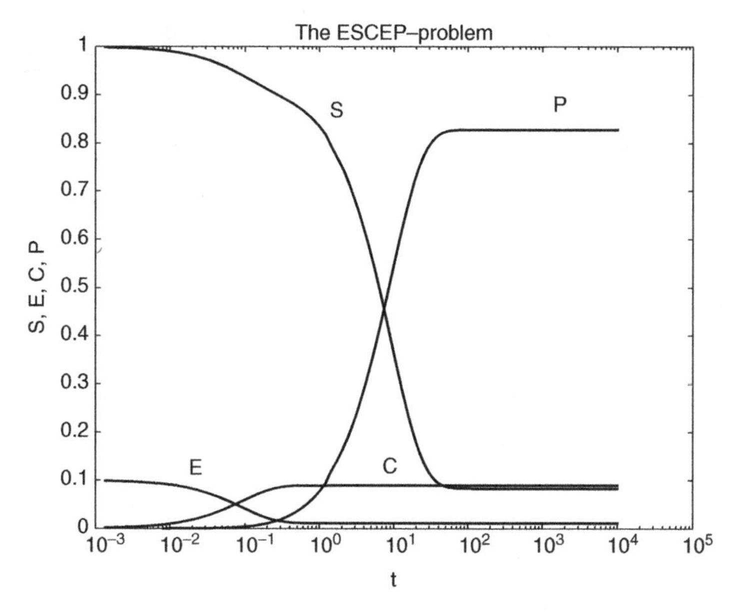

Figure 3.1 Example of solution trajectories

Example 3.2. **(The kinetics of a biochemical reaction system).**
The following reactions and the corresponding ODE system was introduced in Example 2.6:

$$E + S \Longleftrightarrow C \Longleftrightarrow E + P$$

The kinetics of the reaction system is modeled by the ODE system (2.59). In this example, the following values of the parameters are given: $k_1 = 10$, $k_2 = 0.1$, $k_3 = 1$, $k_4 = 10$. With the initial values $E_0 = 0.1$, $S_0 = 1$, $C_0 = 0$, and $P_0 = 0$ we get the following trajectories. Note that it is appropriate for this problem to use logarithmic scale on the t-axis (Figure 3.1).

Example 3.3. **(A problem with an oscillating solution: Van der Pol's equation).**
Van der Pol's equation (see Exercise 2.3.5) occurs, e.g., in modeling of nonlinear electrical circuits:

$$\frac{d^2y}{dt^2} + \epsilon(y^2 - 1)\frac{dy}{dt} + y = 0, \quad y(0) = 1, \quad \frac{dy}{dt}(0) = 0 \tag{3.4}$$

The numerical solution of this ODE problem is described in Section 3.3. The results are preferably visualized as a phase portrait. If the parameter ϵ is given the values $\epsilon = 0.1, 1, 10, 100$, the following phase portraits are drawn in a *parameter study*

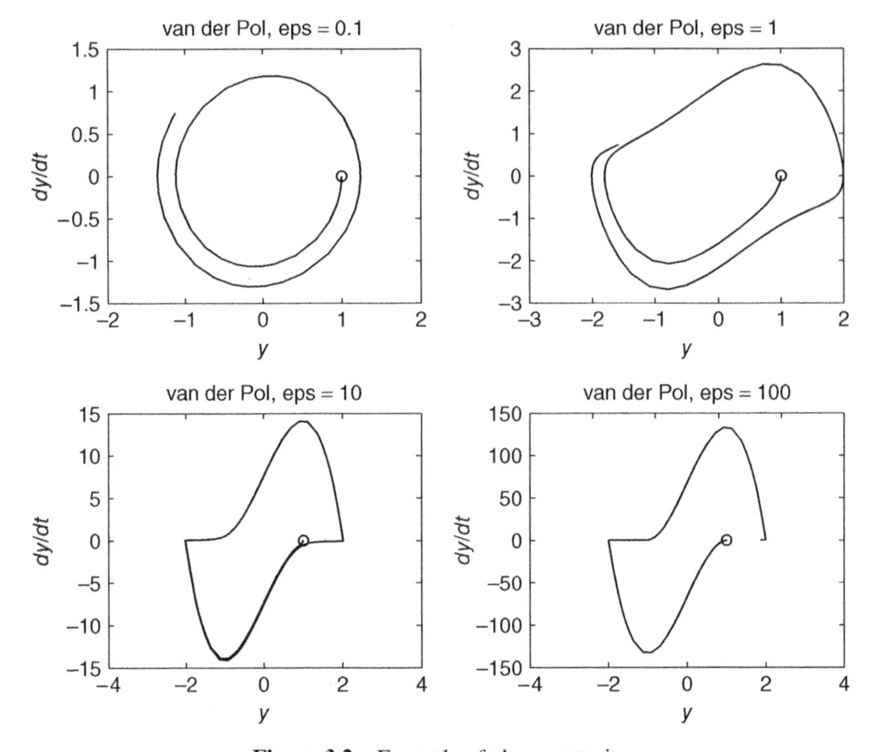

Figure 3.2 Example of phase portraits

(Figure 3.2). Obviously the solutions are oscillating. After an initial transient, the phase portrait approaches a closed curve, a *limit cycle*. In the graph, the point *o* represents the initial value point.

3.2 BASIC PRINCIPLES OF NUMERICAL APPROXIMATION OF ODEs

The solution of an ODE problem (3.1) is a vector-valued function $\mathbf{u}(t)$. An approximate numerical solution can be obtained with a *discretization* method, such as the *finite difference method* (FDM) or an *ansatz* method, e.g., the *finite element method* (FEM) or the *finite volume method* (FVM). In this chapter, the FDM is presented.

With an FDM, we obtain a sequence of points $(t_1, \mathbf{u}_1), (t_2, \mathbf{u}_2), \ldots, (t_N, \mathbf{u}_N)$. These points approximate the exact solution, i.e., $\mathbf{u}_k \approx \mathbf{u}(t_k)$. Examples of well-known FDMs are Euler's method and Runge–Kutta's method.

With an ansatz method, a function $\mathbf{u}_h(x) \approx \mathbf{u}(x)$ is obtained. The approximating function has the form $\mathbf{u}_h(t) = \sum c_j \varphi_j(x)$, where $\varphi_j(x)$ are given basis functions and c_j are coefficients chosen in such a way that the approximation is as accurate as possible in some sense. The FEM is an example of such a method, see Sections 4.3, 6.5, and 7.4.

The FVM is a method for solving certain PDEs approximated by ODE systems. Similar to the FDM approximate values, $\mathbf{u}_1, \mathbf{u}_2, \dots, \mathbf{u}_N$ are calculated at discrete space points $x_1, x_2, \dots, x_N$, see Section 8.3.

When different methods for the IVP are discussed, we can without loss of generality restrict ourselves to the *scalar* case, i.e.,

$$\frac{du}{dt} = f(t, u), \quad u(0) = u_0, \quad 0 < t \le t_{end} \tag{3.5}$$

The exact solution of this problem is denoted by $u(t)$. All numerical formulas in this chapter will be given for the scalar case, unless otherwise is stated.

3.3 NUMERICAL SOLUTION OF IVPs WITH EULER's METHOD

The basic idea in an FDM is that the derivative in (3.5) at time t_k is approximated by a *difference formula* using a few u values $\dots u_{k-1}, u_k, u_{k+1} \dots$ in the neighborhood of u_k. The difference h_k between two consecutive t points, i.e., $h_k = t_{k+1} - t_k$, is called the *stepsize*. Most numerical methods use *variable stepsize*, i.e., h_k takes different values as the solution is computed along the t-axis. Methods where variable stepsize is implemented are also called adaptive methods.

However, *constant stepsize* h can also be used, especially when we want to have better control of the truncation errors (approximation errors) of the u_k values. In that case, the t_k points are *equidistantly spaced* along the t-axis.

The discretized formulation of the ODE problem changes the differential equation into a *difference equation* and the numerical solution is computed as a sequence of u_k values; the u_k values are generated by a *recursion formula*. Linear difference equations are fundamental for understanding numerical stability of FDMs. A short overview is given in Appendix A.2.

The derivative in (3.5) can be approximated by various difference formulas, e.g., the *Euler forward* (3.6) or the *Euler backward* (3.7) approximations:

$$\frac{du}{dt}(t_k) = \frac{u(t_{k+1}) - u(t_k)}{h} + O(h) \tag{3.6}$$

$$\frac{du}{dt}(t_k) = \frac{u(t_k) - u(t_{k-1})}{h} + O(h) \tag{3.7}$$

Both formulas are of first order, i.e., the truncation error is approximately proportional to h provided h is sufficiently small. Leonard Euler was a Swiss mathematician active in the 18th century and one of the most productive ever. He was one of the first to write a textbook in analysis: "Introductio in analysin infinitorum" from 1748.

Another possible approximation would be to use the *central difference* approximation formula, which is of second order:

$$\frac{du}{dt}(t_k) = \frac{u(t_{k+1}) - u(t_{k-1})}{2h} + O(h^2) \tag{3.8}$$

It may seem that the formula (3.8) is better to use instead of (3.6) and (3.7) as it is of second order and therefore more accurate. There are disadvantages, however, using (3.8) for the IVP, related to *numerical instability*, see Section 3.4.2. For certain problems, however, (3.8) works well, see Section 3.5.

Consider first the constant stepsize case. Applying (3.6) to the IVP (3.5) gives Euler's *explicit* method (also called Euler's forward method)

$$\begin{cases} u_k = u_{k-1} + hf(t_{k-1}, u_{k-1}), & u_0 = u(0) \\ t_k = t_{k-1} + h, \quad t_0 = 0, & k = 1, 2, \ldots, N \end{cases} \tag{3.9}$$

When (3.7) is used, we get Euler's *implicit* method (also called Euler's backward method).

$$\begin{cases} u_k = u_{k-1} + hf(t_k, u_k), & u_0 = u(0) \\ t_k = t_{k-1} + h, \quad t_0 = 0, & k = 1, 2, \ldots, N \end{cases} \tag{3.10}$$

It can be shown that both methods give *first-order accuracy* in the u_k values, i.e., the truncation error, for IVPs also called the global error, e_k, fulfills

$$e_k = u(t_k) - u_k = O(h) \tag{3.11}$$

For both methods, we see that when (t_{k-1}, u_{k-1}) is given as input, the next point (t_k, u_k) is computed as output, hence Euler's explicit and implicit methods are both *one-step* methods.

However, in explicit Euler, u_k is computed just by inserting (t_{k-1}, u_{k-1}) into the right hand side, while in implicit, Euler u_k is computed by solving a nonlinear equation. For that reason, we call (3.9) an *explicit* method, while (3.10) is an *implicit* method.

As an implicit method will be more complicated to compute, one may immediately ask: Why use implicit Euler when the computational work is so much larger? The answer is that for certain problems called *stiff* problems, implicit Euler is much more advantageous for *numerical stability* reasons. This will be shown in Section 3.3.4.

The two Euler methods usually have too low order of accuracy to be of practical importance. However, they are very useful for demonstration of principles of the FDM when applied to an IVP. We will therefore show some basic properties using these two methods before we continue to higher order methods in Section 3.4.

***Exercise* 3.3.1.** Show that the difference approximations (3.6) and (3.7) both have first-order accuracy.

***Exercise* 3.3.2.** Show that the difference approximation (3.8) is of second order.

Exercise 3.3.3. Set up the recursion formula when Euler's explicit method (3.9) is applied to the *scalar model problem* for IVPs

$$\frac{du}{dt} = \lambda u, \quad u(0) = u_0$$

where λ is a constant complex parameter. Show that $u_k = (1 + h\lambda)^k u_0$.

Exercise 3.3.4. The following method is sometimes used for solving problems in Newtonian mechanics. For the ODE system

$$\frac{dv}{dt} = a(x, v), \qquad \frac{dx}{dt} = v$$

the Euler–Cromer method (see also Section 3.5) is defined as

$$v_{k+1} = v_k + ha(x_k, v_k), \qquad x_{k+1} = x_k + hv_{k+1}$$

What is the order of this method? Try an analytical approach or make a numerical experiment (see Section 3.3.1).

3.3.1 Euler's Explicit Method: Accuracy

We first show how explicit Euler is used to solve Van der Pol's equation (3.4)

$$\frac{d^2y}{dt^2} + \epsilon(y^2 - 1)\frac{dy}{dt} + y = 0, \quad y(0) = 1, \quad \frac{dy}{dt}(0) = 0$$

This second-order ODE is written as a system of two first-order ODEs by introducing a vector $\mathbf{u}$ with the components $u_1 = y, u_2 = dy/dt$:

$$\begin{cases} \dfrac{du_1}{dt} = u_2, & u_1(0) = 1 \\ \dfrac{du_2}{dt} = -\epsilon(u_1^2 - 1)u_2 - u_1, & u_2(0) = 0 \end{cases}$$

When the explicit Euler method is applied to this system, we get the *recursion* formulas:

$$\begin{cases} u_{1,k} = u_{1,k-1} + hu_{2,k-1}, & u_{1,0} = 1 \\ u_{2,k} = u_{2,k-1} + h(-\epsilon(u_{1,k-1}^2 - 1)u_{2,k-1} - u_{1,k-1}), & u_{2,0} = 0 \\ t_k = t_{k-1} + h, & t_0 = 0 \end{cases}$$

Recall that $u_{1,k} \approx u_1(t_k) = y(t_k)$. How does the global error $e_{1,k} = u_1(t_k) - u_{1,k}$ depend on the stepsize h? The numerical experiment presented in the example below gives an answer to that question.

Example 3.4. Numerical experiment to determine the order of accuracy

For Van der Pol's equation (3.4), let $\epsilon = 1$ and calculate $y(1)$ with explicit Euler. Discretize the t-axis on the interval $[0, 1]$ with constant stepsize h, using N steps. Table 3.1 shows the result obtained with the explicit Euler method.

As we observe in the table, the convergence is very slow. When h is halved, the global error $u_{1,N} - y(1)$ is also halved, hence $e_{1,N} \approx Ch$

From a logarithmic plot, the order of accuracy can be estimated graphically. Taking the logarithm of the general error relation $e \approx Ch^p$ gives $\log\ e \approx \log\ C + p \log\ h$, which is a straight line in a loglog diagram. The slope of that line is p. In Figure 3.3, the slope is 1, hence the order of explicit Euler is $p = 1$.

The error defined by (3.11) is called the *global error*. Another concept important for error estimation and variable stepsize control in adaptive methods is the *local error*, $l(t, h)$, which is defined as the residual obtained when the exact solution is inserted into the explicit Euler method

$$l(t_k, h) = u(t_k) - u(t_{k-1}) - hf(t_{k-1}, u(t_{k-1})) \tag{3.12a}$$

Hence, the local error can be regarded as the truncation error committed in *one* step. With Taylor's expansion, we see that

$$l(t_k, h) = \frac{h^2}{2}\frac{d^2u}{dt^2}(\tau_k), \quad t_{k-1} < \tau_k < t_k \tag{3.12b}$$

In Figure 3.4, it is shown how the global and local errors propagate when one step is taken from t_{k-1} to t_k.

TABLE 3.1 Numerical Solution of Van der Pol's Equation Using Explicit Euler

N	h	$u_{1,N}$	$u_{1,N} - y(1)$
16	1/16	0.5231	0.0254
32	1/32	0.5100	0.0123
64	1/64	0.5037	0.0060
128	1/128	0.5006	0.0029
256	1/256	0.4991	0.0014
512	1/512	0.4984	0.0007
1024	1/1024	0.4980	0.0003
2048	1/2048	0.4978	0.0001

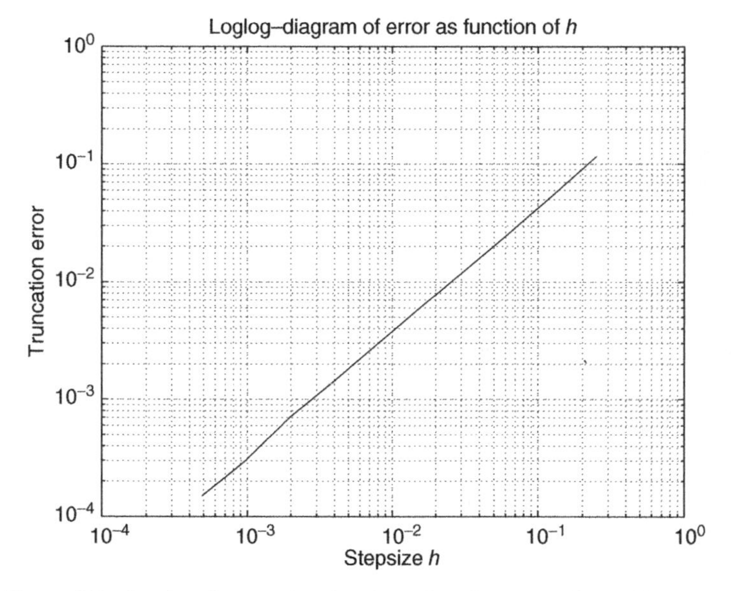

Figure 3.3 Loglog diagram showing the order of accuracy for explicit Euler

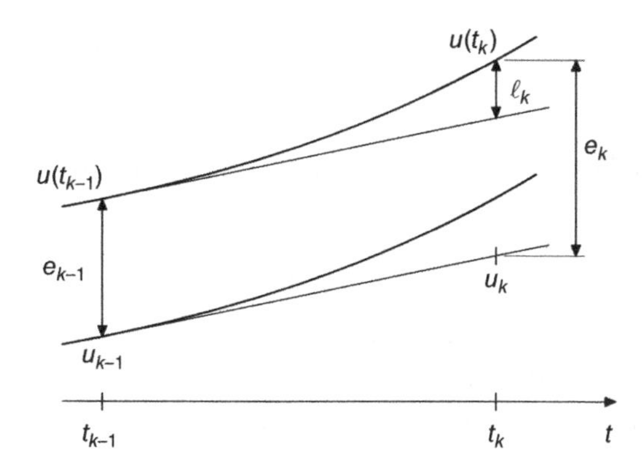

Figure 3.4 Graphical definition of the local and global errors

The global error e_k can be regarded as the accumulated effect of all local errors from t_1 up to t_k. However, it is not true in general that the global error is the sum of the local errors, as the following analysis shows:

Euler's explicit method can be written:

$$0 = -u_k + u_{k-1} + hf(t_{k-1}, u_{k-1})$$

If this equality is added to (3.12) we obtain, after using Taylor's expansion on the second argument u in $f(t, u)$, a global error recursion relation

$$e_k = e_{k-1} + h\frac{\partial f}{\partial u}(t_{k-1}, u_{k-1} + \theta(u(t_{k-1}) - u_{k-1}))e_{k-1} + l(t_k, h), \quad e_0 = 0 \tag{3.13}$$

where $0 < \theta < 1$. From this equality, we see that only if $\partial f/\partial u \equiv 0$, i.e., if $f(t, u)$ does not depend on u, the global error is the sum of the local errors. If this is not the case, assume that

$$\frac{\partial f}{\partial u} \leq \mu \Rightarrow 1 + h\frac{\partial f}{\partial u} \leq 1 + h\mu \leq e^{h\mu}$$

in the neighborhood of the solution, where μ *can* be a negative number. Insert the expression (3.12b) into the global error formula (3.13) and we get

$$e_k \leq (1 + h\mu)e_{k-1} + \frac{h^2}{2}\frac{d^2u}{dt^2}(\tau_k)$$

With the assumption $h\mu > -1$ and

$$\max_{1\leq k\leq n} |\frac{d^2u}{dt^2}(\tau_k)| \leq M$$

we obtain the inequality

$$|e_k| \leq (1 + h\mu)|e_{k-1}| + \frac{h^2}{2}M$$

which gives the following estimate for the global error ($e_0 = 0$)

$$|e_k| \leq \frac{h^2}{2}M\frac{e^{t_k\mu} - 1}{h\mu} = O(h) \tag{3.14}$$

Hence, the global error for Euler's explicit method is of first order. We see from (3.12b) and (3.14) that the local error is one order larger than the global error. In general, this is true for most FDM. If the local error $l(t_k, h)$ is $O(h^{p+1})$, the global error e_k is $O(h^p)$.

3.3.2 Euler's Explicit Method: Improving the Accuracy

Instead of decreasing the stepsize h in explicit Euler to get better accuracy, the method can be improved by calculating the solution with two stepsizes, h and $h/2$, combined with *extrapolation*. Take a step from t_{k-1} to $t_k = t_{k-1} + h$ in the following way:

1. Take one Eulerstep of size h: $u_k^* = u_{k-1} + hf(t_{k-1}, u_{k-1})$
2. Take two Eulersteps of size $h/2$:

$$u_{k-\frac{1}{2}} = u_{k-1} + \frac{h}{2}f(t_{k-1}, u_{k-1}), \quad u_k^{**} = u_{k-\frac{1}{2}} + \frac{h}{2}f(t_{k-\frac{1}{2}}, u_{k-\frac{1}{2}})$$

3. The value u_k is obtained by extrapolation:

$$u_k = 2u_k^{**} - u_k^*$$

The method described in the steps 1–3 above can be shown to be a second-order method. With a numerical experiment on Van der Pol's equation (3.4), this statement can be verified empirically (Table 3.2):

TABLE 3.2 Numerical Solution of Van der Pol's Equation Using Runge's Second-Order Method

N	h	$u_{1,N}$	$u_{1,N} - y(1)$
4	1/4	0.4991	0.0015
8	1/8	0.4980	0.0004
16	1/16	0.4977	0.0001

As is seen in Table 3.2, the errors decrease quadratically, i.e., $e_N = O(h^2)$. Hence, the convergence of improved Euler is much faster than explicit Euler and therefore more efficient as fewer steps have to be taken to achieve a certain accuracy. The improved Euler method is equivalent to a *Runge–Kutta* method (more about these methods in Section 3.4) known as *Runge's second-order method*:

$$k_1 = f(t_{k-1}, u_{k-1}),\ k_2 = f(t_{k-1} + h/2, u_{k-1} + hk_1/2),\ u_k = u_{k-1} + hk_2$$

Exercise 3.3.5. Show that improved Euler is a second-order method. Hint: use the following error expansion valid for explicit Euler: $e_k = c_1 h + c_2 h^2 + c_3 h^3 + \ldots$ and extrapolation according to 3).

Exercise 3.3.6. Verify that improved Euler and Runge's second-order method are the same one-step method.

Exercise 3.3.7. The following method, Heun's method, is another variant of explicit Euler.

$$u_k^* = u_{k-1} + hf(t_{k-1}, u_{k-1}), \quad u_k = u_{k-1} + \frac{h}{2}(f(t_{k-1}, u_{k-1}) + f(t_k, u_k^*))$$

Verify, by solving Van der Pol's equation numerically (write a program) that this method is of second order.

Exercise 3.3.8. Verify that the local error (3.12b) for explicit Euler is given by

$$l(t_k, h) = \frac{h^2}{2}\frac{d^2u}{dt^2}(t_k) + O(h^3)$$

Hence, the local error is $O(h^2)$, while the global error is $O(h)$. Write a program to make a numerical experiment on the scalar test equation defined in Exercise 3.3.3. Let $\lambda = -1$, $u_0 = 1$, $t_0 = 0$, $t_{end} = 1$, and $h = 0.05$. Compute the global and the local errors (neglect the $O(h^3)$ term) and write a table showing k, t_k, u_k, $l(t_k, h) = h^2 u''(t_k)/2$, and $e_k = u(t_k) - u_k$.

3.3.3 Euler's Explicit Method: Stability

In the previous part of this chapter, it is shown how explicit Euler behaves when the stepsize h is small. But what happens if large steps are taken?

Recall first the example of instability in Section 1.3, where the explicit Euler gives an erroneous result with increasing amplitudes in the numerical solution of the vibration equation, although the analytic solution has decreasing amplitudes. What is the reason for this difference in numerical and analytical behavior?

Example 3.5. The following example from *chemical kinetics*, known as *Robertson's problem*, shows that explicit Euler gives wrong results unless the stepsize is small enough.

The following three irreversible reactions are given:

$$A \to B, \qquad B + C \to A + C \qquad 2B \to B + C$$

The following ODE system can be formulated based on the mass action law:

$$\begin{cases} \dfrac{du_1}{dt} = -k_1 u_1 + k_2 u_2 u_3, & u_1(0) = 1 \\ \dfrac{du_2}{dt} = k_1 u_1 - k_2 u_2 u_3 - k_3 u_2^2, & u_2(0) = 0 \\ \dfrac{du_3}{dt} = k_3 u_2^2, & u_3(0) = 0 \end{cases} \tag{3.15}$$

where $k_1 = 0.04$, $k_2 = 10^4$, $k_3 = 3 \cdot 10^7$ are the rate constants of the reactions and u_1, u_2, and u_3 are scaled concentrations of A, B, and C.

If we use the explicit Euler method on this problem, we get the following numerical result, shown in Table 3.3, for $\mathbf{u}(1)$ using different constant stepsizes h:

TABLE 3.3 Stability Depending on the Stepsize

N	h	$\mathbf{u}(1)$
10	0.1	*NaN*
100	0.01	*NaN*
1000	0.001	*NaN*
10^4	10^{-4}	*ok*

In Table 3.3, NaN (Not a Number) means that the method has generated numbers that are too large to be stored in the computer memory, also called overflow. The result of the calculation is acceptable (but not very accurate) when $h = 10^{-4}$. However, the solution is interesting to study on the time interval $[0, 1000]$. Using the stepsize $h = 10^{-4}$ on the whole time interval means that we have to take 10^7 time steps, which is very inefficient.

The solution of the problem is shown in the graph in Figure 3.5 drawn in both linear and logarithmic scales:

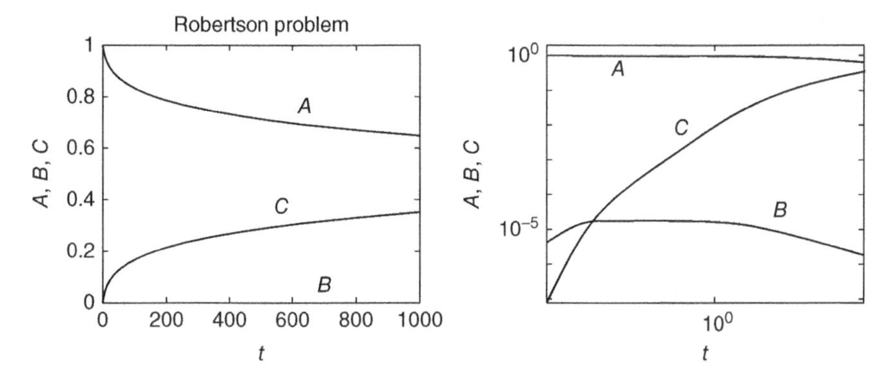

Figure 3.5 Solution of Robertson's problem, in linear and logarithmic diagrams

From the behavior of the solution curves, there seems to be no need for very small stepsizes; the solution is very smooth after a short transient phase in the time interval $(0, 10^{-2})$, where $u_2(t)$ passes a maximum.

Two relevant questions in connection to this problem are

- Why are such small stepsizes needed when explicit Euler is used on this problem?
- Are there other methods better suited to solve this problem?

These questions and their answers are fundamental for understanding the *numerical stability* problems when solving both ODE and PDE problems by discretization.

To investigate the *numerical stability* of methods, some results for LCC systems (see Section 2.2) are needed.

We start with the *scalar model problem* introduced in Exercise 3.3.3. Then we generalize to the LCC problem and finally to a nonlinear system. This is the same increase of complexity as in the analytic stability analysis in Section 2.3.

$$\frac{du}{dt} = \lambda u, \quad \rightarrow \quad \frac{d\mathbf{u}}{dt} = \mathbf{A}\mathbf{u}, \quad \rightarrow \quad \frac{d\mathbf{u}}{dt} = \mathbf{f}(\mathbf{u}) \tag{3.16}$$

Recall that the scalar model problem has the analytic solution $u(t) = e^{\lambda t}u_0$, which is stable for $t \geq 0$ if $Re(\lambda) \leq 0$.

Explicit Euler applied to the scalar model problem gives the following recursion formula (see Exercise 3.3.3):

$$u_k = u_{k-1} + h\lambda u_{k-1} = (1 + h\lambda)u_{k-1} = (1 + h\lambda)^2 u_{k-2} = \cdots = (1 + h\lambda)^k u_0$$

The u_k-sequence is *numerically stable* (bounded) if

$$|1 + h\lambda| \leq 1 \tag{3.17}$$

If λ is real and negative, the inequality can be written:

$$-1 \leq 1 + h\lambda \leq 1 \Rightarrow h \leq \frac{2}{-\lambda} = \frac{2}{|\lambda|} = h_{max} \tag{3.18}$$

If λ is a complex number, the geometrical correspondence to the inequality (3.17) is a disc in the complex $h\lambda$-plane with radius 1 and center in $h\lambda = -1$. We denote this region in the complex $h\lambda$-plane by S_{EE}, the *stability region* for Euler's explicit formula (Figure 3.6).

Hence, Euler's explicit method gives stable solutions to the scalar model problem if

$$h\lambda \in S_{EE} \tag{3.19}$$

It is important at this stage to observe and understand the difference between analytic stability and numerical stability of the model problem $\dot{u} = \lambda u$.

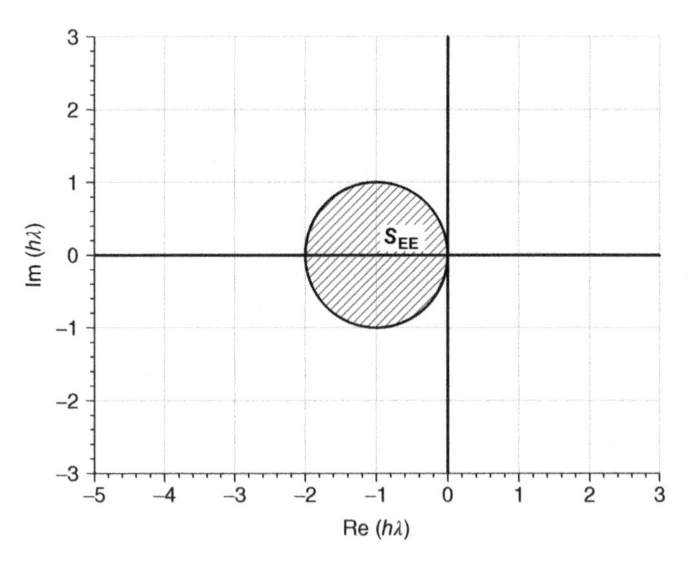

Figure 3.6 Stability region of Euler's explicit method

- Analytic stability applies to the differential equation $\dot{u} = \lambda u$ and the stability criterion is formulated for λ only: $Re(\lambda) \leq 0$.
- Numerical stability applies to the difference equation obtained with the method, i.e., for explicit Euler $u_k = u_{k-1} + h\lambda u_{k-1}$ and the stability criterion is formulated for the product $h\lambda$ as follows: $h\lambda \in S_{EE}$. Observe that if $Re(\lambda) > 0$, the analytic solution of the model problem is unstable and the solution is unbounded as $t \to \infty$. In this case, the explicit Euler method also gives unbounded numerical solutions as $h\lambda$ is situated in the right half plane.

For an LCC system of ODEs

$$\frac{d\mathbf{u}}{dt} = A\mathbf{u}, \quad \mathbf{u}(0) = \mathbf{u}_0$$

we know that the analytic solution is stable for $t \geq 0$ if $Re(\lambda_i) \leq 0$ (assuming there are no multiple eigenvalues with $Re(\lambda_i) = 0$).

Applying the explicit Euler method to this system gives as numerical solution a vector sequence $\mathbf{u}_k$:

$$\mathbf{u}_k = \mathbf{u}_{k-1} + hA\mathbf{u}_{k-1} = (I + hA)\mathbf{u}_{k-1} \tag{3.20}$$

The discussion of numerical stability is easier if we assume that the matrix A is diagonalizable, i.e.,

$$A = S\Lambda S^{-1} \quad \text{or} \quad S^{-1}AS = \Lambda$$

where the columns of S are the eigenvectors of A and Λ is a diagonal matrix with the eigenvalues of A in the diagonal, see Section 2.2.

Define another vector sequence $\mathbf{z}_k$ through the relation $\mathbf{u}_k = S\mathbf{z}_k$. Insert this into (3.20):

$$S\mathbf{z}_k = S\mathbf{z}_{k-1} + hAS\mathbf{z}_{k-1} \tag{3.21}$$

Multiply both sides by S^{-1}

$$\mathbf{z}_k = \mathbf{z}_{k-1} + hS^{-1}AS\mathbf{z}_{k-1} = (I + h\Lambda)\mathbf{z}_{k-1} \tag{3.22}$$

In the sequence (3.22) of vectors $\mathbf{z}_k$, the ith component $z_k^{(i)}$ is *uncoupled* from the other components as $I + h\Lambda$ is diagonal

$$z_k^{(i)} = (1 + h\lambda_i)z_{k-1}^{(i)} \tag{3.23}$$

Hence, the stability criterion for the explicit Euler solution is the same as for the scalar model equation:

$$|1 + h\lambda_i| \leq 1, \qquad i = 1, 2, \ldots, n \tag{3.24}$$

The geometrical interpretation of these inequalities is that *all products* $h\lambda_i$ must be situated in the stability region S_{EE}, i.e.,

$$h\lambda_i \in S_{EE}, \qquad i = 1, 2, \ldots, n \tag{3.25}$$

In the special case that all λ_i are real and $\lambda_n \leq \lambda_{n-1} \leq \cdots \leq \lambda_1 < 0$, we get the stability condition:

$$h \leq \frac{2}{|\lambda_n|} = h_{max} \tag{3.26}$$

i.e., the stepsize must be adjusted to the *fastest* timescale of the system. This may seem contradictory: The fastest component $e^{\lambda_n t}$ decreases very quickly to zero and its value is very soon completely negligible. All the same, this component puts the severest restriction to the stepsize!

For a nonlinear autonomous system of ODEs

$$\frac{d\mathbf{u}}{dt} = \mathbf{f}(\mathbf{u}), \qquad \mathbf{u}(0) = \mathbf{u}_0$$

the numerical stability analysis can be performed in a small neighborhood of a point $(a, \mathbf{b})$ on a solution trajectory $(t, \mathbf{u}(t))$. We want to see how stepsizes can be chosen at this point. To apply the results from the previous LCC problem, approximate the nonlinear ODE system with an LCC system of ODEs. Let $\mathbf{u} = \mathbf{b} + \delta\mathbf{u}$, where $\delta\mathbf{u}(t)$ is a perturbation function. Insert this $\mathbf{u}$ into the ODE:

$$\frac{d\delta\mathbf{u}}{dt} = \mathbf{f}(\mathbf{b} + \delta\mathbf{u}) = \mathbf{f}(\mathbf{b}) + \frac{\partial\mathbf{f}}{\partial\mathbf{u}}(\mathbf{b})\delta\mathbf{u} + hot \tag{3.27}$$

Denote $\mathbf{f}(\mathbf{b})$ by $\mathbf{c}$ and the jacobian evaluated at $\mathbf{b}$ by J. If higher order terms are neglected, the system of ODEs for the perturbation can be written

$$\frac{d\delta\mathbf{u}}{dt} = J\delta\mathbf{u} + \mathbf{c} \tag{3.28}$$

The constant $\mathbf{c}$ does not affect the stability analysis (see Exercise 3.3.10). Therefore, we have the (approximate) result that the explicit Euler method is stable if

$$h\lambda_i(J) \in S_{EE} \tag{3.29}$$

Let us test this result on Robertson's problem introduced in the beginning of this section. The jacobian of the system is

$$J = \begin{pmatrix} -k_1 & k_2u_3 & k_2u_2 \\ k_1 & -k_2u_3 - 2k_3u_2 & -k_2u_2 \\ 0 & 2k_3u_2 & 0 \end{pmatrix}$$

TABLE 3.4 Eigenvalues of the Jacobian of (3.15) Along the Solution Trajectory

t	λ_1	λ_2	λ_3
1	0	−0.16	−1090
10	0	−0.06	−1210
100	0	−0.008	−1950
1000	0	−0.0004	−3670

From the solution trajectory, we choose the solution points $\mathbf{u}(1)$, $\mathbf{u}(10)$, $\mathbf{u}(100)$, and $\mathbf{u}(1000)$ and compute the eigenvalues of the jacobian in these points. The result is shown in Table 3.4.

As all eigenvalues are real and nonpositive, the stability criterion (3.26) applies, so we conclude that a maximal stepsize h_{max} for Euler's explicit method would be $h_{max} = 2/3670 \approx 5 \cdot 10^{-4}$ for this problem in accordance with the numerical experiment in the beginning of this section. Using an adaptive variant of the method will not increase the efficiency as the stepsize must be kept very small on the whole time interval.

Robertson's problem is an example of a *stiff* ODE system. Such a system is characterized by the location in the complex plane of the eigenvalues $\lambda_i, i = 1, 2, \dots, n$ of the jacobian. The system is stiff if

- λ_i are situated in the left half complex plane, i.e., $Re(\lambda_i) \leq 0$.
- $Re(\lambda_i)$ are of very different size, i.e., the system contains time constants of widely varying sizes.

The conclusion is that Euler's explicit method is not suited for stiff ODE problems.

Exercise 3.3.9. Show the instability of explicit Euler in Example 1.1 by verifying that the values $h\lambda_i$ are not inside the stability region of explicit Euler for the stepsize used, $h = 0.1$. Which is the largest stepsize giving decreasing oscillations for this problem?

Exercise 3.3.10. Verify that the constant vector $\mathbf{c}$ in (3.28) does not affect the stability analysis, i.e., the stability results are the same as for the homogeneous ODE system. Hint: Let $\delta\mathbf{v} = \delta\mathbf{u} + J^{-1}\mathbf{c}$ and formulate (3.28) for $\delta\mathbf{v}$.

Exercise 3.3.11. Write a program to compute and plot the stability region S_H for Heun's method presented in Exercise 3.3.7.

3.3.4 Euler's Implicit Method

As Euler's explicit method is very inefficient on stiff problems, the question is if there are other methods that are better suited for these problems.

Apply Euler's implicit method on the scalar model problem:

$$u_k = u_{k-1} + h\lambda u_k \tag{3.30}$$

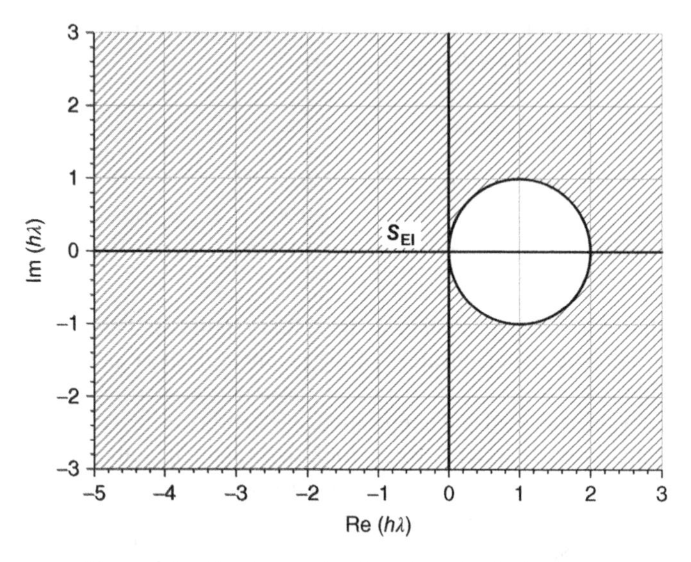

Figure 3.7 Stability region for Euler's implicit method

We solve this equation with respect to u_k and obtain

$$u_k = \frac{1}{1 - h\lambda} u_{k-1} \tag{3.31}$$

The sequence is stable if

$$|1 - h\lambda| \geq 1 \tag{3.32}$$

The stability region of the implicit Euler method is shown in the graph in Figure 3.7.

Hence, the stability region S_{EI} consists of almost the whole complex plane; just the interior of the disc with radius 1 and center in $h\lambda = 1$ is *not* in the stability region. Therefore, *the whole left half plane* belongs to the stability region. This method is appropriate for stiff systems as $h\lambda_i$ belongs to the stability region for all $h > 0$. Such a method is called an *A-stable* method. However, Euler's implicit method is only a first-order method and will therefore be inefficient for *accuracy* reasons: the stepsize h must be small in order to meet a certain error tolerance. In addition, there is another price to pay. For a system of nonlinear ODEs, the following nonlinear algebraic system of equations must be solved with respect to $\mathbf{u}_k$ in *every* time step:

$$\mathbf{u}_k = \mathbf{u}_{k-1} + h\mathbf{f}(t_k, \mathbf{u}_k) \tag{3.33}$$

This can be done with Newton's method (see Section A.1), which, however, consumes much computer time. An efficient method for stiff ODE systems must be based on the following:

- an implicit higher order method,
- variable stepsize technique,
- use of the sparsity (see Section A.5) of the jacobian.

Exercise 3.3.12. Assume that $Re(\lambda) > 0$ in the scalar model problem, i.e., the analytical solution is unstable as $t \to \infty$. Find out if the numerical solution with implicit Euler is unstable or stable for a fixed λ-value as the stepsize h increases from zero to larger values.

3.3.5 The Trapezoidal Method

The trapezoidal method can be regarded as a symmetric combination of Euler's explicit and implicit methods

$$u_k = u_{k-1} + \frac{h}{2}(f(t_{k-1}, u_{k-1}) + f(t_k, u_k)) \tag{3.34}$$

The trapezoidal method is a one-step implicit method and has Second-order accuracy:

$$e_k = u(t_k) - u_k = O(h^2)$$

The stability region S_{TM} of the trapezoidal method is exactly the left half part of the complex plane (Figure 3.8).

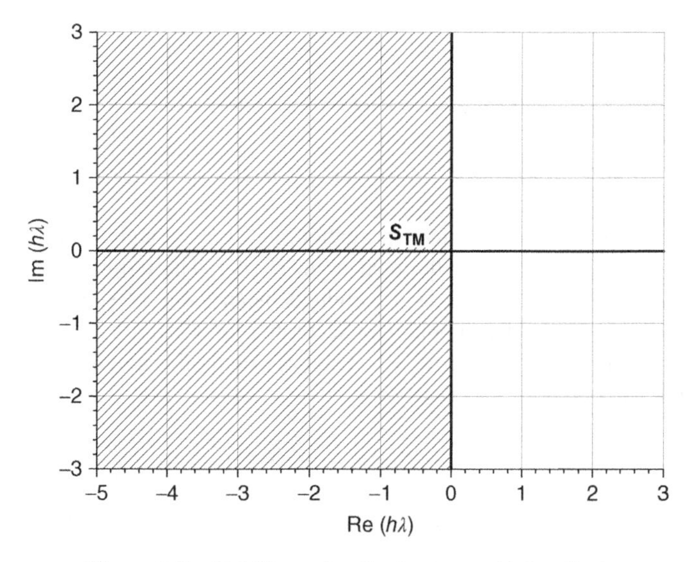

Figure 3.8 Stability region for the trapezoidal method

An advantage with the trapezoidal method is that the region of analytic stability coincides with S_{TM}. However, the solutions of the scalar model equation obtained with this method will contain oscillations that are damped out very slowly when $\lambda << 0$. Therefore, the method is not recommended for very stiff problems, use instead BDF methods (see Section 3.4.2).

Exercise 3.3.13. Write a program that visualizes the numerical solution of the trapezoidal method applied to the scalar model equation for some negative values of λ.

3.4 HIGHER ORDER METHODS FOR THE IVP

3.4.1 Runge–Kutta Methods

The improved Euler method (see Section 3.3.2), also called Runge's second- order method, is an example from the Runge–Kutta family of methods (RK methods). The most well-known member of this family is the *classical* RK4 method:

$$u_k = u_{k-1} + \frac{h}{6}(k_1 + 2k_2 + 2k_3 + k_4) \tag{3.35}$$

$$t_k = t_{k-1} + h, \qquad i = 1, 2, \ldots, N$$

where

$$k_1 = f(t_{k-1}, u_{k-1})$$

$$k_2 = f(t_{k-1} + h/2, u_{k-1} + hk_1/2)$$

$$k_3 = f(t_{k-1} + h/2, u_{k-1} + hk_2/2)$$

$$k_4 = f(t_{k-1} + h, u_{k-1} + hk_3)$$

Like Euler's method, this is a one-step method, i.e., it is enough to know *one* point (t_{k-1}, u_{k-1}) in order to compute the next point (t_k, u_k). The term $(k_1 + 2k_2 + 2k_3 + k_4)/6$ in (3.35) can be regarded as a weighted mean value of slopes evaluated in four different points in order to obtain an accurate representation of the slope of the solution between (t_{k-1}, u_{k-1}) and (t_k, u_k). The RK4 method is a fourth-order method, i.e., the truncation error e_k satisfies

$$e_k = u(t_k) - u_k = O(h^4) \tag{3.36}$$

Carl Runge and Wilhelm Kutta were German mathematicians active in the beginning of the 20th century.

Applying the RK4 method to Van der Pol's equation to compute $y(1)$, just as we did for Euler's method, results in Table 3.5:

TABLE 3.5 Numerical Solution of Van der Pol's Equation Using the RK4 Method

N	h	$u_{1,N}$	$u_{1,N} - y(1)$
1	1	0.5052	0.0076
2	1/2	0.4981	0.0005
4	1/4	0.4976	$<10^{-4}$
8	1/8	0.4976	$<10^{-4}$

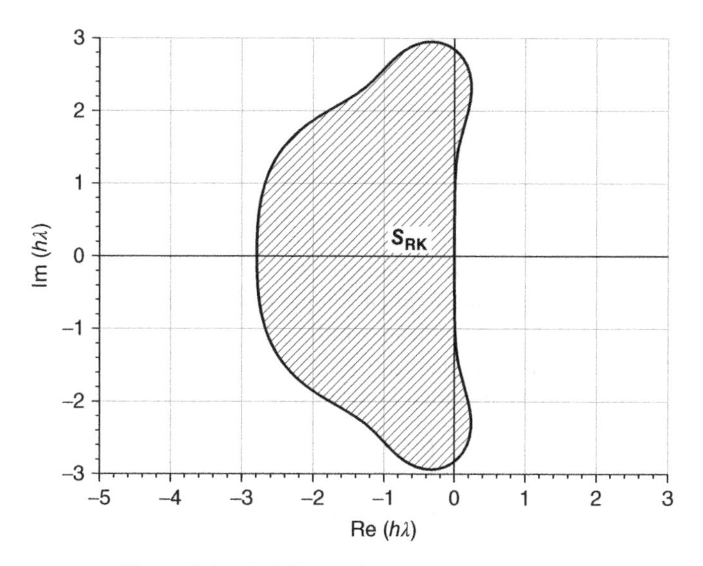

Figure 3.9 Stability region for the RK4 method

Hence, we obtain four digit accuracy for $N = 4$, which is much more efficient than the explicit Euler method.

Just like the explicit Euler method, the RK4 method has a certain stability region, defined by applying the method to the scalar model equation. It is an algebraic exercise to show that the stability region for the RK4 method satisfies the inequality corresponding to the graph in Figure 3.9.

$$|1 + h\lambda + \frac{h^2\lambda^2}{2} + \frac{h^3\lambda^3}{6} + \frac{h^4\lambda^4}{24}| \leq 1 \tag{3.37}$$

As the stability region of RK4 is bounded and the order of accuracy is high, the RK4 method is well suited for *nonstiff* problems.

A more general formula for an explicit RK method is given by the following scheme of an *s-stage* RK method:

$$u_k = u_{k-1} + h(b_1k_1 + b_2k_2 + \cdots + b_sk_s) \tag{3.38}$$

$$t_k = t_{k-1} + h, \qquad k = 1, 2, \dots, N$$

$$k_1 = f(x_{k-1}, u_{k-1})$$

$$k_2 = f(x_{k-1} + c_2 h, u_{k-1} + h a_{21} k_1)$$

....

$$k_s = f(x_{k-1} + c_s h, u_{k-1} + h(a_{s1} k_1 + a_{s2} k_2 + \cdots + a_{ss-1} k_{s-1}))$$

In schematic form, the formula can be described by a matrix tableau, the Butcher tableau, of the form

$$\begin{pmatrix} \mathbf{c} & A \\ & \mathbf{b}^T \end{pmatrix} = \begin{pmatrix} 0 & & & & & \\ c_2 & a_{21} & & & & \\ \cdot & \cdot & \cdot & & & \\ c_s & a_{s1} & a_{s2} & \dots & a_{ss-1} & \\ & b_1 & b_2 & \dots & b_{s-1} & b_s \end{pmatrix} \tag{3.39}$$

The tableau (3.39) represents the formula for an explicit Runge–Kutta method. The matrix A is then lower triangular. Implicit Runge–Kutta methods can also be defined. In that case, the matrix A in the tableau is full.

John Butcher is a numerical analyst from New Zeeland still active.

With this formalism, the modified Euler method can be written as

$$\begin{pmatrix} 0 & & \\ 1/2 & 1/2 & \\ & 0 & 1 \end{pmatrix}$$

and the RK4 method as

$$\begin{pmatrix} 0 & & & & \\ 1/2 & 1/2 & & & \\ 1/2 & 0 & 1/2 & & \\ 1 & 0 & 0 & 1 & \\ & 1/6 & 1/3 & 1/3 & 1/6 \end{pmatrix}$$

Adaptive methods can be constructed from *embedded* RK methods. An embedded RK method uses the same $\mathbf{c}$ and A part of the matrix tableau but consists of two different $\mathbf{b}$ vectors. The first one, denoted by $\mathbf{b}$ (giving a method of order p), is used to compute the next value u_k. The second one denoted $\hat{\mathbf{b}}$ (giving a method of order $p + 1$) computes a $\hat{u}_k$ used just for error estimation:

$$\begin{pmatrix} \mathbf{c} & A \\ & \mathbf{b}^T \\ & \hat{\mathbf{b}}^T \end{pmatrix} = \begin{pmatrix} 0 & & & & & \\ c_2 & a_{21} & & & & \\ \cdot & \cdot & \cdot & & & \\ c_s & a_{s1} & a_{s2} & \dots & a_{ss-1} & \\ & b_1 & b_2 & \dots & b_{s-1} & b_s \\ & \hat{b}_1 & \hat{b}_2 & \dots & \hat{b}_{s-1} & \hat{b}_s \end{pmatrix} \tag{3.40}$$

Assume that the local error (see Section 3.3.1) of the method corresponding to **b** satisfies

$$l(t_k, h) \approx Ch^{p+1} \tag{3.41}$$

Assume also that $\hat{u}_k$ is much more accurate than u_k, so that $\delta = |\hat{u}_k - u_k|$ is an estimate of $l(t_k, h)$ for the step taken from t_{k-1} to t_k.

If *tol* is a given error tolerance value for the absolute error, we can use the following stepsize strategy when taking a step h from t_{k-1} to t_k:

Algorithm 3.1. **(Stepsize control)**

1. *accept* the step (u_k is accepted) if $\delta \leq tol$
   The new step h_{next} should satisfy $tol = Ch_{next}^{p+1}$, hence $h_{next} = h(\frac{tol}{\delta})^{1/(p+1)}$.
2. *reject* the step (u_k is rejected) if $\delta > tol$
   Start again from t_{k-1} with the new stepsize $h_{new} = h/2$, return to 1.

One well-known embedded Runge–Kutta method is the Bogacki–Shampine order 2/3 pair, used in, e.g., ode23 in the MATLAB-library

$$\begin{pmatrix} 0 & & & & \\ 1/2 & 1/2 & & & \\ 3/4 & 0 & 3/4 & & \\ 1 & 2/9 & 3/9 & 4/9 & 0 \\ & 2/9 & 3/9 & 4/9 & 0 \\ & 7/24 & 6/24 & 8/24 & 3/24 \end{pmatrix}$$

Here $\hat{\mathbf{b}}$ corresponds to a second-order method, while **b** is of order three.

Exercise 3.4.1. Formulate the RK method associated with the following tableau

$$\begin{pmatrix} 0 & & & \\ 1 & 1 & & \\ 1/2 & 1/4 & 1/4 & \\ & 1/6 & 1/6 & 4/6 \\ & 1/2 & 1/2 & 0 \end{pmatrix}$$

Apply this method to the test problem $\dot{u} = \lambda u$ and deduce from that result the order of the method giving the solution values and also the order of the method used for error estimation.

Exercise 3.4.2. Show the inequality (3.37) for the stability of the RK4 method.

Exercise 3.4.3. Write a program for plotting the graph of the stability area of the RK4 method.

Exercise 3.4.4. Write a program for simulation of the particle motion problem described in Example 2.3. Use the RK4 method with constant stepsize. The following values of the model parameters can be used: $g = 10\ \mathrm{m/s^2}$, $\|\mathbf{v}_0\| = 10\ \mathrm{m/s}$, $y0 = 2$ m, $c = 0.01\ \mathrm{N\cdot s^2/m}$, $m = 1$ kg, $\alpha = 20°$, and $\alpha = 60°$. Make a plot of the two particle trajectories showing the motion from the initial point to the point where the particle hits the ground.

3.4.2 Linear Multistep Methods

In linear multistep methods, information is used not only from the previous step (t_{k-1}, u_{k-1}) but also from earlier steps (t_{k-2}, u_{k-2}), (t_{k-3}, u_{k-3}), ..., (t_{k-p}, u_{k-p}).

A simple example of a *two-step method* is the *explicit midpoint method*, also known as the leap-frog method based on the central difference approximation (3.5)

$$\begin{cases} u_k = u_{k-2} + 2hf(t_{k-1}, u_{k-1}) \\ t_k = t_{k-1} + h, \quad k = 2, 3, \dots, N \end{cases} \tag{3.42}$$

Inserting $k = 2$ in the formula, we see that the method is not self-starting, i.e., apart from u_0, the initial value, we also need a value of u_1. This could be obtained, e.g., using the explicit Euler method, i.e., $u_1 = u_0 + hf(t_0, u_0)$.

The midpoint method is a second-order method, but unfortunately it is not suited for most kinds of IVPs (see, however, Section 3.5) which the following example shows:

Example 3.6. Consider the explicit midpoint method applied to the scalar model problem. We get the difference equation:

$$u_k = u_{k-2} + 2h\lambda u_{k-1}, \qquad u_0 = 1, \quad u_1 = 1 + h\lambda$$

The results for $h = 0.1, \lambda = -1$ and $\lambda = -2$ are shown in the graphs in Figure 3.10.

The analytic solution is the smooth curve and the numerical solution is the oscillating curve with increasing amplitude. This method is unstable regardless of the stepsize h. It does not help to decrease the stepsize; the oscillations will occur sooner or later. Hence, the explicit midpoint method is not suited for solving IVP in general but can be designed to work for mechanical problems, see Section 3.5. The stability region is derived in Section A.2.

The general definition of a linear p step method is

$$u_k = \alpha_1 u_{k-1} + \alpha_2 u_{k-2} \cdots \alpha_p u_{k-p} + h(\beta_0 f_k + \beta_1 f_{k-1} + \cdots \beta_p f_{k-p}) \tag{3.43}$$

where $f_k = f(t_k, u_k)$. To start such a method, we need apart from u_0 also $u_1, u_2, \dots u_{p-1}$. These values can be calculated with some one-step method, e.g., the RK4 method.

If $\beta_0 = 0$ the method is *explicit*, if $\beta_0 \neq 0$ the method is *implicit*. Some well-known classical multistep methods are, e.g.,

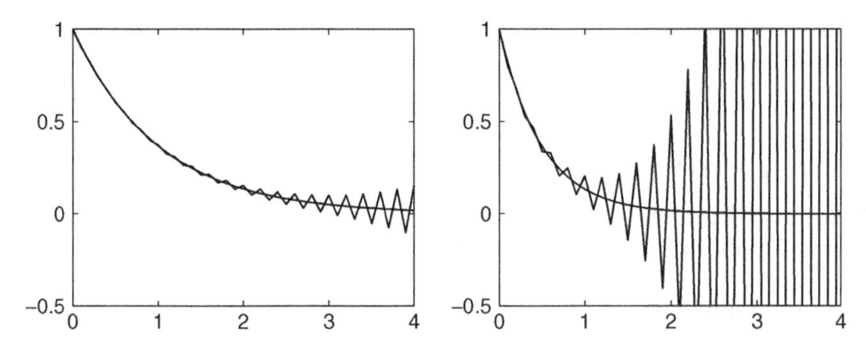

Figure 3.10 Numerical instability for the leap-frog method

- Adams–Bashforth (explicit):

$$u_k = u_{k-1} + h\left(f_{k-1} + \frac{1}{2}\nabla f_{k-1} + \frac{5}{12}\nabla^2 f_{k-1} + \ldots\right) \tag{3.44}$$

- Adams–Moulton (implicit)

$$u_k = u_{k-1} + h\left(f_k + \frac{1}{2}\nabla f_k - \frac{1}{12}\nabla^2 f_k + \ldots\right) \tag{3.45}$$

- Backward Differentiation Formulas (also called BDF-p or Gear's method of order p and well suited for stiff IVPs)

$$\nabla u_k + \frac{1}{2}\nabla^2 u_k + \cdots + \frac{1}{p}\nabla^p u_k = hf_k, \quad p \leq 6 \tag{3.46}$$

where

$$\begin{aligned} \nabla f_k &= f_k - f_{k-1} \\ \nabla^2 f_k &= f_k - 2f_{k-1} + f_{k-2} \\ \nabla^3 f_k &= f_k - 3f_{k-1} + 3f_{k-2} - f_{k-3} \end{aligned} \tag{3.47}$$

J. Adams was a British mathematician active in the middle of the 19th century. F. Bashforth and F. Moulton were scientists active around 1900 and W. Gear is an American numerical analyst who published his method around 1970.

Automatic stepsize control can be designed in a way similar to RK methods. For linear multistep methods, the *local error* $l(t_k, h)$ (see Section 3.3.1) is estimated as follows:

$$l(t_k, h) = Cu^{(p+1)}(t_k)h^{p+1} \tag{3.48}$$

where C is a known constant depending on the method. The derivative can be estimated from some difference approximation, see Appendix A.3. As an example for

Adams–Bashforth method, $p = 1$, (which is in fact explicit Euler) the local error is estimated from

$$l(t_k, h) = \frac{1}{2}\frac{\nabla^2 u_k}{h^2}h^2 = \frac{\nabla^2 u_k}{2} \tag{3.49}$$

and the algorithm for stepsize control now follows Algorithm 3.1.

Exercise 3.4.5. Present in the form (3.43), the first- and second-order methods of Adams–Bashforth, Adams–Moulton, and Gear.

Exercise 3.4.6. Compute the stability region of BDF-2.

Exercise 3.4.7. Compute the solution of the ESCEP problem described in Example 2.6. Use BDF-2 with stepsize $h = 0.01$. (It is of course more effective to use variable stepsize).

3.5 SPECIAL METHODS FOR SPECIAL PROBLEMS

In many physical and chemical problems, it is important in numerical simulations to preserve certain invariants, such as the total energy, the total mass, and/or the positivity of the solution components. Many important classes of models possess invariants that are linear and quadratic expressions in the solution components $u_1, u_2, \ldots u_n$. In numerical calculations, we cannot expect the invariants to be preserved exactly but should be preserved to within rounding errors.

3.5.1 Preserving Linear and Quadratic Invariants

Linear invariants occurring, e.g., in the conservation of mass of a system, are preserved by most methods solving the IVP. For an ODE system, assume that a number of linear relations are time invariant

$$\frac{d\mathbf{u}}{dt} = \mathbf{f}(t, \mathbf{u}),\ \ \mathbf{u}(0) = \mathbf{u}_0, \qquad B\mathbf{u}(t) = B\mathbf{u}_0$$

where B is an $m \times n,\ m < n$ matrix. By differentiating the linear relation with respect to t, we obtain

$$B\frac{d\mathbf{u}}{dt} = B\mathbf{f}(t, \mathbf{u}) = 0$$

When using, e.g., explicit Euler, we get after multiplication by B

$$B\mathbf{u}_k = B\mathbf{u}_{k-1} + hB\mathbf{f}(t_{k-1}, \mathbf{u}_{k-1}) = B\mathbf{u}_{k-1} = \ \ldots\ B\mathbf{u}_1 = B\mathbf{u}_0$$

Hence, the linear relations are preserved along the approximate solution trajectory $\mathbf{u}_1, \mathbf{u}_2, \ldots \mathbf{u}_n$. It is easy to conclude the same property for, e.g., classical Runge–Kutta methods and linear multistep methods. Observe that even if the components of the solution vector $\mathbf{u}$ are computed with a global error of say 10^{-4}, the computed linear invariants will hold an accuracy on rounding error level, typically 10^{-15}.

Example 3.7. In the biochemical system presented in Example 2.6, the matrix S in the right hand side of the ODE system (2.59) has rank 2, hence there are two linear relations between the rows of S, see Exercise 2.5.7.

$$B\mathbf{f}(\mathbf{c}) = \begin{pmatrix} 1 & 0 & 1 & 0 \\ 0 & 1 & 1 & 1 \end{pmatrix} \begin{pmatrix} -1 & 1 & 1 & -1 \\ -1 & 1 & 0 & 0 \\ 1 & -1 & -1 & 1 \\ 0 & 0 & 1 & -1 \end{pmatrix} \mathbf{r}(\mathbf{c}) = 0$$

for all vectors c, hence $B\mathbf{c} = B\mathbf{c}_0$.

In general, quadratic invariants are not preserved, but there is one method, a variant of the trapezoidal method, called the *implicit midpoint method*

$$\mathbf{u}_k = \mathbf{u}_{k-1} + h\mathbf{f}\left(t_{k-1} + \frac{h}{2}, \frac{\mathbf{u}_{k-1} + \mathbf{u}_k}{2}\right) \tag{3.50}$$

that preserves the relation $\mathbf{u}^T C\mathbf{u} = \mathbf{u}_0^T C\mathbf{u}_0$, where C is a symmetric matrix. By time differentiation of the quadratic relation and dividing by 2, we get

$$\mathbf{u}^T C\frac{d\mathbf{u}}{dt} = \mathbf{u}^T C\mathbf{f}(t, \mathbf{u}) = 0$$

Multiply (3.50) first by $\mathbf{u}_{k-1}^T C$, then by $\mathbf{u}_k^T C$, add these two equalities and you find that $\mathbf{u}_k^T C\mathbf{u}_k = \mathbf{u}_{k-1}^T C\mathbf{u}_{k-1}$, hence the quadratic relation is preserved.

Exercise 3.5.1. Given the parameter representation of a circle of radius 1

$$x = cos(\varphi), \qquad y = sin(\varphi), \quad 0 \le \varphi \le 2\pi$$

and the corresponding ODE system where $u_1 = x$ and $u_2 = y$.

$$\frac{d\mathbf{u}}{d\varphi} = \begin{pmatrix} 0 & -1 \\ 1 & 0 \end{pmatrix} \mathbf{u}, \quad \mathbf{u}(0) = \begin{pmatrix} 1 \\ 0 \end{pmatrix}$$

Write a program showing graphs of the phase portraits (u_1, u_2) obtained with the (i) explicit Euler, (ii) implicit Euler, and (iii) implicit midpoint method. Conclude that explicit Euler gives a growing spiral, implicit Euler gives a shrinking spiral, and the implicit midpoint method gives an ellipse. Try also to verify this by mathematical proofs.

3.5.2 Preserving Positivity of the Numerical Solution

In the numerical simulation of physical and chemical systems, there is a demand for positivity for certain solution components. A concentration or a pressure, e.g., can never take negative values. Apart from being unphysical, negativity may also make the ODE system unstable and totally ruin the whole solution. A simple remedy would be, e.g., to set a negative concentration component to zero, which, however, will destroy the mass balances (linear invariants) of the system.

Preserving numerical positivity when solving an ODE system with solution components $u_i(t) > 0$ as long as $u_i(0) \geq 0$ is harder. One way can be to utilize the special form of the right hand side of the ODE system. As an example, take the system (2.49) being a model for chemical kinetics. We can write this system in the alternative form

$$\frac{d\mathbf{c}}{dt} = K(\mathbf{c})\mathbf{c}, \quad \mathbf{c}(0) = \mathbf{c}_0 \tag{3.51}$$

where $K(\mathbf{c})$ is an $N \times N$ matrix. Using implicit Euler on this usually stiff ODE system gives the algebraic equation system

$$\mathbf{c}_k = \mathbf{c}_{k-1} + hK(\mathbf{c}_k)\mathbf{c}_k \rightarrow (I - hK(\mathbf{c}_k))\mathbf{c}_k = \mathbf{c}_{k-1}$$

For the special matrices $K(\mathbf{c})$ occurring in chemical kinetics, it can be shown that if $\mathbf{c}_{k-1} > 0$ (all components of $\mathbf{c}_{k-1}$ are positive), the inverse of $I - hK(\mathbf{c}_{k-1})$ has all elements positive, thus $\mathbf{c}_k > 0$, and positivity will be preserved if the initial value $\mathbf{c}_0$ is nonnegative. Now, if the nonlinear equation system is solved with an iteration method of type

$$(I - hK(\mathbf{c}_k^{(i)}))\mathbf{c}_k^{(i+1)} = \mathbf{c}_{k-1}$$

then positivity is preserved. On the other hand that iteration scheme does not give quadratic convergence. Newton's method (see Appendix A.1) does but will not always produce positive iterates, at least not for chemical kinetics ODEs.

3.5.3 Methods for Newton's Equations of Motion

For conservative mechanical systems, the total energy is preserved. One very important application is *molecular dynamics*, where the physical movements of interacting molecules are simulated by numerically solving Newton's equation of motion (2.43). The Hamiltonian (2.46) for such a system of n particles can be written as the sum of the kinetic energy T and the potential energy U

$$H = T + U = \sum_{i=1}^{n} \frac{\|\mathbf{p}_i\|^2}{2m_i} - \sum_{j=1,i\neq j}^{n} \frac{\gamma m_i m_j}{\|\mathbf{r}_i - \mathbf{r}_j\|} \tag{3.52}$$

Newton's equation will be

$$m_i \ddot{\mathbf{r}}_i = - \sum_{j=1,i\neq j} \frac{\gamma m_i m_j (\mathbf{r}_i - \mathbf{r}_j)}{\|\mathbf{r}_i - \mathbf{r}_j\|^3}, \quad i = 1, 2, \ldots, n$$

Another important application of a mechanical system is a system of masses and springs introduced in (2.44) (with $D_v = 0$ and $\mathbf{F} = 0$) and for which Newton's equations take the form

$$M\ddot{\mathbf{x}} + K\mathbf{x} = 0, \quad \mathbf{x}(0) = \mathbf{x}_0, \quad \dot{\mathbf{x}}(0) = \mathbf{v}_0$$

For this ODE system, the hamiltonian H (quadratic form) is invariant

$$H = \frac{1}{2}\dot{\mathbf{x}}^T M\dot{\mathbf{x}} + \frac{1}{2}\mathbf{x}^T K\mathbf{x} \tag{3.53}$$

For Newton's equation of motion, special numerical methods have been developed and much used, e.g., the leap-frog method and Verlet's method, which are both of second-order accuracy, but do not belong to the Runge–Kutta or linear multistep methods. One reason for using special methods for Newton's equation is that Runge–Kutta and linear multistep methods do not preserve the total energy of a conservative mechanical system but allow the energy to drift away from its constant value as time increases. The Verlet method has its name from the French physicist Loup Verlet, who rediscovered the method for computer simulation of molecular dynamics in the 1960s.

If we write Newton's equation of motion on scalar form as a first-order system

$$\dot{r} = v, \quad r(0) = r_0 \qquad \ddot{r} = a(r), \quad v(0) = v_0 \tag{3.54}$$

where $a(r)$ depends only on the position r, Verlet's method for the position and the velocity is

$$\begin{cases} v_k = (r_{k+1} - r_{k-1})/2h \\ r_{k+1} = 2r_k - r_{k-1} + h^2 a(r_k) \end{cases} \tag{3.55}$$

As is seen from (3.55), the derivative approximations are based on the symmetric central difference formulas (3.8) and (4.24). The method has second-order accuracy. It is also explicit and needs only one evaluation of the right hand side function $a(r)$ per time step. If just the positions $r_0, r_1, \dots\ r_k, \dots$ are wanted only the second of the two formulas are needed. This method also goes under the name Störmer's method, after the Norwegian mathematician Carl Störmer, active in the first part of the 20th century.

Verlet's method is not self-starting meaning that the initial conditions $r_0 = r(0)$ and $v_0 = v(0)$ are not enough to start the recursion. Here r_1 is also needed and can be computed from some approximation of the first step taken, e.g.,

$$r_1 = r_0 + hv_0 + \frac{h^2}{2}a(r_0) \tag{3.56}$$

The order of computation will then be $r_0, v_0, r_1, r_2, v_1, r_3, v_2$, etc.

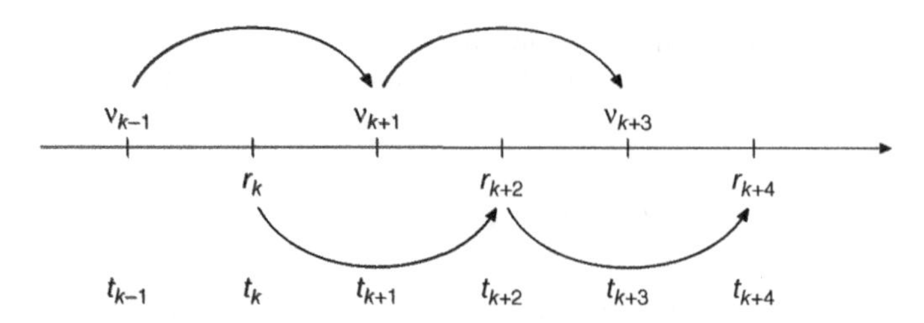

Figure 3.11 Steps taken by different variables on a staggered grid

The leap-frog method is based on the following representation of Newton's equations of motion: $\dot{v} = a(r), \quad v(0) = v_0, \qquad \dot{r} = v, \quad r(0) = r_0$

$$\begin{cases} v_{k+1} = v_{k-1} + 2ha(r_k) \\ r_{k+2} = r_k + 2hv_{k+1} \end{cases} \tag{3.57a}$$

which can also be written with midpoint values

$$\begin{cases} v_{k+1/2} = v_{k-1/2} + ha(r_k) \\ r_{k+1} = r_k + hv_{k+1/2} \end{cases} \tag{3.57b}$$

This method is also based on the symmetric approximation of the derivatives. It has a strong resemblance with the Euler–Cromer method (see Exercise 3.3.4), which, however, is based on the Euler forward approximation giving only first- order accuracy, while the leap-frog method is of second order. In the beginning of Section 3.4.2, the leap-frog method was tested on the scalar model problem $\dot{u} = \lambda u$ and found to be useless when, e.g., $\lambda = -1$. However, for hamiltonian systems in mechanics, it turns out to be both effective and reliable, just like Verlet's method. Here we also need a way to start the method, e.g., with (3.56). The order of computation will be r_0, v_0, r_1, v_2, r_3, v_4, r_5, etc. From this sequence and Figure 3.11, we see that velocities v are evaluated at a grid *staggered* with respect to the positions r. In general, a staggered grid method means that different variables are computed on different grids, see further Section 8.2.7.

3.6 THE VARIATIONAL EQUATION AND PARAMETER FITTING IN IVPs

The variational equation (VE) was introduced in Chapter 2 as a tool for analytic *stability analysis* of solution curves and critical points. The aim was to see if a perturbed curve in the neighborhood of a given solution trajectory will *converge* to the given curve. Hence, analytical stability is investigated by perturbing the *initial values* in the problem (see Section 2.3.1).

Another version of the VE occurs in *sensitivity analysis*, i.e., when we want to compute how sensitive a solution is with respect to *parameters* in the right hand side. Sensitivity analysis is based on the IVP formulation (3.2)

$$\frac{d\mathbf{u}}{dt} = \mathbf{f}(t, \mathbf{u}, \mathbf{p}), \qquad \mathbf{u}(0) = \mathbf{u}_0(\mathbf{p}) \tag{3.58}$$

To investigate the sensitivity, assume that the solution trajectory $\mathbf{u}(t, \mathbf{p})$ has been computed for a given value of the parameter vector $\mathbf{p}$ and the task is to study a neighboring solution trajectory $\mathbf{u}(t, \mathbf{p} + \delta\mathbf{p})$. Taylor expansion gives

$$\mathbf{u}(t, \mathbf{p} + \delta\mathbf{p}) = \mathbf{u}(t, \mathbf{p}) + \frac{\partial \mathbf{u}}{\partial \mathbf{p}}(t, \mathbf{p})\delta\mathbf{p} + hot \tag{3.59}$$

Hence, we need the *sensitivity matrix* $S(t, \mathbf{p})$ defined as

$$S(t, \mathbf{p}) = \frac{\partial \mathbf{u}}{\partial \mathbf{p}} \tag{3.60}$$

$S(t, \mathbf{p})$ is obtained as the solution of the matrix differential equation derived by differentiating (3.58) with respect to $\mathbf{p}$:

$$\frac{dS}{dt} = \frac{\partial \mathbf{f}}{\partial \mathbf{u}} S + \frac{\partial \mathbf{f}}{\partial \mathbf{p}}, \qquad S(0, \mathbf{p}) = \frac{\partial \mathbf{u}_0}{\partial \mathbf{p}} \tag{3.61}$$

The sensitivity matrix can be used for *parameter estimation*. Assume we want to fit the parameter vector $\mathbf{p}$ in (3.58) to measured data $(t_k, \mathbf{u}_k^*)$, $k = 1, 2, \dots, M$ of the state variable $\mathbf{u(t)}$. The residual vector at each measurement point is defined as

$$\mathbf{r}_k = \mathbf{u}(t_k, \mathbf{p}) - \mathbf{u}_k^* \approx 0, \quad k = 1, 2, \dots, M$$

Also define the global sensitivity matrix $\mathbf{S(p)}$, the global measurement vector $\mathbf{u}^*$ and the global residual vector $\mathbf{r(p)}$ according to

$$\mathbf{S(p)} = \begin{pmatrix} S(t_1, \mathbf{p}) \\ S(t_2, \mathbf{p}) \\ \cdot \\ \cdot \\ S(t_M, \mathbf{p}) \end{pmatrix}, \quad \mathbf{u}^* = \begin{pmatrix} \mathbf{u}_1^* \\ \mathbf{u}_2^* \\ \cdot \\ \cdot \\ \mathbf{u}_M^* \end{pmatrix}, \quad \mathbf{r} = \begin{pmatrix} \mathbf{r}_1 \\ \mathbf{r}_2 \\ \cdot \\ \cdot \\ \mathbf{r}_M \end{pmatrix}$$

Gauss–Newton's method can be used (see Section A.1). Assume $\mathbf{p}^{(i)}$ is an iterate in a sequence $\mathbf{p}^{(0)}, \mathbf{p}^{(1)}, \dots, \mathbf{p}^{(i)}$. Given a start parameter vector $\mathbf{p}^{(0)}$, iterate according to

$$\begin{cases} \mathbf{S}(\mathbf{p}^{(i)})\delta\mathbf{p}^{(i)} = -\mathbf{r}^{(i)} \\ \mathbf{p}^{(i+1)} = \mathbf{p}^{(i)} + \delta\mathbf{p}^{(i)} \end{cases}$$

until the correction $\|\delta\mathbf{p}^{(i)}\|$ is small enough.

Example 3.8. Consider the following system of ODEs containing three parameters k_1, k_2, and k_3:

$$\frac{dy_1}{dt} = k_1 y_1 - k_2 y_1 y_2, \quad y_1(0) = 1.0$$

$$\frac{dy_2}{dt} = k_2 y_1 y_2 - k_3 y_2, \quad y_2(0) = 0.3$$

Assume we have measurements of y_1 and y_2 according to the matrix

$$\mathbf{Y}_{msr} = \begin{pmatrix} 1 & 1.1 & 1.3 & 1.1 & 0.9 & 0.7 & 0.5 & 0.6 & 0.7 & 0.8 & 1 \\ 0.3 & 0.35 & 0.4 & 0.5 & 0.5 & 0.4 & 0.3 & 0.25 & 0.25 & 0.3 & 0.35 \end{pmatrix}$$

at time points $t_k = k \cdot 0.5, k = 0, 1, 2, \ldots 10$. We want to estimate the unknown parameters k_1, k_2, and k_3 from the measurements using the algorithm described in Section 3.6. For that we need the following jacobians:

$$J(\mathbf{y}) = \begin{pmatrix} k_1 - k_2 y_2 & -k_2 y_1 \\ k_2 y_2 & k_2 y_1 - k_3 \end{pmatrix}, \quad \frac{\partial \mathbf{f}}{\partial \mathbf{p}} = \begin{pmatrix} y_1 & -y_1 y_2 & 0 \\ 0 & y_1 y_2 & -y_2 \end{pmatrix}$$

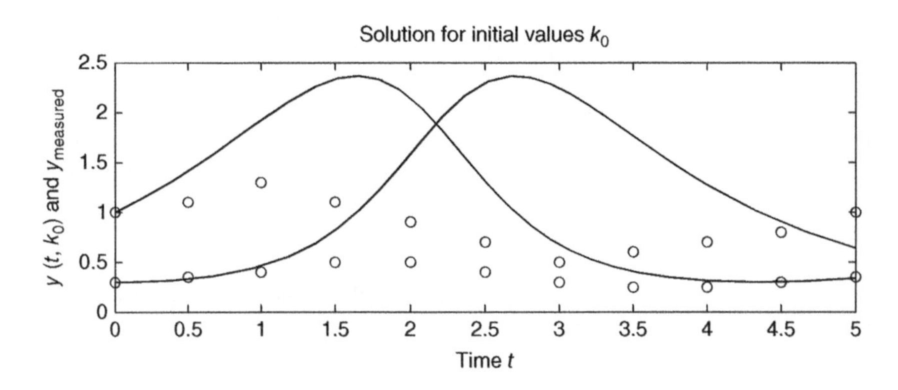

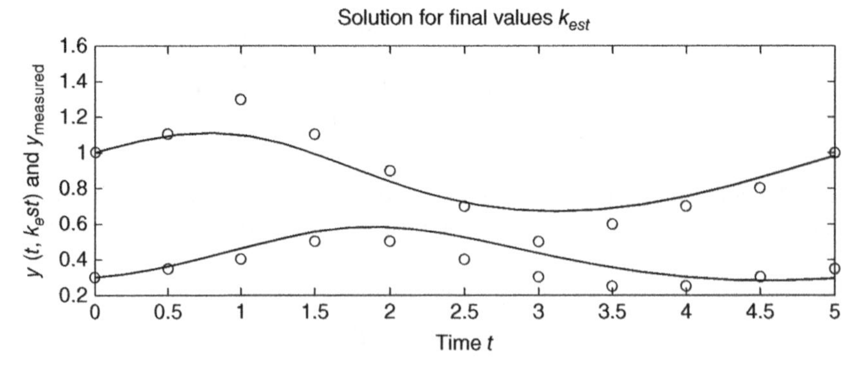

Figure 3.12 Measurements and solution curves before and after the Gauss–Newton method

The initial guess for the parameter vector is $\mathbf{k}_0 = (1, 1, 1)^T$. After a few iterations with the Gauss–Newton method, the iterations have converged (Figure 3.12) to $\mathbf{k}_{est} = (0.86, 2.08, 1.81)$.

BIBLIOGRAPHY

1. G. Dahlquist, Å. Björck, "Numerical Methods", Chapter 8, Dover, 2003
2. G. Golub, J. Ortega, "Scientific Computing and Differential Equations", Chapter 2, Academic Press, 1992
3. C. Moler, "Numerical Computing with MATLAB", Chapter 7, SIAM, 2004

   A textbook with many physical applications

4. A.L. Garcia, "Numerical Methods for Physics", 2nd ed, chapter 2-3, Prentice Hall, 2000

   Extensive treatment of the IVP is found in the textbooks

5. U. Ascher, L. Petzold, "Computer Methods for ODEs and DAEs", SIAM, 1998
6. J.C. Butcher, "Numerical Methods for Ordinary Differential Equations", Wiley, 2003
7. E. Hairer, S. Nörset, G. Wanner, "Solving Ordinary Differential Equations I: Non-stiff problems", Springer 2009
8. E. Hairer, G. Wanner, "Solving Ordinary Differential Equations II: Stiff and Differential-Algebraic Problems", Springer 2010
9. A. Iserles, "A First Course in the Numerical Analysis of Differential Equations", Cambridge University Press, 1996
10. L. Shampine, "Numerical Solution of ODEs", Chapman and Hall, 1994

4

NUMERICAL METHODS FOR BOUNDARY VALUE PROBLEMS

In this chapter, numerical methods for boundary value problems (BVPs) are described. Most of the methods are presented for second-order ordinary differential equations (ODEs) as such problems are very common in applications. However, generalization to higher order ODEs and other types of BVP formulations are also presented in various application problems. The independent variable is changed from t to x, as many BVPs are formulated for 1D space-dependent problems.

A nonlinear second-order ODE

$$\frac{d^2u}{dx^2} = f\left(x, u, \frac{du}{dx}\right) \tag{4.1}$$

has a general solution that depends on two arbitrary constants C_1 and C_2

$$u = u(x, C_1, C_2) \tag{4.2}$$

In general, we cannot find the analytic solution. Hence, there is no explicit expression of the solution in the form (4.2) in which values of the constants C_1 and C_2 can be inserted. To obtain a unique solution, we therefore need two conditions. In Chapter 3, these conditions were given as initial conditions. In this chapter, we show how appropriate *boundary conditions* (BCs) define a solution that can be computed approximately with numerical methods.

Just like an initial value problem (IVP), a BVP usually contains a number of parameters occurring both in the right hand side function $f(x, u, u')$ and/or in the BCs. As a consequence, parameter studies are needed for BVPs, too. Hence, it is necessary to have accurate, efficient, and robust methods.

Introduction to Computation and Modeling for Differential Equations, Second Edition. Lennart Edsberg.
© 2016 John Wiley & Sons, Inc. Published 2016 by John Wiley & Sons, Inc.

In case the ODE is linear, e.g.,

$$\frac{d^2u}{dx^2} + a(x)\frac{du}{dx} + b(x)u = c(x) \tag{4.3}$$

the arbitrary constants C_1 and C_2 enter linearly in the general solution

$$u(x) = u_{part}(x) + C_1\Phi_1(x) + C_2\Phi_2(x)$$

where $u_{part}(x)$ is a particular solution and $\Phi_1(x)$, $\Phi_2(x)$ are two linearly independent solutions of the homogeneous ODE (the ODE with $c(x) \equiv 0$). However, even for linear problems of type (4.3), analytic solutions cannot be found in general.

BCs can be formulated in various ways. As was stated in Chapter 2, a unique solution is determined by specifying *initial conditions* which can be regarded as a special case of BCs. An initial value formulation of (4.1) is obtained by rewriting the second-order ODE as a system of two first-order ODEs (see Section 2.1) and then selecting two initial values $u(x_0)$ and $u'(x_0)$. Let $y_1 = u$ and $y_2 = u'$

$$\frac{dy_1}{dx} = y_2, \qquad y_1(x_0) = u(x_0) \tag{4.4a}$$

$$\frac{dy_2}{dx} = f(x, y_1, y_2), \quad y_2(x_0) = u'(x_0) \tag{4.4b}$$

Frequent forms of BCs to a second-order ODE are named after Peter Dirichlet and Carl Neumann, German mathematicians active in the middle of the 19th century.

1. *Dirichlet's BC* specifies the solution $u(x)$ at an endpoint of an x interval (a, b), say

$$u(a) = u_a \quad \text{and/or} \quad u(b) = u_b \tag{4.5}$$

   where u_a and/or u_b are known values.
2. *Neumann's BC* specifies the derivative $u'(x)$ at an endpoint

$$\frac{du}{dx}(a) = u'_a \quad \text{and/or} \quad \frac{du}{dx}(b) = u'_b \tag{4.6}$$

   where u'_a and/or u'_b are known values.
3. *Robin's BC* or *mixed BC* or *generalized Neumann's BC* specifies a linear combination of $u(x)$ and $u'(x)$ at an endpoint, e.g.,

$$\frac{du}{dx}(a) = \alpha_1 u(a) + \beta_1 \quad \text{and/or} \quad \frac{du}{dx}(b) = \alpha_2 u(b) + \beta_2 \tag{4.7}$$

   where α_1, α_2, β_1, and/or β_2 are known values.

Any combination of two of the BCs (4.5), (4.6), and (4.7) can be given as BCs to (4.1). They can be compactly written in the form

$$\gamma_1 \frac{du}{dx}(a) = \alpha_1 u(a) + \beta_1, \qquad \gamma_2 \frac{du}{dx}(b) = \alpha_2 u(b) + \beta_2 \tag{4.8}$$

where the parameters γ_i, α_i, and β_i, $i = 1, 2, 3$ are chosen appropriately. The BCs (4.8) are *linear*.

Yet another type of BC is a *periodic BC* that is formulated when the solution is to be continued periodically outside the interval (a, b), e.g.,

$$u(a) = u(b), \qquad \frac{du}{dx}(a) = \frac{du}{dx}(b) \tag{4.9}$$

4.1 APPLICATIONS

To illustrate various formulations of BVPs, a number of application problems are presented in the examples below. Some of them are solved numerically later in this chapter. The mathematical modeling background to some examples is presented in Chapter 9.

Example 4.1. **(Stationary flow of a hot fluid in a pipe).**

Consider a pipe of length L (m) with a cylindrical cross section of radius R (m). A hot fluid is transported through the pipe. Exchange of heat with the environment takes place through the wall of the pipe (Figure 4.1). This process can be modeled with the following BVP:

$$-\frac{d}{dz}\left(\kappa \frac{dT}{dz}\right) + \rho C v \frac{dT}{dz} + \frac{2h}{R}(T - T_{out}) = 0, \quad 0 < z < L \tag{4.10a}$$

$$T(0) = T_0, \quad -\kappa \frac{dT}{dz}(L) = k(T(L) - T_{out}) \tag{4.10b}$$

This BVP is a model of the thermal energy balance of the fluid in the pipe. In Chapter 9, Example 9.6, this ODE is derived from physical assumptions, where the three terms in the ODE (4.10a) represent, respectively,

1. heat conduction,
2. heat convection, and
3. heat loss through the wall.

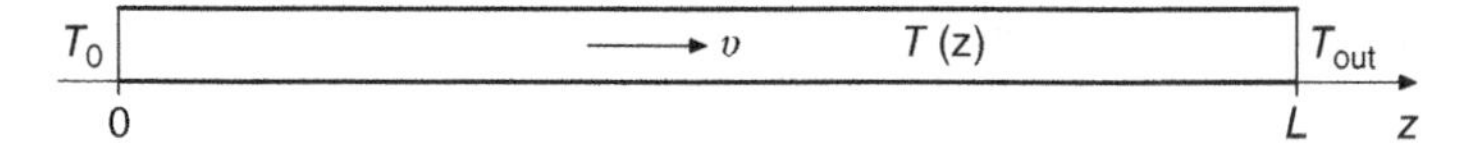

Figure 4.1 Steady heat transport by a fluid through a pipe

There can also be terms modeling heat sources. Such a term is represented in (4.10a) as a source term $f(T)$ in the ODE.

The BCs (4.10b) correspond to constant temperature T_0 at the inlet of the pipe and, at the outlet, heat flux is proportional to the difference between the fluid temperature at the outlet $T(L)$ and the ambient temperature T_{out}.

In (4.10), z (m) is the length coordinate, $T = T(z)$ (K) the temperature along the pipe, κ [J/(K· m· s)] the heat conduction coefficient of the fluid, v (m/s) the flow velocity, ρ (kg/ m^3) the density of the fluid, C [J/(K· kg)] the heat capacity, and h and k [J/(K· m^2· s)] are the heat transfer coefficients between the fluid and the environment. Here, h is associated with heat loss through the wall of the pipe and k with the heat transfer at the right end. T_0 (K) is the inlet temperature, T_{out} (K) is the ambient temperature at the wall and outside the endpoint at $z = L$. If all parameters $L, R, \kappa, \rho, C, v, h, k, T_{out}$, and T_0 in the model (4.10) are constant, the BVP is linear with constant coefficients. Even for this simple problem, there are as many as 10 parameters of which the solution $T(z)$ is dependent. In Chapter 9, it is shown how a scaling procedure will reduce the number of parameters to 3.

A nonlinear formulation of the ODE is obtained if some of the parameters, e.g., κ, are temperature dependent, i.e., $\kappa = \kappa(T)$.

The nonlinearity may also occur in a BC, e.g., as a radiation formulation at the right endpoint

$$-\kappa\frac{dT}{dz}(L) = \sigma(T(L)^4 - T_{out}^4)$$

where σ [J/(K^4· m^2· s)] is the heat transfer coefficient for radiation.

Note that this problem will appear in other variants of the model in Chapters 6, 7, and 9.

Example 4.2. **(Concentration profile in a spherical catalyst particle).**

In a spherical porous catalyst particle with radius R (m), the concentration $c(r)$ (mol/ m^3) of a substance diffusing into the particle where it reacts with the catalyst is modeled by the BVP

$$D\left(\frac{d^2c}{dr^2} + \frac{2}{r}\frac{dc}{dr}\right) = kc^p, \quad 0 < r < R \tag{4.11a}$$

$$\frac{dc}{dr}(0) = 0, \qquad c(R) = c_0 \tag{4.11b}$$

where D (m^2/ s) is the diffusion coefficient, kc^p [mol/(m^3· s)] the rate expression of the chemical reaction, and c_0 (mol/ m^3) the concentration of the substance on the surface of the particle (see Figure 4.2). In the rate expression, k is the rate constant and p, a positive integer, is the order of the reaction. Observe that the ODE is not defined at $r = 0$. The form of the ODE valid at this particular point can be derived with l'Hopital's rule (see Exercise 2.1.6 in Chapter 2).

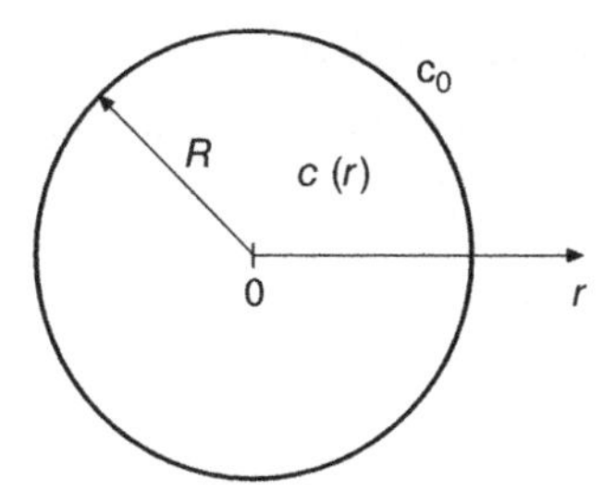

Figure 4.2 Concentration profile in a spherical catalyst particle

Higher order BVPs are also common in applications. A fourth-order ODE describes transversal bending of an elastic beam.

Example 4.3. **(Transversally loaded beam).**
The transversal deformation $u(x)$ (m) of a beam of length L (m) can be modeled by the ODE

$$\frac{d^2}{dx^2}\left(EI(x)\frac{d^2u}{dx^2}\right) = f(x), \quad 0 < x < L \tag{4.12a}$$

where E (N/ m^2] is the elasticity module, $I(x)$ (m^4) the cross section moment of inertia, and $f(x)$ (N/m) the load acting transversally on the beam (Figure 4.3). For this ODE, four BCs are needed to specify a unique solution, e.g.,

$$u(0) = 0, \quad \frac{du}{dx}(0) = 0, \quad \frac{d^2u}{dx^2}(L) = 0, \quad \frac{d^3u}{dx^3}(L) = 0 \tag{4.12b}$$

BVPs can also be formulated as systems of ODEs of first order, e.g.,

$$\begin{aligned} \frac{du}{dx} &= f(x, u, v), \quad u(a) = u_a \\ \frac{dv}{dx} &= g(x, u, v), \quad v(b) = v_b \end{aligned} \tag{4.13}$$

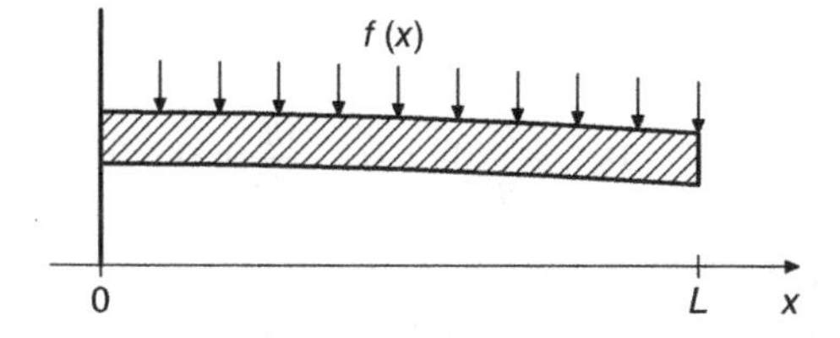

Figure 4.3 Displacement of a loaded beam

Example 4.4. **(Counterflow heat exchanger).**

Consider a heat exchanger consisting of two pipes of length L (m) exchanging energy and with counterflow streams of hot and cold water

$$u_H \frac{dH}{dx} = -a(H - C), \qquad H(0) = H_0, \quad 0 < x < L \tag{4.14a}$$

$$-u_C \frac{dC}{dx} = a(H - C), \qquad C(L) = C_0 \tag{4.14b}$$

where H (K) is the temperature of the hot water, C (K) the temperature of the cold water, u_H and u_C (m/s) the fluid velocities, and a (1/s) a temperature transfer coefficient describing the heat flow between the two pipes (Figure 4.4).

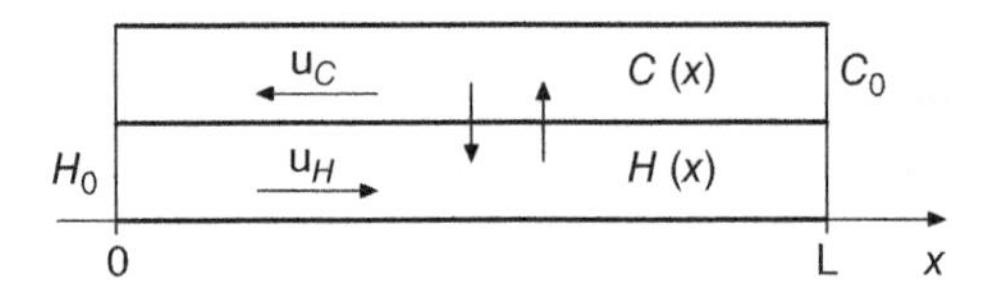

Figure 4.4 Counter flow heat exchange

A general formulation of a two-point BVP is

$$\frac{d\mathbf{u}}{dx} = \mathbf{f}(x, \mathbf{u}), \quad \mathbf{g}(\mathbf{u}(a), \mathbf{u}(b)) = 0 \tag{4.15a}$$

Often in applications $\mathbf{g}$ is linear, in which case the BCs can be written:

$$B_0\mathbf{u}(a) + B_1\mathbf{u}(b) = \mathbf{b} \tag{4.15b}$$

Example 4.5. **(Blasius' boundary layer equation in fluid dynamics).**

The nonlinear third-order ODE

$$2\frac{d^3f}{d\eta^3} + f\frac{d^2f}{d\eta^2} = 0, \quad 0 < \eta < \infty \tag{4.16a}$$

with BCs

$$f(0) = 0, \quad \frac{df}{d\eta}(0) = 0, \quad \frac{df}{d\eta}(\infty) = 1 \tag{4.16b}$$

is a classical problem in fluid dynamics.

From the solution $f(\eta)$, the velocity components $u(x, y)$ (m/s) and $v(x, y)$ (m/s) of the stream close to a plane wall (see Figure 4.5) can be computed:

$$u(x, y) = u_\infty \frac{df}{d\eta}(\eta), \qquad v(x, y) = \sqrt{\frac{\nu u_\infty}{4x}}\left(\eta\frac{df}{d\eta}(\eta) - f(\eta)\right), \qquad \eta = y\sqrt{\frac{u_\infty}{\nu x}}$$

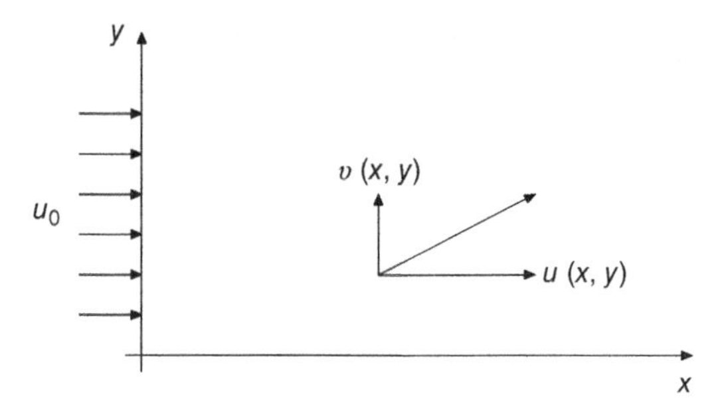

Figure 4.5 Blasius' boundary layer flow

where u_∞ (m/s) is the free stream velocity far from the wall and ν (m^2/s) is the kinematic viscosity. Observe that one of the BCs is given at $\eta = \infty$.

As was pointed out at the end of Section 2.1, uniqueness of a solution to a BVP is more complicated than for an IVP. The following two examples show what can happen.

Example 4.6. (An eigenvalue problem).

The eigenfrequencies of a circular membrane are the frequencies for which the membrane can oscillate without the influence of external forces. The fundamental model for an oscillatory system is the wave equation (see Section 5.1)

$$\frac{\partial^2 u}{\partial t^2} = c^2 \left(\frac{\partial^2 u}{\partial x^2} + \frac{\partial^2 u}{\partial y^2} \right)$$

where u (m) is the deflection of the membrane, t (s) time, and c (m/s) the velocity of the wave.

We look at time harmonic solutions of angular frequency ω,

$$u(x, y, t) = e^{i\omega t} v(x, y)$$

Insert into the wave equation to obtain the Helmholtz equation

$$c^2 \left(\frac{\partial^2 v}{\partial x^2} + \frac{\partial^2 v}{\partial y^2} \right) + \omega^2 v = 0 \tag{4.17}$$

Now let the membrane be fixed along the boundary of the circle. We wish to find values of ω such that the BVP with homogeneous BC $v = 0$ on the boundary has nontrivial solutions. This is an eigenvalue problem for the Laplace equation, see also Problem (5) in Section 5.1. Hermann von Helmholtz was a German scientist active around the end of the 19th century.

For a circular membrane of radius R (m), use polar coordinates (r, φ). Assume there is no angular dependence and we obtain the BVP

$$-\frac{1}{r}\frac{d}{dr}\left(r\frac{dv}{dr} \right) = \lambda v, \quad \frac{dv}{dr}(0) = 0, \;\; v(R) = 0 \tag{4.18}$$

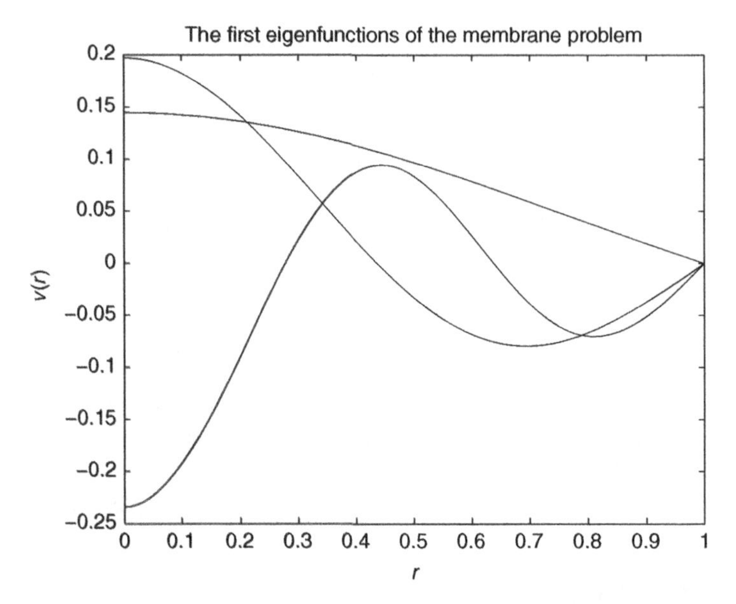

Figure 4.6 Some eigensolutions of the circular membrane problem

where $\lambda = (\omega/c)^2$. This BVP has *infinitely* many solutions, see Figure 4.6 for the eigenfunctions corresponding to the smallest eigenvalues.

***Example 4.7.* (A nonlinear BVP with two solutions).**
Another example of a BVP not having a unique solution is

$$u'' + e^{u+1} = 0, \quad u(0) = u(1) = 0, \quad 0 < x < 1 \tag{4.19}$$

This problem has *exactly two* solutions, plotted in the graph in Figure 4.7.

The two Examples 4.6 and 4.7 show that the question of uniqueness of a solution to a BVP is not as transparent as for an IVP.

4.2 DIFFERENCE METHODS FOR BVPs

The numerical way of treating a BVP is different from the way of solving an IVP. An IVP is solved with a *marching technique* starting from given initial values, then proceeding step by step forward until the final time point is reached.

In a BVP, there are not enough initial values to start the marching technique. Instead, in the discretization of a BVP, a *matching technique* leads to coupled algebraic equations. Hence, a finite difference method (FDM) for numerical solution of a BVP transforms the ODE problem into a system of algebraic equations. The system

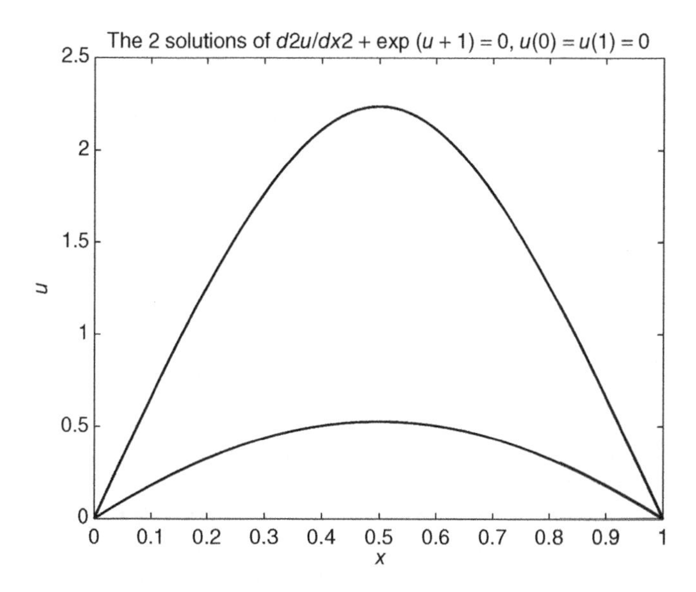

Figure 4.7 Example showing non-uniqueness for a BVP

will be linear or nonlinear depending on whether the BVP is linear or nonlinear. The algebraic equations will generally be sparse and should therefore be solved by sparse direct methods or by iterative methods (see Section A.5).

4.2.1 A Model Problem for BVPs, Dirichlet's BCs

As a model problem for numerical solution of a BVP, we take the elementary linear problem

$$-\frac{d^2u}{dx^2} = f(x), \quad 0 < x < 1 \tag{4.20a}$$

$$u(0) = 0, \qquad u(1) = 0 \tag{4.20b}$$

The general solution of the ODE is

$$u(x) = C_1 + C_2 x - \int_0^x F(s)ds \tag{4.21}$$

where C_1 and C_2 are arbitrary constants and $F(s) = \int_0^s f(t)dt$. Integrating by parts gives

$$\int_0^x F(s)ds = [sF(s)]_0^x - \int_0^x sF'(s)ds = \int_0^x (x-s)f(s)ds$$

Inserting the BCs (4.20b) gives

$$C_1 = 0, \qquad C_2 = \int_0^1 (1-s)f(s)ds$$

and the solution can be written

$$u(x) = x \int_0^1 (1 - s)f(s)ds - \int_0^x (x - s)f(s)ds \tag{4.22a}$$

or

$$u(x) = \int_0^1 G(x, s)f(s)ds$$

where

$$G(x, s) = \begin{cases} s(1 - x), & 0 \leq s \leq x \\ x(1 - s), & x \leq s \leq 1 \end{cases} \tag{4.22b}$$

$G(x, s)$ is called the *Green's function* after the British mathematician George Green active in the beginning of the 19th century (Figure 4.8).

Observe that if $f(x) \geq 0$, $0 \leq x \leq 1$, then the solution $u(x) \geq 0$.

The principle of the *FDM* (see Section 3.2) is to discretize the continuous variable x and to replace the derivatives by difference approximations. The following algorithm transforms the ODE problem into an algebraic system of equations:

DG = Discretize the interval to a Grid,
DD = Discretize the Differential equation,
DB = Discretize the BCs.

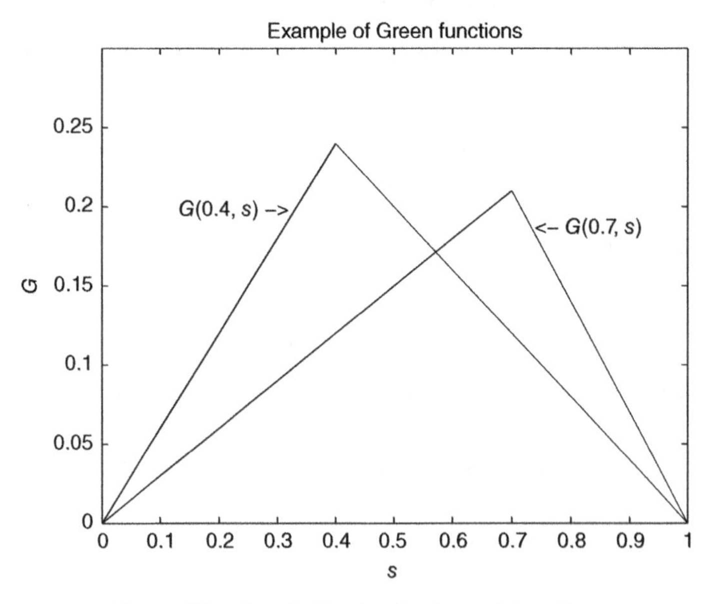

Figure 4.8 Green's function for the model problem

This algorithm is now illustrated on the model problem (4.20).

DG—discretize the x interval and introduce a numbering of the grid points:

$$\begin{array}{ccccccccccc} 0 & & & & h & & & & 1 & \\ \hline x_0 & x_1 & & x_{i-1} & & x_i & & x_N & x_{N+1} & x \end{array}$$

We call this basic grid **G**. If the stepsize h is constant, then the discretization above gives the following relation between h and N, the number of *inner* points in the interval [0, 1]:

$$h(N+1) = 1$$

DD—choose difference approximations for the derivatives. In Chapter 3, we have already introduced a second-order formula for the first derivative, the central difference approximation:

$$\frac{du}{dx}(x_i) = \frac{u(x_{i+1}) - u(x_{i-1})}{2h} + O(h^2) \tag{4.23}$$

For the second derivative, there is a similar second-order formula:

$$\frac{d^2u}{dx^2}(x_i) = \frac{u(x_{i+1}) - 2u(x_i) + u(x_{i-1})}{h^2} + O(h^2) \tag{4.24}$$

Use (4.24) to discretize (4.20a) at an arbitrary inner gridpoint x_i:

$$-\frac{u_{i+1} - 2u_i + u_{i-1}}{h^2} = f(x_i) \tag{4.25}$$

This relation is valid at all the *inner points*, i.e., $i = 1, 2, \dots, N$.

DB—as u is given at the two endpoints also being gridpoints, the Dirichlet BCs (4.20b) can be represented exactly:

$$u_0 = 0, \quad u_{N+1} = 0 \tag{4.26}$$

The BCs (4.26) are inserted into the first and last equation of (4.25). The result of the whole discretization procedure is the following system of N unknowns $u_1, u_2, \dots, u_N$ and N equations:

$$\begin{aligned} -\frac{u_2 - 2u_1}{h^2} &= f(x_1) \\ -\frac{u_{i+1} - 2u_i + u_{i-1}}{h^2} &= f(x_i), \quad i = 2, 3, \dots, N-1 \\ -\frac{-2u_N + u_{N-1}}{h^2} &= f(x_N) \end{aligned} \tag{4.27}$$

The system is written in matrix form:

$$\mathbf{Au} = \mathbf{f} \tag{4.28}$$

where

$$A = \begin{pmatrix} 2 & -1 & 0 & 0 & \dots \\ -1 & 2 & -1 & 0 & \dots \\ & & \dots & & \\ \dots & 0 & -1 & 2 & -1 \\ \dots & 0 & 0 & -1 & 2 \end{pmatrix}, \qquad \mathbf{f} = h^2 \begin{pmatrix} f(x_1) \\ f(x_2) \\ \cdot \\ f(x_{N-1}) \\ f(x_N) \end{pmatrix} \tag{4.29}$$

The $N \times N$ matrix A is *symmetric, positive definite*, and *tridiagonal*. This will not be the case for BVPs in general; the matrix can be both unsymmetric and have another structure than tridiagonal. In most cases, however, the matrix has some *banded* structure. A in (4.29) can also be written $A = tridiag(-1, 2, -1)$.

The matrix A in (4.29) plays an important role for insight into the numerical properties of methods for both ODEs and partial differential equations (PDEs). In particular, the eigenvalues λ_j are important. They can be computed exactly (see Section A.2):

$$\lambda_j = 4 \sin^2 \left(\frac{\pi j}{2(N+1)} \right), \quad j = 1, 2, \dots, N \tag{4.30}$$

We see that all eigenvalues are positive and that

$$\lambda_{\min} \approx \pi^2/(N+1)^2, \qquad \lambda_{\max} \approx 4 \tag{4.31}$$

It can be shown that A^{-1} has only positive elements. Hence, if $f(x) \geq 0$ on the interval $0 \leq x \leq 1$, then the discrete solution $\mathbf{u}$ will be positive just as the analytic solution (4.22a).

The tridiagonality of A can be utilized computationally in two ways:

- Computer storage: if only the nonzero diagonals of A are stored, much less storage is needed than storing a full matrix.
- Computer time: the number of floating point operations (flops) needed to solve $A\mathbf{u} = \mathbf{b}$ is much less for a tridiagonal matrix than for a full matrix.

In fact, if A is treated as a full matrix, N^2 elements must be stored, while if A is stored in a sparse way, only $3N - 2$ elements need to be stored. In addition, if A is treated as a full matrix, Gaussian elimination needs about $N^3/3$ flops to solve $A\mathbf{u} = \mathbf{b}$, while a tridiagonal version needs only about $5N$ flops (see Section A.5).

These observations concerning computing efficiency for 1D problems become more important when solving BVPs in 2D or 3D (see Chapter 7).

As the accuracy of the difference approximation (4.24) is of second order, it should be expected that the accuracy of the numerical solution $u_i = u_i(h)$ also is of second order. The following numerical experiment verifies this statement. For a mathematical proof, see Section 4.2.3.

TABLE 4.1 Numerical Solution of (4.20) Using Central Difference Approximations

N	h	$u_i(h), x = 0.5$	$e(h)$
3	1/4	0.1067	$5.37 \cdot 10^{-3}$
7	1/8	0.1026	$1.31 \cdot 10^{-3}$
15	1/16	0.1017	$3.26 \cdot 10^{-4}$
31	1/32	0.1014	$8.14 \cdot 10^{-5}$
63	1/64	0.1013	$2.03 \cdot 10^{-5}$
127	1/128	0.1013	$5.09 \cdot 10^{-6}$

Example 4.8. The following numerical experiment verifies that the discretization of (4.20) gives second-order approximation to the solution. With $f(x) = \sin(\pi x)$, the analytic solution is $u(x) = \sin(\pi x)/\pi^2$. The following result was obtained for constant stepsizes $h = 1/2, 1/4, 1/8, \ldots$. In Table 4.1, the truncation error $e(h) = u_i(h) - u(x_i)$ at the point $x_i = 0.5$ for the different stepsizes is shown. A loglog graph of $e(x_i)$ as function of h shows that the order of accuracy is two.

Exercise 4.2.1. Calculate the matrix A and the vector $\mathbf{b}$ in the case that the stepsize $h_i = x_{i+1} - x_i$ is not constant. All approximations should have second-order accuracy.

4.2.2 A Model Problem for BVPs, Mixed BCs

What modifications must be made if the BCs in (4.20) are changed? We illustrate by treating the following BVP:

$$-\frac{d^2u}{dx^2} = f(x), \quad 0 < x < 1 \qquad \frac{du}{dx}(0) = 0, -\frac{du}{dx}(1) = u(1) - 1 \tag{4.32}$$

Introduce an equidistant grid with the following numbering of the points:

0 h 1

x_1 x_2 x_{i-1} x_i x_{N-1} x_N x

We call this grid **G0**. Use Euler's method to approximation to the BCs:

$$\frac{u_2 - u_1}{h} = 0, \qquad -\frac{u_N - u_{N-1}}{h} = u_N - 1 \tag{4.33}$$

The disadvantage with this discretization, however, is that the first-order approximations in the BCs will give only first-order accuracy in *all* points of the grid (although the ODE is discretized to second order).

Exercise 4.2.2. Verify first-order accuracy at $x = 0.5$ of the problem given in (4.32) when using the Euler approximations (4.33) in the BCs by writing a program similar to the one in Example 4.8.

To obtain second-order accuracy, there are various possibilities:

G1—modify the grid and the indexing by incorporating two outer points (ghost points) $x_0 = -h$ and $x_{N+1} = 1 + h$. The inner points are now $x_1 = 0, x_2 = h, \dots , x_N = 1$.

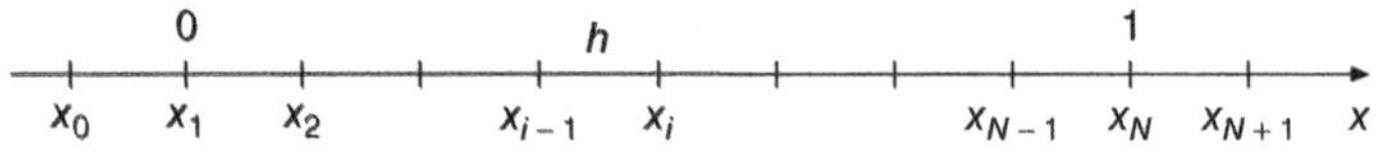

G2—modify the grid so that an endpoint with a derivative condition is located in the middle between two gridpoints. The Euler approximation $u'(0) = (u_1 - u_0)/h$ is a symmetric approximation for this grid and has therefore second-order accuracy.

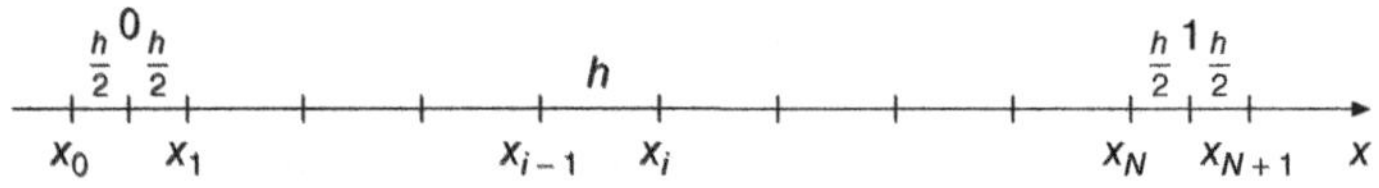

G3—use the original grid **G0** but apply second-order *unsymmetric* difference approximation for the BCs, e.g.,

$$\frac{du}{dx}(0) = \frac{-u(2h) + 4u(h) - 3u(0)}{2h} + O(h^2) \tag{4.34a}$$

and, at the other end of the interval,

$$\frac{du}{dx}(1) = \frac{3u(1) - 4u(1-h) + u(1-2h)}{2h} + O(h^2) \tag{4.34b}$$

With this grid and these difference approximations, no points outside the interval [0, 1] are needed. Hence, we can use the grid **G0**.

We illustrate a second-order method for solving (4.32) using the grid modification **G1** given above. Following the three steps **DG**,**DD**, and **DB** from above:

DG—use the grid **G1**, but observe that now $h(N-1) = 1$.

DD—the ODE is now valid also at the inner points $x_1 = 0$ and $x_N = 1$.

$$-\frac{u_{i+1} - 2u_i + u_{i-1}}{h^2} = f(x_i), \quad i = 1, 2, \dots , N \tag{4.35a}$$

DB—the BC approximations are

$$\frac{u_2 - u_0}{2h} = 0, \qquad -\frac{u_{N+1} - u_{N-1}}{2h} = u_N - 1 \tag{4.35b}$$

Hence, we get $N+2$ equations for the $N+2$ unknowns, $u_0, u_1, \dots, u_{N+1}$ and the matrix A is modified to an $(N+2)\times(N+2)$ matrix. The system of equations is

$$A = \begin{pmatrix} 1 & 0 & \mathbf{-1} & 0 & \dots & \\ -1 & 2 & -1 & 0 & \dots & \\ 0 & -1 & 2 & -1 & \dots & \\ & & & & & \\ \dots & 0 & -1 & 2 & -1 & \\ \dots & 0 & \mathbf{-1} & 2h & 1 & \end{pmatrix} \qquad \mathbf{f} = h^2 \begin{pmatrix} 0 \\ f(x_1) \\ f(x_2) \\ \\ f(x_{N-1}) \\ f(x_N) \\ 0 \end{pmatrix} + 2h \begin{pmatrix} 0 \\ 0 \\ 0 \\ \\ 0 \\ 0 \\ 1 \end{pmatrix}$$

The matrix A corresponds to the difference approximations of u' and u''. The right hand side vector $\mathbf{f}$ consists of one term originating from $f(x)$ and the other term from the BCs.

We see that the matrix is almost tridiagonal; the two boldface $\mathbf{-1}$ elements in the first and last row make the matrix *pentadiagonal*. However, by eliminating u_0 and u_{N+1}, i.e., solving the first and last equation with respect to u_0 and u_{N+1}, respectively,

$$u_0 = u_2 \qquad u_{N+1} = u_{N-1} - 2h(u_N - 1) \tag{4.36}$$

and inserting these expressions into the equations for $i = 1$ and $i = N$

$$\begin{aligned} -\frac{-2u_1 + 2u_2}{h^2} &= f(x_1) \\ -\frac{2u_{N-1} - (2+2h)u_N}{h^2} &= f(x_N) + \frac{2}{h} \end{aligned} \tag{4.37}$$

gives a tridiagonal linear system of N equations for the N unknowns $(u_1, \dots, u_N)$ with the tridiagonal $N \times N$ matrix:

$$A = \begin{pmatrix} 2 & -2 & 0 & \dots & & \\ -1 & 2 & -1 & \dots & & \\ & & & & & \\ \dots & 0 & -1 & 2 & -1 \\ \dots & 0 & 0 & -2 & 2+2h \end{pmatrix} \tag{4.38}$$

The reduction to tridiagonal form, however, is not necessary from computational point of view. On the contrary, it is easier to put the BCs (4.35b) as the first and last equations in $A\mathbf{u} = \mathbf{b}$ and the equations (4.35a) in between.

Exercise 4.2.3. Write programs for solution of (4.32) using the two grids **G2** and **G3** above. Make the discretization, set up the equations, formulate on matrix form, and solve the linear system of equations giving the approximate solution. Verify second-order accuracy.

Exercise 4.2.4. Use Taylor expansion to show that the difference formulas in (4.24) and (4.34) are of second order.

Exercise 4.2.5. If the grid **G2** is used, how do we compute $u(0)$ and $u(1)$ with formulas giving second-order accuracy from the computed values $u_0, u_1, \dots, u_{N+1}$?

4.2.3 Accuracy

In order to discuss the truncation errors in the numerical solution of (4.20), we define the *local discretization error*, the residual obtained when the exact solution is inserted into the discretization (4.25)

$$l(x,h) = -\frac{u(x+h) - 2u(x) + u(x-h)}{h^2} - f(x) \tag{4.39}$$

Using Taylor expansion, it is easy to show that

$$l(x,h) = \frac{1}{12}u^{(4)}(x+\theta h)h^2, \qquad 0 < \theta < 1, \tag{4.40}$$

i.e., the local error is of second order.

Define the *global error* at the point x_i as

$$e(x_i, h) = u(x_i) - u_i \tag{4.41}$$

where $u(x)$ is the analytic solution of (4.20) and u_i is the numerical solution obtained from (4.28).

At the point $x = x_i$ the following two relations are valid:

$$\begin{aligned} l(x_i,h) &= -\frac{u(x_i+h) - 2u(x_i) + u(x_i-h)}{h^2} - f(x_i) \\ 0 &= -\frac{u_{i+1} - 2u_i + u_{i-1}}{h^2} - f(x_i) \end{aligned} \tag{4.42}$$

Subtract the second equality from the first:

$$l(x_i,h) = -\frac{e(x_{i+1},h) - 2e(x_i,h) + e(x_{i-1},h)}{h^2} \tag{4.43}$$

As there is no error in the BCs, we also have

$$e(x_0,h) = e(x_{N+1},h) = 0 \tag{4.44}$$

Hence, the global error satisfies the following linear system of equations:

$$A\mathbf{e} = h^2\mathbf{l} \rightarrow \mathbf{e} = h^2A^{-1}\mathbf{l} \tag{4.45}$$

where **l** and **e** are the vectors of local and global errors at the discretization points, respectively. Taking the norm of both sides, we get

$$\|\mathbf{e}\| \le h^2 \|A^{-1}\| \, \|\mathbf{l}\| \tag{4.46}$$

where the norm is chosen as the max-norm, i.e., $\|.\| = \|.\|_\infty$. The matrix A^{-1} can be computed and its max-norm shown to be

$$\|A^{-1}\|_\infty \le \frac{N^2}{4} \tag{4.47}$$

For the global error, we thus obtain second-order accuracy

$$\max_i |e(x_i, h)| = \|\mathbf{e}\|_\infty \le \frac{1}{(N+1)^2} \frac{N^2}{4} \|\mathbf{l}\|_\infty < \frac{h^2}{48} \max_{0<x<1} |u^{(4)}(x)| = O(h^2)$$

4.2.4 Spurious Solutions

In the previous section in this chapter, it was shown how the FDM behaves when the stepsize h is small. But what happens if large steps are taken?

For the discussion of this question, we change the model problem (4.20) to contain a first derivative term. This model equation is known as the *advection–diffusion equation* and is a slight modification of the model problem (with $f(x) = 0$)

$$-\epsilon \frac{d^2u}{dx^2} + \frac{du}{dx} = 0, \quad 0 < x < 1, \quad u(0) = 0, u(1) = 1 \tag{4.48}$$

The first term in (4.48) corresponds to diffusion transport, while the second term describes convection transport (see also Chapter 9). The parameter ϵ is a positive constant. The general solution of this LCC problem is

$$u(x) = C_1 + C_2 e^{x/\epsilon}$$

Imposing BCs gives

$$u(x) = \frac{e^{x/\epsilon} - 1}{e^{1/\epsilon} - 1} \tag{4.49}$$

If $\epsilon >> 1$, $u(x) \approx x$, which is the straight line between (0,0) and (1,1).

If $0 < \epsilon << 1$, however, $u(x) \approx e^{-(1-x)/\epsilon}$, which is almost equal to zero everywhere except in the neighborhood of $x = 1$ where $u(1) = 1$. Hence, the solution $u(x)$ almost makes a jump at $x = 1$. We say that the problem has a *boundary layer* at $x = 1$ and the problem is an example of a *singular perturbation problem.*

For a problem having a solution with this almost discontinuous behavior in a very small part on the interval [0, 1], we might expect numerical difficulties unless the stepsize is appropriately adjusted. Discretize the ODE using the FDM with second-order difference approximations on the grid **G**

$$-\epsilon \frac{u_{i+1} - 2u_i + u_{i-1}}{h^2} + \frac{u_{i+1} - u_{i-1}}{2h} = 0, \quad i = 1, 2, \ldots N \tag{4.50}$$

This difference equation with the BCs $u_0 = 0$, $u_{N+1} = 1$ can be solved according to the method in Section A.2 and the solution is

$$u_i = \frac{1 - \left(\frac{1+Pe}{1-Pe}\right)^i}{1 - \left(\frac{1+Pe}{1-Pe}\right)^{N+1}}, \quad i = 0, 1, \dots, N+1 \tag{4.51}$$

where the numerical Peclet number Pe (Jean Peclet was a French physicist active in the first half of the 19th century) is defined as

$$Pe = \frac{h}{2\epsilon} \tag{4.52}$$

We note that if $Pe > 1$, the solution u_i is oscillatory, which is not the case in the analytical solution (Figure 4.9). The amplitude of the spurious oscillations increases with the stepsize h. This numerical phenomenon might be regarded as a kind of instability, but, unlike the IVP, the amplitude cannot increase unboundedly, as the largest step h that can be taken is the size of the whole interval, which is finite in a numerical computation.

If $Pe < 1$, there will be no oscillations (see Figure 4.9). However, if ϵ is very small, the stepsize h must also be very small to meet the condition $Pe < 1$. On the other hand, for a very small ϵ, the analytic solution is almost zero in the whole interval $[0, 1]$ except in the boundary layer where the solution steeply goes to one. This numerical phenomenon resembles stiffness introduced for IVPs in Chapter 3. For stiff IVPs, implicit methods can be used to avoid very small stepsizes. So, what is the remedy here?

Approximate the first derivative with a backward difference. This leads to the FDM

$$-\epsilon \frac{u_{i+1} - 2u_i + u_{i-1}}{h^2} + \frac{u_i - u_{i-1}}{h} = 0, \quad i = 1, 2, \dots, N \tag{4.53}$$

This approximation of the first derivative is also called the *upwind difference* and is of great importance in Chapter 8 on hyperbolic PDEs. This method postpones the oscillations but has the disadvantage of giving only first-order accuracy. What this FDM actually does is to increase the ϵ value in (4.48). As the upwind difference can be written

$$\frac{u_i - u_{i-1}}{h} = \frac{u_{i+1} - u_{i-1}}{2h} - \frac{h}{2} \frac{u_{i+1} - 2u_i + u_{i-1}}{h^2} \tag{4.54}$$

we obtain the following FDM equivalent to (4.53)

$$-\epsilon_h \frac{u_{i+1} - 2u_i + u_{i-1}}{h^2} + \frac{u_{i+1} - u_{i-1}}{2h} = 0, \quad i = 1, 2, \dots N \tag{4.55}$$

where $\epsilon_h = \epsilon(1 + Pe)$. Hence, in the original ODE (4.48), the diffusion parameter ϵ has been increased: we have added a term called *artificial diffusion* (or *artificial*

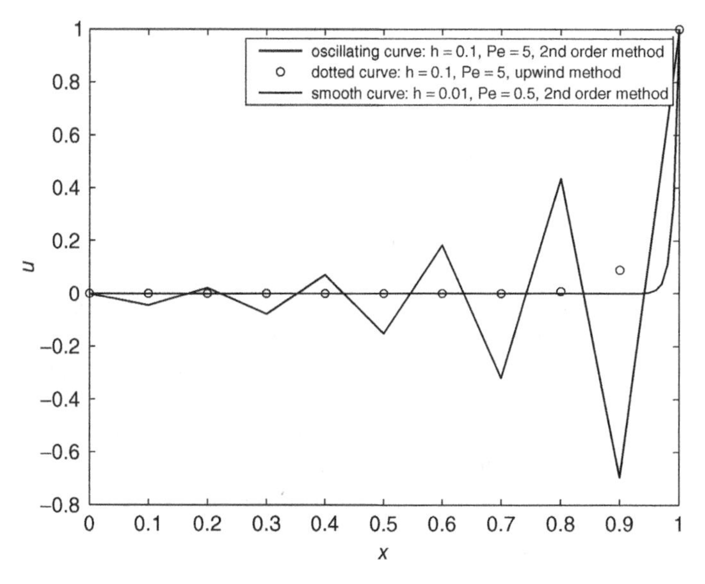

Figure 4.9 Spurious oscillations in the advection-diffusion Equation

viscosity) to the problem and the discretization (4.55) corresponds to the perturbed BVP

$$-\epsilon_h \frac{d^2u}{dx^2} + \frac{du}{dx} = 0, \quad 0 < x < 1, \quad u(0) = 0, u(1) = 1 \tag{4.56}$$

4.2.5 Linear Two-Point BVPs

The BVP (4.3), now formulated in a way suited for many applications is

$$-\frac{d}{dx}(\kappa(x)\frac{du}{dx}) + p(x)\frac{du}{dx} + q(x)u = f(x), \quad a < x < b \tag{4.57a}$$

with BCs

$$\kappa(a)\frac{du}{dx}(a) = \alpha_1 u(a) + \beta_1 \tag{4.57b}$$

$$\kappa(b)\frac{du}{dx}(b) = \alpha_2 u(b) + \beta_2 \tag{4.57c}$$

represent a linear BVP with linear Neumann or mixed BCs, depending on the values of α_i, and β_i, $i = 1, 2$. A Dirichlet condition at, e.g., the left boundary is implemented by setting $u(a) = u_a$ instead of (4.57b). With the grid **G1** and the central difference approximations of the derivatives, we obtain

$$\kappa(x_1)\frac{u_2 - u_0}{2h} - \alpha_1 u_1 = \beta_1 \tag{4.58a}$$

$$-\frac{1}{h}\left(\kappa(x_{i+1/2})\frac{u_{i+1}-u_i}{h} - \kappa(x_{i-1/2})\frac{u_i-u_{i-1}}{h}\right)$$

$$+p(x_i)\frac{u_{i+1}-u_{i-1}}{2h} + q(x_i)u_i = f(x_i), i = 1, \ldots, N \tag{4.58b}$$

$$\kappa(x_N)\frac{u_{N+1}-u_{N-1}}{2h} - \alpha_2 u_N = \beta_2 \tag{4.58c}$$

where x_0, x_{N+1} are ghostpoints, $x_1 = a, x_N = b$, and $h(N-1) = b - a$. This is a linear algebraic system $A\mathbf{u} = \mathbf{b}$ with a pentadiagonal matrix.

For a Dirichlet BC at, say, the left boundary point, i.e., $u(a) = u_a$, equation (4.58a) takes the form $u_1 = u_a$ and no ghost point x_0 is needed.

The system can be reduced to a tridiagonal system if u_0 and u_{N+1} are eliminated, but this is not necessary, as the reduction to a tridiagonal system is not easier from a programming point of view. The matrix A and the right hand side $\mathbf{b}$ is easier to set up if we start from (4.58). The generation of the linear system of these equations can be easily automatized if we skip the elimination of ghost points.

Example 4.9. **(Interface problem in heat conduction)**

In Example 4.1, we have assumed that the heat conduction coefficient κ is constant for the fluid. However, for heat conduction in a long metal rod composed of two different materials, the heat conductivity will be different in the two regions $(0, \xi)$ and (ξ, L), where ξ is the interface point. The following mathematical model consisting of two coupled BVPs can be formulated assuming there are no convection terms

$$-\frac{d}{dz}\left(\kappa_1\frac{dT_1}{dz}\right) + \frac{2h}{R}(T_1 - T_{out}) = 0, \quad 0 < z < \xi \tag{4.59a}$$

with the BC at the left endpoint $T_1(0) = T_0$

$$-\frac{d}{dz}\left(\kappa_2\frac{dT_2}{dz}\right) + \frac{2h}{R}(T_2 - T_{out}) = 0, \quad \xi < z < L \tag{4.59b}$$

with the BC at the right endpoint $-\kappa\frac{dT_2}{dz}(L) = k(T_2(L) - T_{out})$. At the interface point, the temperature and the heat flux are continuous, i.e.,

$$T_1(\xi) = T_2(\xi) \qquad -\kappa_1\frac{dT_1}{dz}(\xi) = -\kappa_2\frac{dT_2}{dz}(\xi) \tag{4.59c}$$

Hence in the interface point, we have a discontinuity in the first derivative and the second derivative does not exist at that point. The formulation of this problem with the finite element method (FEM) is found in Example 4.12.

4.2.6 Nonlinear Two-Point BVPs

The problem (4.1), i.e., a nonlinear second-order ODE

$$\frac{d^2u}{dx^2} = f\left(x, u, \frac{du}{dx}\right) \quad a < x < b \tag{4.60a}$$

with nonlinear BCs:

$$g_1\left(a, u(a), \frac{du}{dx}(a)\right) = 0 \tag{4.60b}$$

$$g_2\left(b, u(b), \frac{du}{dx}(b)\right) = 0 \tag{4.60c}$$

gives a nonlinear BVP.

With the grid **G1**, the following equations are obtained after discretization:

$$g_1\left(a, u_1, \frac{u_2 - u_0}{2h}\right) = 0$$

$$\frac{u_{i+1} - 2u_i + u_{i-1}}{h^2} - f\left(x_i, u_i, \frac{u_{i+1} - u_{i-1}}{2h}\right) = 0, \quad i = 1, \dots, N$$

$$g_2\left(b, u_N, \frac{u_{N+1} - u_{N-1}}{2h}\right) = 0 \tag{4.61}$$

This can be written as a nonlinear system of algebraic equations

$$\mathbf{F}(\mathbf{u}) = 0$$

where

$$\mathbf{F}(\mathbf{u}) = \begin{pmatrix} g_1(a, u_0, u_1, u_2) \\ \cdots \\ \dfrac{u_{i+1} - 2u_i + u_{i-1}}{h^2} - f(x_i, u_{i-1}, u_i, u_{i+1}) \\ \cdots \\ g_2(b, u_{N-1}, u_N, u_{N+1}) \end{pmatrix} \quad \mathbf{u} = \begin{pmatrix} u_0 \\ u_1 \\ \vdots \\ u_N \\ u_{N+1} \end{pmatrix} \tag{4.62}$$

This system can be solved with Newton's method, see Section A.1. The jacobian is almost tridiagonal, hence in each Newton-iteration a banded system is to be solved.

Exercise 4.2.6. Discretize the BVP in Example 4.1 with the FDM. Use the following values of the parameters in the problem: $L = 10$, $\kappa = 0.5$, $h = 5$, $k = 10$, $\rho = 1$, $C = 1$, $R = 1$, $T_0 = 400$, and $T_L = 300$

a) Let $v = 0$ (no convection term, only conduction). Plot the discretized solution T_i approximating $T(z_i)$ using the grid size $N = 10, 20, 40, 80$ in the same graph. Note the convergence of the solution curves.

b) Vary ν in a parameter study using the grid size $N = 40$. Choose the values $\nu = 0.1, 0.5, 1, 10$ and plot the solutions in the same graph. Note the spurious oscillations when $\nu = 10$! Investigate if the oscillations decrease if the grid size N is increased.

Exercise 4.2.7. Discretize the BVP in Example 4.2 with the FDM. Use the parameter values $D = 1$, $k = 1$, $R = 1$, and $c_0 = 1$.

a) First solve the problem for $p = 1$ giving a linear BVP. Write a program solving the linear system of equations and choose in the discretization suitable grid size N. Plot the numerical solution for different values of N showing the convergence.
b) Change the reaction order parameter to $p = 2$. The BVP is now nonlinear. Formulate the nonlinear algebraic system of equation (4.62) for this problem and find the jacobian of the system. Write a program implementing Newton's method for this problem, see Section A.1.

Exercise 4.2.8. The fourth-order ODE in Example 4.3 with modified BCs can be written as a system of two second-order ODEs:

$$\frac{d^2M(x)}{dx^2} = f(x), \quad M(L) = 0$$

$$\frac{d^2u}{dx^2} = \frac{M(x)}{EI(x)}, \quad u(0) = 0, \quad u'(0) = 0, \quad u(L) = 0$$

The function $M(x)$ is the moment acting on the beam. Choose simple values for the parameters: $L = 1$, $E = 1$, $I = 1$, and $f = 1$.

a) Using the FDM, formulate the two algebraic systems of equations corresponding to the two BVPs. They are coupled through the discretized moment vector $\mathbf{M}$.
b) Write a program that solves the systems of equations in a) and plots the numerical solution u_i in a graph. Choose suitable values of the grid parameter N.

4.2.7 The Shooting Method

In the shooting method, the BVP is treated as an IVP in an iterative way. We illustrate the shooting method on the nonlinear formulation (4.60a)

$$\frac{d^2u}{dx^2} = f\left(x, u, \frac{du}{dx}\right), \quad a < x < b$$

with Dirichlet's BCs

$$u(a) = \alpha, \quad u(b) = \beta \tag{4.63}$$

We write this BVP as an IVP

$$\frac{dy_1}{dx} = y_2, \quad y_1(a) = u(a) \tag{4.64a}$$

$$\frac{dy_2}{dx} = f(x, y_1, y_2), \quad y_2(a) = \frac{du}{dx}(a) \tag{4.64b}$$

In the formulation above, $u(a)$ is known $u(a) = \alpha$, but $u'(a)$ is unknown. If we introduce the unknown slope at $x = a$ as a parameter s, the shooting parameter, we can illustrate geometrically in Figure 4.10 the meaning of changing s:

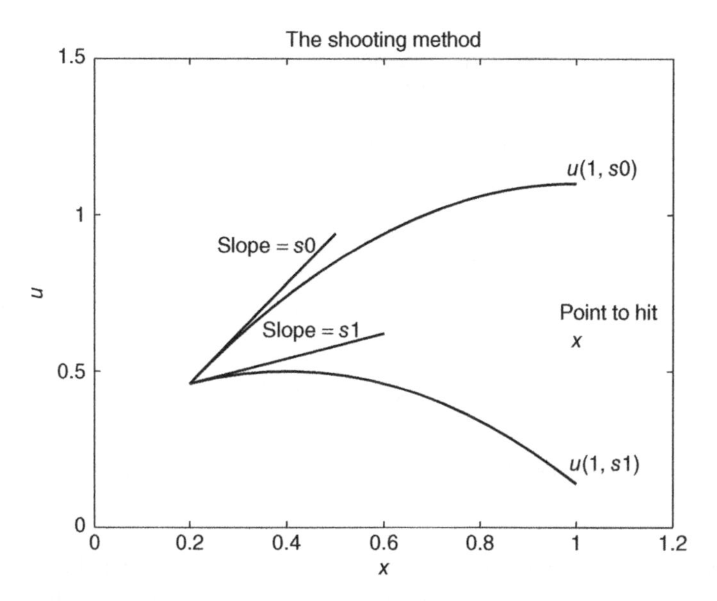

Figure 4.10 Graph showing the idea behind the shooting method

The computational task is to device a systematic method that computes s so that the BC at the right endpoint is met. When s is varied, the value of $u(b)$ at the right endpoint $x = b$ is changed, i.e., $u(b)$ is a function of s: $u(b) = u(b, s)$. We want to find the value of s for which $u(b, s) = \beta$, i.e., we have to solve the following equation with respect to s:

$$u(b, s) - \beta = 0 \tag{4.65}$$

This is a nonlinear scalar equation in s. It has to be solved numerically as the analytical solution $u(x, s)$ is not known. This excludes Newton's method as in general we cannot form $\partial u/\partial s$ needed in Newton's method.

The method to be used here is instead the *secant method*. The secant method applied to the general problem $F(z) = 0$ is

$$z_{i+1} = z_i - \frac{F(z_i)(z_i - z_{i-1})}{F(z_i) - F(z_{i-1})}, \quad i = 1, 2, \ldots \tag{4.66}$$

where z_0 and z_1 are given initial guesses. Hence, by applying the secant method to the BVP (4.64) with $F(s) = u(b, s) - \beta$ we have the shooting method.

Example 4.10. **(Blasius' boundary layer equation in fluid dynamics).**
The nonlinear third-order ODE in Example 4.5

$$2\frac{d^3f}{d\eta^3} + f\frac{d^2f}{d\eta^2} = 0, \quad 0 < \eta < \infty$$

with BCs

$$f(0) = 0, \quad \frac{df}{d\eta}(0) = 0, \quad \frac{df}{d\eta}(\infty) = 1$$

is formulated as a first-order system of ODEs. Let $u_1 = f, u_2 = F', u_3 = F''$

$$\frac{du_1}{d\eta} = u_2, \quad u_1(0) = 0$$
$$\frac{du_2}{d\eta} = u_3, \quad u_2(0) = 0$$
$$\frac{du_3}{d\eta} = -\frac{1}{2}u_1u_3, \quad u_3(0) = s$$

where s, the shooting parameter, is to be computed so that $u_2(s, L) = 1$, where L is a large number.

Exercise 4.2.9. Write a program solving the BVP in Example 4.10. Try different values of L (say L = 10,20,40) and check the convergence of the $f(\eta)$ curve.

4.3 ANSATZ METHODS FOR BVPs

Ansatz methods applied to BVPs will result in systems of algebraic equations to be solved just as the FDM. They are linear or nonlinear depending on whether the BVP is linear or nonlinear.

As a model problem for ansatz methods, we use the same BVP as before (4.20), i.e.,

$$-\frac{d^2u}{dx^2} = f(x), \quad 0 < x < 1, \quad u(0) = 0, \quad u(1) = 0$$

The solution $u(x)$ is approximated by an *ansatz* $u_h(x)$:

$$u(x) \approx u_h(x) = \sum_{j=1}^{N} c_j\varphi_j(x) \tag{4.67}$$

where $\varphi_j(x)$ are given *basis functions* and c_j are *coefficients* to be determined so that $u_h(x)$ is a "good" approximation in some sense. We assume here that the basis functions satisfy the BCs, i.e.,

$$\varphi_j(0) = 0, \quad \varphi_j(1) = 0, \quad j = 1, 2, \ldots, N \tag{4.68}$$

Hence, the ansatz solution $u_h(x)$ satisfies the BCs of the model problem, i.e., $u_h(0) = 0, u_h(1) = 0$.

The subscript h in $u_h(x)$ refers to the fact that the basis functions are defined from some sort of discretization of the interval (0, 1).

4.3.1 Starting with the ODE Formulation

Insert the ansatz into the BVP and we obtain a residual function $r(x)$:

$$r(x) = \frac{d^2 u_h}{dx^2} + f(x) \neq 0 \tag{4.69}$$

We want the residual function to be small in some sense.

If $r(x) \equiv 0$ for all x in the interval $0 \leq x \leq 1$, then $u_h(x)$ is the exact solution of (4.20). Of course we cannot expect $u_h(x)$ to be exact in a general case, but a small residual can be achieved in various ways:

1. $r(x_i) = 0, i = 1, 2, \ldots N$, i.e., the residual is zero in points $x_i \in [0, 1]$, the *collocation* method
2. $r(x)$ is orthogonal to $\varphi_i(x)$, i.e., $\int_0^1 r(x)\varphi_i(x)dx = 0,\ i = 1, 2, \ldots, N$, *Galerkin's method*

Boris Galerkin was a Russian engineer and mathematician active in the first half of the 20th century.

Galerkin's method is demonstrated here. Insert $r(x)$ into the integral conditions:

$$\int_0^1 \left(\sum_{j=1}^{N} c_j \frac{d^2 \varphi_j}{dx^2} + f(x) \right) \varphi_i dx = 0, \quad i = 1, 2, \ldots, N \tag{4.70}$$

As c_j are independent of x, we can move them outside the integral:

$$\sum_{j=1}^{N} c_j \int_0^1 \frac{d^2 \varphi_j}{dx^2} \varphi_i dx + \int_0^1 f(x)\varphi_i dx = 0 \tag{4.71}$$

Now use integration by parts on the first integral

$$\int_0^1 \frac{d^2 \varphi_j}{dx^2} \varphi_i dx = \left[\frac{d\varphi_j}{dx} \varphi_i \right]_0^1 - \int_0^1 \frac{d\varphi_i}{dx} \frac{d\varphi_j}{dx} dx \tag{4.72}$$

Because of the BCs, the first term on the right hand side is equal to zero and we get

$$\sum_{j=1}^{N} c_j \int_0^1 \frac{d\varphi_i}{dx}\frac{d\varphi_j}{dx}dx = \int_0^1 f(x)\varphi_i dx \tag{4.73}$$

which is a linear system of equations

$$\mathbf{Ac} = \mathbf{f}, \quad A_{ij} = \int_0^1 \frac{d\varphi_i}{dx}\frac{d\varphi_j}{dx}dx, \quad f_i = \int_0^1 f(x)\varphi_i dx \tag{4.74}$$

where the matrix A is $N \times N$, symmetric, and positive definite. Observe that so far we cannot say anything about the sparsity of A. That depends on the choice of basis functions $\varphi_i(x), i = 1, 2, \ldots, N$, see Section 4.3.3.

4.3.2 Starting with the Weak Formulation

4.3.2.1 The Model Problem The analytical solution (4.22) to the ODE problem (4.20) is referred to as the *strong (or classical) solution,* meaning that $u(x)$ must be twice piecewise continuously differentiable and fulfill the BCs to satisfy the BVP.

There is, however, an alternative formulation, the *weak (or variational) formulation,* based on an integral relation to which the BCs are coupled. This formulation requires $u(x)$ to be only once *piecewise continuously differentiable* and is called the *weak solution.*

The weak solution is obtained by multiplying the ODE by a *test function* $v(x)$ fulfilling the BCs $v(0) = v(1) = 0$, integrate over the interval $(0, 1)$ and then use integration by parts

$$-\int_0^1 u''(x)v(x)dx = -[u'(x)v(x)]_0^1 + \int_0^1 u'(x)v'(x)dx = \int_0^1 f(x)v(x)dx$$

The outintegrated term vanishes as $v(0) = v(1) = 0$ and the original ODE problem (4.20) can now be stated in weak form: Find $u(x) \in U$ such that

$$\int_0^1 u'(x)v'(x)dx = \int_0^1 f(x)v(x)dx, \quad \text{for all } v(x) \in U \tag{4.75}$$

Both $u(x)$ and $v(x)$ belong to the same *function space*, $U \equiv PC^1_{[0,1]}(0, 1)$, which means all functions that are once piecewise continuously differentiable on the interval $(0, 1)$ and zero at the boundary points $[0]$ and $[1]$. We use the notation PC as C in most textbooks is reserved for functions that are continuously differentiable up to a certain order.

If the BVP (4.20) is given with another BC

$$-\frac{d^2u}{dx^2} = f(x), \qquad u(0) = 0, \quad u'(1) = 0 \tag{4.76}$$

the weak formulation must be modified. As before multiply with a test function $v(x)$, integrate, and then integrate by parts. The outintegrated term is

$$-u'(1)v(1) + u'(0)v(0) = u'(0)v(0) = 0 \qquad if \quad v(0) = 0$$

and the weak formulation will be: Find $u(x) \in U$ such that

$$\int_0^1 u'(x)v'(x)dx = \int_0^1 f(x)v(x)dx, \qquad \text{for all } v(x) \in U$$

where $U = PC^1_{[0]}(0, 1)$. Hence, the BC $u(0) = 0$ is imposed into the function space U, while the BC $u'(1) = 0$ comes out from the weak formulation. BCs that are imposed explicitly to the function space U (here $u(0) = 0$) are called *essential BCs*, while BCs that comes out from the weak formulation are called *natural BCs*. Most often Dirichlet BCs are essential, while Neumann's and mixed BCs are natural.

Comment. It should be noted that there is also a third way of formulating (4.20) called *Ritz' method* or the *minimization formulation*

$$min_{v(x)\in U} \int_0^1 (\frac{1}{2}(v'(x))^2 - f(x)v(x))dx$$

where $U \equiv PC^1_{[0,1]}(0, 1)$. The function giving the minimum value to the integral expression is denoted by $u(x)$ and is the same function that solves the variational problem (4.75). This technique, however, belongs to variational calculus and is not taken up here. End of comment.

It may seem like magic that the weak formulation works, i.e., that (4.20) and (4.75) give the same solution and that a method can be based on a relation that is satisfied for infinitely many functions! The following analogy with a linear system of algebraic equations shows that "although this be madness, yet there is method in it" (Shakespeare, Hamlet, Act 2). Assume that A is an arbitrary $N \times N$ matrix and let $\mathbf{u} \in R^N$ be the solution of

$$A\mathbf{u} = \mathbf{b}$$

Multiply both sides by $\mathbf{v}^T$, an arbitrary (row) vector in R^N

$$\mathbf{v}^T(A\mathbf{u} - \mathbf{b}) = 0 \quad \text{for all } \mathbf{v} \in R^N \tag{4.77}$$

and we obtain the weak formulation. On the other hand, if (4.77) is true, choose from R^N N linearly independent vectors, put them as columns in the $N \times N$ matrix W, and we obtain

$$W^T(A\mathbf{u} - \mathbf{b}) = 0$$

As W is nonsingular, we can multiply by the inverse of W^T giving

$$A\mathbf{u} - \mathbf{b} = 0 \rightarrow A\mathbf{u} = \mathbf{b}$$

Now, back to the weak formulation (4.75). A well-known approximation method is Galerkin's method already mentioned in Section 4.3.1. This method is based on projection of $u(x)$ onto a finite dimensional linear subspace U_h of U. Assume this subspace is built up by basis functions $\varphi_i(x)$, where $\varphi_i(0) = 0$, $\varphi_i(1) = 0$, $i = 1, 2, \dots, N$. We now look for an ansatz approximation $u_h(x)$ that belongs to the subspace U_h

$$u_h(x) = \sum_{j=1}^{N} c_j \varphi_j(x)$$

and satisfies the weak formulation in the subspace, i.e., find $u_h(x) \in U_h$ such that

$$\int_0^1 \frac{du_h}{dx}\frac{dv_h}{dx}dx = \int_0^1 f(x)v_h dx, \quad \text{for all } v_h \in U_h \tag{4.78}$$

In Galerkin's method, the basis functions $\varphi_i(x)$ are used as test functions $v_h(x)$ and we obtain

$$\sum_{j=1}^{N} c_j \int_0^1 \frac{d\varphi_i}{dx}\frac{d\varphi_j}{dx}dx = \int_0^1 f(x)\varphi_i dx, \quad i = 1, 2, \dots N \tag{4.79}$$

which is the same linear system of equations as (4.74), where we started with the differential equation. The trick in both cases is to use partial integration to get rid of the second derivative.

4.3.2.2 A Linear Two-Point BVP Now, let us generalize Galerkin's method to the linear two-point BVP defined by (4.57). This generalization is helpful to have as a background to Chapter 7 where elliptic PDEs are studied. As before, multiply (4.57a) with a test function $v(x)$

$$\int_a^b \left(-\frac{d}{dx}\left(\kappa(x)\frac{du}{dx}\right)v(x)dx + p(x)\frac{du}{dx}v(x) + q(x)u(x)v(x)\right)dx = \int_a^b f(x)v(x)dx \tag{4.80}$$

Use integration by parts on the first term

$$-\int_a^b \frac{d}{dx}\left(\kappa(x)\frac{du}{dx}\right)v(x)dx = -\left[\kappa(x)\frac{du}{dx}v(x)\right]_a^b + \int_a^b \kappa(x)\frac{du}{dx}\frac{dv}{dx}dx \tag{4.81}$$

Using (4.57b) and (4.57c), the outintegrated term can be written

$$\kappa(a)\frac{du}{dx}(a)v(a) - \kappa(b)\frac{du}{dx}(b)v(b) = (\alpha_1 u(a) + \beta_1)v(a) - (\alpha_2 u(b) + \beta_2)v(b) \tag{4.82}$$

With the BCs given here, we cannot restrict $u(x)$ or $v(x)$ to be zero at $x = a$ or $x = b$. Instead the values of $u(x)$ and $v(x)$ at the boundary points will come out from the following weak formulation: Find $u(x) \in U$ such that

$$\int_a^b (\kappa(x)\frac{du}{dx}\frac{dv}{dx} + p(x)\frac{du}{dx}v(x) + q(x)u(x)v(x))dx + (\alpha_1 u(a)v(a) - \alpha_2 u(b)v(b))$$

$$= \int_a^b f(x)v(x)dx - (\beta_1 v(a) - \beta_2 v(b)) \quad \text{for all } v(x) \in U \tag{4.83}$$

where U is $PC^1(0,1)$, i.e., all functions once piecewise continuously differentiable on $(0,1)$. Hence, the BCs in (4.57) are not explicitly given but comes out as a consequence of the weak formulation, hence the BCs are *natural*. Note that terms containing $u, du/dx$ and $v, dv/dx$ occur in the left hand side as quadratic terms, while terms containing only v are collected in the right hand side.

Example 4.11. The weak formulation of problem (4.32) is formulated as follows. Starting from (4.78), the outintegrated term is

$$-u'(1)v(1) + u'(0)v(0) = (u(1) - 1)v(1)$$

Hence, the weak formulation has the form: Find $u(x) \in U$ such that

$$\int_0^1 u'v'dx + u(1)v(1) = \int_0^1 f(x)v(x)dx + v(1), \qquad \text{for all } v(x) \in U$$

where $U = PC^1(0,1)$. Both BCs are natural.

Now, use Galerkin's method on the weak formulation (4.83), i.e., let

$$u_h(x) = \sum_{j=1}^{N} c_j\varphi_j(x), \quad v_h(x) = \varphi_i(x), \ i = 1, 2, \ldots N$$

The coefficients c_j are obtained from a linear system of algebraic equations

$$(A\kappa + P + Q + B)\mathbf{c} = \mathbf{f} \tag{4.84}$$

where the elements of the $N \times N$ matrices $A\kappa$, P, Q, B are

$$A\kappa_{ij} = \int_a^b \kappa(x)\frac{d\varphi_i}{dx}\frac{d\varphi_j}{dx}dx, \quad P_{ij} = \int_a^b p(x)\varphi_i\frac{d\varphi_j}{dx}dx$$

$$Q_{ij} = \int_a^b q(x)\varphi_i(x)\varphi_j(x)dx, \quad B_{ij} = \alpha_1\varphi_i(a)\varphi_j(a) - \alpha_2\varphi_i(b)\varphi_j(b) \tag{4.85}$$

and the right hand side is the $N \times 1$ vector $\mathbf{f}$ with the elements

$$f_i = \int_a^b f(x)\varphi_i(x)dx - \beta_1\varphi_i(a) + \beta_2\varphi_i(b) \tag{4.86}$$

If $\kappa(x) > 0$, $q(x) > 0$, $\alpha_1 > 0$, $\alpha_2 < 0$, and $p(x) \equiv 0$, the linear system of equations (4.84) is *symmetric* and positive definite. This special case is very common in applications and is found in, e.g., heat conduction, structural mechanics, diffusion, and electromagnetics. Linear systems of equations with a symmetric and positive system matrix can be solved with, e.g., Cholesky's method or some conjugate gradient-based method, see Section A.5.

If $p(x) \neq 0$, the linear equation system (4.84) is *unsymmetric* and other methods must be used, e.g., Gaussian elimination or the iterative GMRES method, see Section A.5.

Example 4.12. The interface problem in Example 4.10 is formulated in weak form. Let $T(z) = u(z) + T_0$. With this transformation, we obtain the essential BC $u(0) = 0$. Find $u(z) \in U$ such that

$$\int_0^L \kappa \frac{du}{dz}\frac{dv}{dz} dz + \int_0^L \frac{2h}{R} u(z)v(z)dz + (u(L) + T_0 - T_{out})v(L)$$

$$= \int_0^L \frac{2h}{R} T_{out} v(z)dz - (T_0 - T_{out})v(L)$$

for all $v \in U$, where $U = PC^1_{[0]}(0, 1)$. The parameter κ takes different values in the two regions, i.e., $\kappa = \kappa_1$, $0 < z < \xi$, and $\kappa = \kappa_2$, $\xi < z < L$. Observe that the interface BCs (4.59c) are automatically "built in" in the weak formulation.

In this section, we have stated the weak formulation and Galerkin's method with *integral expressions*. A general, abstract way of formulation is based on *Sobolev spaces*, see reference 4 in the bibliography.

Exercise 4.3.1. Given the BVP $-u' = f(x)$, $u'(0) = 0$, $u(1) = 0$ $0 < x < 1$ present the weak formulation of the problem. Which BC is natural and which is essential?

Exercise 4.3.2. Given the BVP $-u' = f(x)$, $u(0) = a$, $u'(1) = b$ $0 < x < 1$. Here the BC's are *inhomogeneous*. Present the weak formulation of this problem. Hint: Introduce the help function $w(x)$ through $u(x) = w(x) + a + bx$, where $w(0) = 0$, $w'(1) = 0$

Exercise 4.3.3. Given the BVP $-u' = f(x)$, $u(0) = a$, $u(1) = b$ $0 < x < 1$. Present the weak formulation of this problem. Hint: Find a transformation similar to the one in Exercise 4.3.2.

4.3.3 The Finite Element Method

This section is not an extensive treatment of the FEM for solving BVPs in 1D. Instead it is intended to be an overview of a method that can be generalized to problems in 2D (see Section 7.4).

So far we have not said anything about the basis functions. Obviously they influence the structure of the matrix A. When the FDM (4.27) is used, the matrix A is a tridiagonal matrix. We should come down to something similar for the ansatz method, if the methods are to be comparable with respect to computer storage and computer time needed to solve $A\mathbf{c} = \mathbf{b}$.

If we choose the ansatz solution to be a polynomial of degree N, e.g.,

$$\tilde{u}(x) = \sum_{j=1}^{N-1} c_j x^j (x-1) \tag{4.87}$$

where the basis functions are $\varphi_j(x) = x^j(x-1)$ fulfilling the BCs. It is easy to show that this choice of basis functions will give a *full* matrix A, i.e., most elements $A_{i,j} \neq 0$.

Assume we want our ansatz function to be a *piecewise* polynomial of degree one, i.e., a function that is put together from straight lines in a continuous way at the *nodes* x_i, $i = 1, 2, \dots, N$ as in the graph in Figure 4.11, where the stepsize h_i can be variable or constant, i.e., $\sum_{i=0}^{N} h_i = 1$ or $h(N+1) = 1$.

The basis functions of the ansatz $u_h(x) = \sum_{j=1}^{n} c_j\varphi_j$ will also be piecewise polynomials. If we want $c_j = u_h(x_j)$, we see that φ_j must fulfill

$$\varphi_j(x_i) = \begin{cases} 1, & \text{if } i = j \\ 0, & \text{if } i \neq j \end{cases} \tag{4.88}$$

Hence, the ansatz function $u_h(x)$ will be a piecewise linear interpolation polynomial through the points $(x_i, u_h(x_i))$. A graph of this basis function looks like the one shown in Figure 4.12.

The derivative $\varphi_j'(x)$ of the basis function has the graph shown in Figure 4.13.

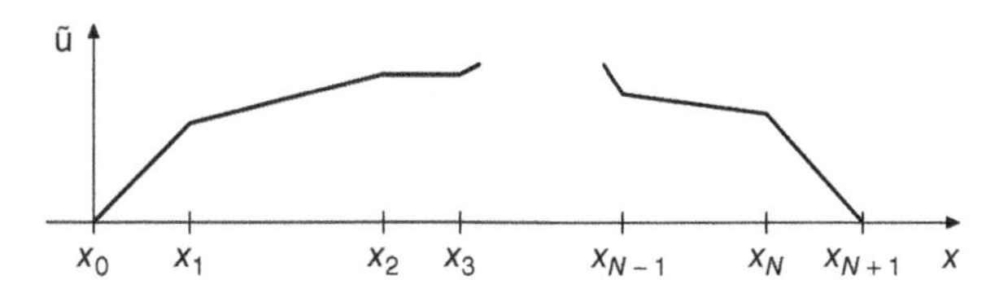

Figure 4.11 A piecewise linear ansatz function

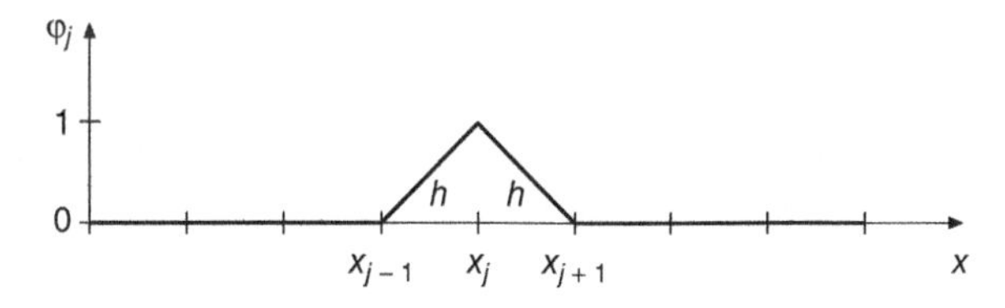

Figure 4.12 A piecewise linear basis function, a roof function

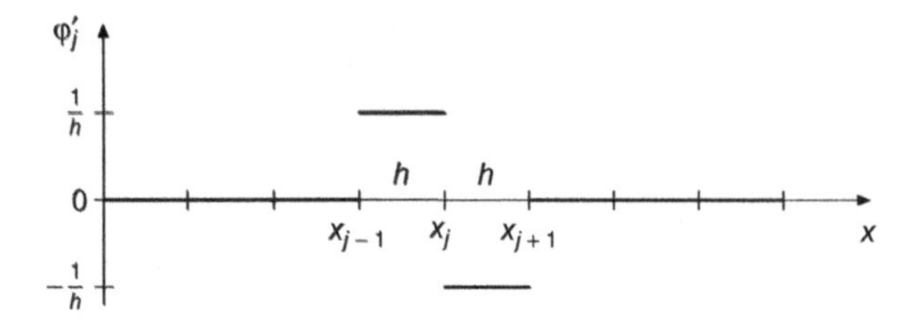

Figure 4.13 The derivative of a roof function

A few simple calculations using (4.74) give as result

$$A_{ij} = -\frac{1}{h}, \quad j = i-1, j = i+1 \quad A_{ii} = \frac{2}{h} \tag{4.89}$$

Hence, the $N \times N$ matrix A, the 'stiffness matrix' is the same as in (4.29) (apart from the factor $1/h$).

$$A = \frac{1}{h}\begin{pmatrix} 2 & -1 & 0 & 0 & \dots \\ -1 & 2 & -1 & 0 & \dots \\ & & \dots & & \\ \dots & 0 & -1 & 2 & -1 \\ \dots & 0 & 0 & -1 & 2 \end{pmatrix}$$

Stiffness is here related to a material property and not the concept "stiff" in connection with certain ODE problems (see Section 3.3.3). The right hand side vector **f** is often called the *load vector*.

The computation of the right hand side **f** where

$$f_i = \int_{x_i-h}^{x_i+h} f(x)\varphi_i(x)dx$$

can be done approximately with, e.g., the trapezoidal method:

$$f_i \approx hf(x_i) \tag{4.90}$$

Hence, for the model problem (4.20), we get exactly the same system of linear equations (4.29) for the FDM and the FEM. This, however, is not true in general.

Exercise 4.3.4. Modify the calculations leading to A and **f** in the case that the stepsize is not constant, i.e., $h_i = x_{i+1} - x_i$.

Exercise 4.3.5. Given the BVP (4.57) with $\kappa(x) = 1$, $p(x) = 1$, $q(x) = 1$, compute the elements $A\kappa_{i,j}$, $P_{i,j}$, and $Q_{i,j}$ in (4.84).

Exercise 4.3.6. Write a program to solve the BVP (4.32) with the FEM. Compare with the FDM solution of the same problem.

BIBLIOGRAPHY

1. C. Moler, "Numerical Computing with MATLAB", Chapter 8, SIAM, 2004
2. G. Golub, J. Ortega, "Scientific Computing and Differential Equations", Chapter 3, Academic Press, 1992
3. H.B. Keller, "Numerical Methods for Two-point Boundary Value Problems", Blaisdell, London, 1968
4. K. Eriksson, D. Estep, P. Hansbo, C. Johnson, "Computational Differential Equations", Studentlitteratur, 1996

5

PARTIAL DIFFERENTIAL EQUATIONS

This chapter is not intended to be an extensive treatment of mathematical properties of partial differential equations (PDEs) but rather a survey of important definitions, concepts, and results. For a more careful description, the reader may consult some of the mathematical textbooks on PDEs referenced in the end of this chapter.

A general formulation of a system of PDEs expressed in vector form is

$$\frac{\partial \mathbf{u}}{\partial t} = \mathbf{f}\left(t, x, y, z, \mathbf{u}, \frac{\partial \mathbf{u}}{\partial x}, \frac{\partial \mathbf{u}}{\partial y}, \frac{\partial \mathbf{u}}{\partial z}, \frac{\partial^2 \mathbf{u}}{\partial x^2}, \frac{\partial^2 \mathbf{u}}{\partial y^2}, \frac{\partial^2 \mathbf{u}}{\partial z^2}, \dots\right) \tag{5.1}$$

The variables t, x, y, z, the *independent* variables, are defined in some region, bounded or unbounded and the variable $\mathbf{u}$, the *dependent* variable, is a solution of (5.1) if it satisfies the PDEs for all t, x, y, z in the region. Among the independent variables, t is used to denote *time* and x, y, z denote the *space* variables. If time is among the independent variables, the PDE problem is called a *time-dependent* or an *evolution* problem, if not the problem is called an *equilibrium* or a *steady-state* problem.

The solution of a PDE is called a *field*, a function that depends on time and space or just space. The field is *scalar* valued if the PDE (5.1) is scalar and *vector* valued if (5.1) is a system of PDEs.

A problem formulated in three space dimensions x, y, z is called a *3D problem* (three dimension). The mathematical and numerical treatment of a PDE is often easier when formulated in 2Ds x, y or 1D x.

As was illustrated in Chapter 1, a PDE has infinitely many solutions. To obtain a unique solution, we need to have *initial* and *boundary* conditions. It depends on the character of the PDE what conditions should be given in order to have a *well-posed* problem, i.e., a problem the solution of which is *stable* with respect to perturbations in boundary and initial data.

Introduction to Computation and Modeling for Differential Equations, Second Edition. Lennart Edsberg.
© 2016 John Wiley & Sons, Inc. Published 2016 by John Wiley & Sons, Inc.

5.1 CLASSICAL PDE PROBLEMS

The classical PDEs presented in most textbooks are given below as problems (1–5). They are all scalar linear PDEs with constant coefficients. They will be used as *model problems* for numerical methods presented in later chapters.

1. Parabolic PDE in 1D (heat equation or diffusion equation):

$$\frac{\partial u}{\partial t} = \kappa \frac{\partial^2 u}{\partial x^2}, \qquad 0 < x < 1, \quad 0 < t \le t_{end} \tag{5.2}$$

   where κ is a constant positive parameter.
   Initial condition: $u(x, 0) = u_0(x)$, a known function of x.
   Boundary conditions: $u(0, t) = \alpha(t), u(1, t) = \beta(t)$, where $\alpha(t)$ and $\beta(t)$ are known functions of t.

2. Hyperbolic PDE in 1D (wave equation):

$$\frac{\partial^2 u}{\partial t^2} = c^2 \frac{\partial^2 u}{\partial x^2}, \qquad 0 < x < 1, \quad 0 < t \le t_{end} \tag{5.3}$$

   where c is a constant parameter.
   Initial conditions: $u(x, 0) = u_0(x), \frac{\partial u}{\partial t}(x, 0) = v_0(x)$
   Boundary conditions: $u(0, t) = \alpha(t), u(1, t) = \beta(t)$

3. Hyperbolic first-order PDE in 1D (advection equation):

$$\frac{\partial u}{\partial t} + a \frac{\partial u}{\partial x} = 0, \qquad 0 < x < 1, \quad 0 < t \le t_{end} \tag{5.4}$$

   where a is a constant positive parameter.
   Initial condition: $u(x, 0) = u_0(x)$
   Boundary condition: $u(0, t) = \alpha(t)$

4. Elliptic PDE in 2D (Poisson's equation on the unit square)

$$\frac{\partial^2 u}{\partial x^2} + \frac{\partial^2 u}{\partial y^2} = f(x, y), \quad (x, y) \in \Omega = \{(x, y) : 0 < x, y < 1\} \tag{5.5}$$

   Boundary condition: $u = 0$ on $\partial\Omega$, the boundary of Ω
   The important special case when $f(x, y) \equiv 0$ in (5.5) is called *Laplace's equation*.

5. Eigenvalue problem in 2D: (Helmholtz' equation on the unit square)

$$\frac{\partial^2 u}{\partial x^2} + \frac{\partial^2 u}{\partial y^2} = \lambda u, \quad (x, y) \in \Omega = \{(x, y) : 0 < x, y < 1\} \tag{5.6}$$

   Boundary condition: $u = 0$ on $\partial\Omega$.

The equations above are special cases of a general linear second-order PDE:

$$a\frac{\partial^2 u}{\partial x^2} + 2b\frac{\partial^2 u}{\partial x \partial y} + c\frac{\partial^2 u}{\partial y^2} + d_1\frac{\partial u}{\partial x} + d_2\frac{\partial u}{\partial y} + eu = f(x, y) \tag{5.7}$$

which is called, assuming that at least one of a, b, c is $\neq 0$

$$\begin{aligned} &\text{hyperbolic} \quad \text{if} \quad b^2 - ac > 0 \\ &\text{parabolic} \quad \text{if} \quad b^2 - ac = 0 \\ &\text{elliptic} \quad \text{if} \quad b^2 - ac < 0 \end{aligned}$$

If $a = b = c = 0$, but $d_1 \neq 0$ and $d_2 \neq 0$, the PDE is *hyperbolic*.

In a linear PDE of type (5.7), the coefficients a, b, c, d_1, d_2, e are constant or depend on (x, y). For a time-dependent PDE, change y to t. If $f(x, y) \equiv 0$, (5.7) is *homogenous*, otherwise *inhomogeneous*.

Example 5.1. **(The Euler–Tricomi PDE).**

$$\frac{\partial^2 u}{\partial x^2} = x\frac{\partial^2 u}{\partial y^2}$$

This is an example where the coefficient depends on (x, y). The PDE is hyperbolic when $x > 0$ and elliptic when $x < 0$. Francesco Tricomi was an Italian mathematician active in the middle of the 20th century.

Example 5.2. **(The Prandtl–Glauert equation).**

Compressible, inviscid flow over a thin airfoil can be modeled in 2D by the PDE

$$(1 - M_\infty^2)\frac{\partial^2 \Phi}{\partial x^2} + \frac{\partial^2 \Phi}{\partial y^2} = 0$$

where M_∞ is the Mach number and Φ is the velocity potential. This PDE is hyperbolic if $M_\infty > 1$ (supersonic flow) and elliptic if $M_\infty < 1$ (subsonic flow). Ludwig Prandtl was a German and Hermann Glauert was a British physicist active in the first half of the 20th century.

Example 5.3. A fundamental solution of the wave equation (5.3) is obtained from the ansatz (compare with Example 4.6)

$$u(x, t) = e^{i\omega t} v(x)$$

Insert this ansatz into (5.3) and the following ODE for $v(x)$ is obtained

$$\frac{d^2 v}{dx^2} + k^2 v = 0$$

where $k = \omega/c$ is called the *wave number*. A particular solution of (5.3) can then be written

$$u(x, t) = e^{i(\omega t \pm kx)} = e^{i\omega(t \pm cx)}$$

which is a *traveling wave*, a very common model for, e.g., energy transfer in physics, such as mechanical, electromagnetic, or quantum mechanical waves.

In an extended definition, the space-derivative terms are modified in a way that admits *nonlinear* versions of the generic PDEs:

1. Nonlinear heat (or diffusion) equation in 1D:

$$\frac{\partial u}{\partial t} = \frac{\partial}{\partial x}\left(\kappa(u)\frac{\partial u}{\partial x}\right) \tag{5.8}$$

2. Nonlinear wave equation in 1D:

$$\frac{\partial^2 u}{\partial t^2} = c^2(u)\frac{\partial^2 u}{\partial x^2} \tag{5.9}$$

3. Nonlinear advection equation in 1D:

$$\frac{\partial u}{\partial t} + a(u)\frac{\partial u}{\partial x} = 0 \tag{5.10}$$

4. Nonlinear Laplace's equation in 2D:

$$\frac{\partial}{\partial x}\left(\kappa_x(u)\frac{\partial u}{\partial x}\right) + \frac{\partial}{\partial y}\left(\kappa_y(u)\frac{\partial u}{\partial y}\right) = 0 \tag{5.11}$$

Analytic solutions to PDEs can seldom be found but exist for the linear model problems. Usually analytic solutions are complicated consisting of, e.g., infinite series expansions or integral expressions.

For Cauchy formulations (initial values are given for $-\infty < x < \infty$) of the time-dependent problems (1–3) analytic solutions can be formulated. For Laplace's equation, analytic solutions exist when the region Ω has a simple geometry, e.g., a square or a circle.

Augustin Louis Cauchy, Pierre Simon de Laplace, and Simeon Denis Poisson were French mathematicians active in the beginning of the 19th century.

1. The heat equation, with Cauchy conditions

$$\frac{\partial u}{\partial t} = \kappa\frac{\partial^2 u}{\partial x^2},\quad u(x, 0) = u_0(x), -\infty < x < \infty,\ t > 0$$

The solution can be written as an integral

$$u(x, t) = \frac{1}{2\sqrt{\pi\kappa t}}\int_{-\infty}^{+\infty} e^{-(x-s)^2/4\kappa t}u_0(s)ds \tag{5.12}$$

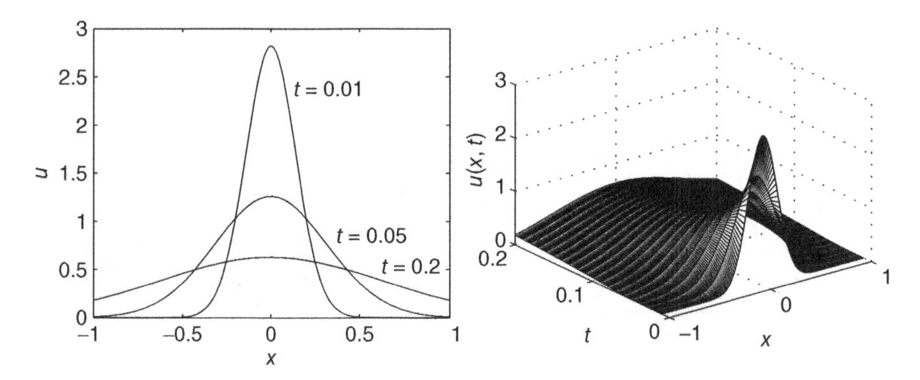

Figure 5.1 The fundamental solution of the heat equation, $\kappa = 1$

provided the integral exists for the given function $u_0(x)$. In the special case, $u_0(x) = \delta(x)$, i.e., the Dirac delta-function at $x = 0$, we obtain the fundamental solution (Figure 5.1)

$$u(x,t) = \frac{1}{2\sqrt{\pi\kappa t}} e^{-x^2/4\kappa t} \tag{5.13}$$

The problem is well posed if $t > 0$ and ill posed if $t < 0$.

2. The wave equation with Cauchy conditions

$$\frac{\partial^2 u}{\partial t^2} = c^2 \frac{\partial^2 u}{\partial x^2}, \; u(x,0) = u_0(x), \; \frac{\partial u}{\partial t}(x,0) = v_0(x), -\infty < x < \infty$$

has the following solution, also called d'Alambert's solution

$$u(x,t) = \frac{1}{2}(u_0(x-ct) + u_0(x+ct)) + \frac{1}{2c}\int_{x-ct}^{x+ct} v_0(s)ds \tag{5.14}$$

For the initial values $u_0(x) = e^{-x^2}$, $v_0(x) = 0$ and the parameter value $c = 1$, the solution is (see Figure 5.2)

$$u(x,t) = \frac{1}{2}\left(e^{-(x-t)^2} + e^{-(x+t)^2}\right)$$

This problem is well posed in both t directions, i.e., both $t > 0$ and $t < 0$.

3. The advection equation and its solution are presented in Chapter 1.
4. Laplace's equation on the unit square

$$\frac{\partial^2 u}{\partial x^2} + \frac{\partial^2 u}{\partial y^2} = 0, \quad (x,y) \in \Omega = \{(x,y) : 0 < x, y < 1\}$$

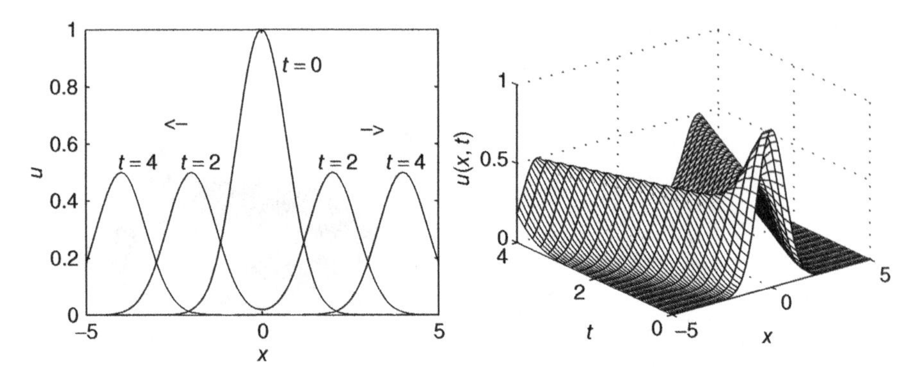

Figure 5.2 D'Alembert's solution of the wave equation, c = 1

with the following boundary conditions (one of which is nonhomogeneous)

$$u(0, y) = u(1, y) = 0, \ 0 < y < 1, \quad u(x, 0) = 0, u(x, 1) = g(x), \ 0 \le x \le 1$$

has the following infinite series solution obtained with the analytic method *separation of variables*

$$u(x, t) = \sum_{k=1}^{\infty} c_k \sin(k\pi x) \sinh(k\pi y) \tag{5.15}$$

where c_k are coefficients given by integral expressions. With the principle of *superposition*, the analytic solution of Laplace's equation with all four boundaries having nonhomogeneous boundary conditions can be written as the sum of four sums of type (5.15).

5.2 DIFFERENTIAL OPERATORS USED FOR PDES

The PDEs of science and engineering are formulated in 3D and often complicated to present in component form. Therefore, it is convenient to use *differential operators* to make the notation more compact. The frequently used differential operators are the *gradient*, the *divergence*, the *rotation* (*curl*), and the *laplacian*.

Let $\Phi = \Phi(x, y, z)$ be a scalar field, $\mathbf{F} = (P(x, y, z), Q(x, y, z), R(x, y, z))^T$ be a vector field, and (x, y, z) the usual *cartesian coordinates*. The differential operator ∇ (nabla) is defined as

$$\nabla = \left(\frac{\partial}{\partial x}, \frac{\partial}{\partial y}, \frac{\partial}{\partial z} \right)^T$$

The differential operators are defined as

$$grad\ \Phi = \nabla \Phi = \left(\frac{\partial \Phi}{\partial x}, \frac{\partial \Phi}{\partial y}, \frac{\partial \Phi}{\partial z} \right)^T \tag{5.16}$$

$$div\ \mathbf{F} = \nabla \cdot \mathbf{F} = \frac{\partial P}{\partial x} + \frac{\partial Q}{\partial y} + \frac{\partial R}{\partial z} \tag{5.17}$$

$$curl\ \mathbf{F} = \nabla \times \mathbf{F} = \begin{pmatrix} \frac{\partial R}{\partial y} - \frac{\partial Q}{\partial z} \\ \frac{\partial P}{\partial z} - \frac{\partial R}{\partial x} \\ \frac{\partial Q}{\partial x} - \frac{\partial P}{\partial y} \end{pmatrix} \tag{5.18}$$

$$div\ grad\ \Phi = \Delta\Phi = \frac{\partial^2 \Phi}{\partial x^2} + \frac{\partial^2 \Phi}{\partial y^2} + \frac{\partial^2 \Phi}{\partial z^2} \tag{5.19}$$

where Δ is known as the *laplacian*, which can also be written as $\nabla \cdot \nabla$ or ∇^2.

Observe that the notations ∇ and ∇^2 are also used as operators for backward differentiation approximations, see Section 3.4.

In 2D, the corresponding operators have to be modified. Let $\Phi = \Phi(x, y)$ be a scalar field and $\mathbf{F} = (P(x, y), Q(x, y))^T$ be a vector field. Then,

$$grad\ \Phi = \nabla\Phi = \left(\frac{\partial \Phi}{\partial x}, \frac{\partial \Phi}{\partial y}\right)^T \tag{5.20}$$

$$div\ \mathbf{F} = \nabla \cdot \mathbf{F} = \frac{\partial P}{\partial x} + \frac{\partial Q}{\partial y} \tag{5.21}$$

$$curl\ \mathbf{F} = \nabla \times \mathbf{F} = \frac{\partial Q}{\partial x} - \frac{\partial P}{\partial y} \tag{5.22}$$

$$div\ grad\ \Phi = \Delta\Phi = \frac{\partial^2 \Phi}{\partial x^2} + \frac{\partial^2 \Phi}{\partial y^2} \tag{5.23}$$

Two important results from vector analysis are

If *curl* $\mathbf{F} = 0$, there exists a potential function Φ satisfying $\mathbf{F} = \nabla\Phi$.

If *div* $\mathbf{F} = 0$, there exists a vector potential $\mathbf{A}$ satisfying $\mathbf{F} = curl\ \mathbf{A}$.

The operators *grad*, *div*, and *curl* have physical interpretations. The gradient $\nabla\Phi$ is a measure of the *steepness* of Φ is in the x,y, and z directions, the *div*-operator can be interpreted as the *source strength* of the field $\mathbf{F}$ and the *curl*-operator corresponds to the *rotation strength* of a velocity field $\mathbf{v}$. For further discussion on these three operators, see Section A.4.

All definitions (5.16)–(5.23) are formulated in *cartesian coordinates* (x, y, z). In applications, however, it is sometimes suitable to use

cylindrical coordinates (r, φ, z): $x = r\cos(\varphi)$, $y = r\sin(\varphi)$, $z = z$

spherical coordinates (r, θ, φ): $x = r\sin(\theta)\cos(\varphi)$, $y = r\sin(\theta)\sin(\varphi)$, $z = r\cos(\theta)$.

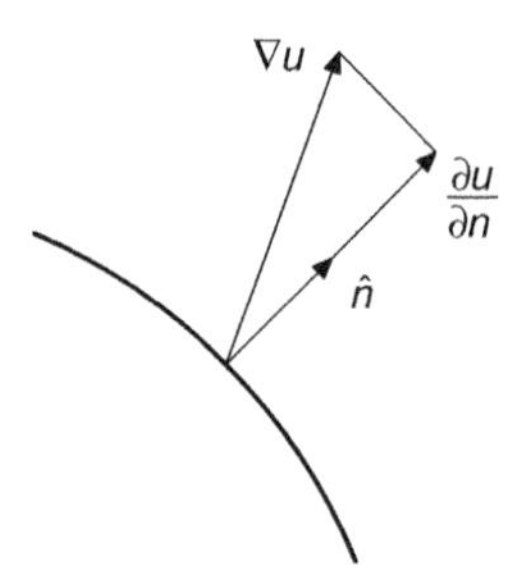

Figure 5.3 The geometrical meaning of the normal derivative

We refer to mathematical textbooks for deriving the transformations leading to the forms of the operators expressed in these coordinates. For use in coming chapters, however, the laplacian in 3D is presented for the three coordinate systems:

$$\Delta\Phi = \frac{\partial^2\Phi}{\partial x^2} + \frac{\partial^2\Phi}{\partial y^2} + \frac{\partial^2\Phi}{\partial y^2} \tag{5.24a}$$

$$\Delta\Phi = \frac{1}{r}\frac{\partial}{\partial r}\left(r\frac{\partial\Phi}{\partial r}\right) + \frac{1}{r^2}\frac{\partial^2\Phi}{\partial\varphi^2} + \frac{\partial^2\Phi}{\partial z^2} \tag{5.24b}$$

$$\Delta\Phi = \frac{1}{r^2}\frac{\partial}{\partial r}\left(r^2\frac{\partial\Phi}{\partial r}\right) + \frac{1}{r^2\sin(\theta)}\frac{\partial}{\partial\theta}\left(\sin(\theta)\frac{\partial\Phi}{\partial\theta}\right) + \frac{1}{r^2\sin^2(\theta)}\frac{\partial^2\Phi}{\partial\varphi^2} \tag{5.24c}$$

Observe that the coefficients depend on r and r, θ respectively, hence they are *variable*, but the PDEs are still linear.

Another differential operator used in 2D and 3D boundary conditions is the *normal derivative*. It is defined as the projection of the gradient in a boundary point onto the outward normal unit vector $\hat{\mathbf{n}}$ from that point (see Figure 5.3).

$$\frac{\partial u}{\partial n} = (\nabla u)\cdot\hat{\mathbf{n}} \tag{5.25}$$

Exercise 5.2.1. Given the following PDEs. Investigate for each one if it is parabolic, hyperbolic, or elliptic.

$$\text{a) } \frac{\partial u}{\partial t} = \frac{\partial^2 u}{\partial x^2} + u, \text{ b) } \frac{\partial^2 u}{\partial x^2} + \frac{\partial^2 u}{\partial y^2} = -1, \text{ c) } \frac{\partial u}{\partial t} + u\frac{\partial u}{\partial x} = 0, \text{ d) } \frac{\partial u}{\partial t} + \frac{\partial u}{\partial x} = \frac{\partial^2 u}{\partial x^2} + u$$

Exercise 5.2.2. Given the PDE problem

$$\frac{\partial u}{\partial t} + \frac{\partial u}{\partial x} = \frac{\partial^2 u}{\partial x^2}, \quad u(0,t) = 1, u(1,t) = 0, \; u(x,0) = u_0(x)$$

Show that the first-order space-derivative term in the PDE can be eliminated by choosing α appropriately in the transformation $u = ve^{\alpha x}$. Formulate the PDE for the variable v and the corresponding initial and boundary conditions.

Exercise 5.2.3. Verify that Laplace's equation in 2D polar coordinates (r, φ), where $x = r\cos\varphi, y = r\sin\varphi$, is

$$\frac{\partial^2 u}{\partial r^2} + \frac{1}{r}\frac{\partial u}{\partial r} + \frac{1}{r^2}\frac{\partial^2 u}{\partial \varphi^2} = 0$$

Show that that the sum of the first two terms can be written

$$\frac{1}{r}\frac{\partial}{\partial r}\left(r\frac{\partial u}{\partial r}\right)$$

Exercise 5.2.4. Show that the first term in the spherical laplacian can be written

$$\frac{1}{r^2}\frac{\partial}{\partial r}\left(r^2\frac{\partial u}{\partial r}\right) = \frac{\partial^2 u}{\partial r^2} + \frac{2}{r}\frac{\partial u}{\partial r} = \frac{1}{r}\frac{\partial^2(ru)}{\partial r^2}$$

Exercise 5.2.5. Spherical waves from a point source are modeled by the PDE

$$\frac{\partial^2 u}{\partial t^2} = c^2\left(\frac{\partial^2 u}{\partial r^2} + \frac{2}{r}\frac{\partial u}{\partial r}\right)$$

Show that this PDE is equivalent to the PDE

$$\frac{\partial^2(ru)}{\partial t^2} = c^2\frac{\partial^2(ru)}{\partial r^2}$$

that has the general solution

$$u(r,t) = \frac{1}{r}(f(r - ct) + g(r + ct))$$

where $f(r)$ and $g(r)$ are arbitrary twice differentiable functions.

Exercise 5.2.6. Show that the fundamental solution (5.13) satisfies the heat equation.

Exercise 5.2.7. Show that d'Alembert's solution (5.14) satisfies the wave equation. Show it first for $v_0 \equiv 0$.

Exercise 5.2.8. Use separation of variables to show that the solution of problem (4) has the series solution (5.15). Derive the integral expressions for the coefficients c_k.

Exercise 5.2.9. Show that $u(x,y) = \sin(i\pi x)\sin(j\pi y),\ i = 1,2,\ldots\ j = 1,2,\ldots$ are solutions (eigenfunctions) to model problem (5). Calculate the corresponding eigenvalues.

Exercise 5.2.10. Show the following differential operator relations:

a) $div\ (\nabla u) = \Delta u$
b) $div\ (curl\ \mathbf{u}) = 0$, i.e., the divergence of a rotation equals zero.
c) $curl\ (\nabla u) = 0$, i.e., the rotation of a gradient equals zero.
d) $div\ (\rho\mathbf{u}) = \rho div\ \mathbf{u} + \nabla\rho \cdot \mathbf{u}$
e) $(\mathbf{u} \cdot \nabla)\mathbf{u} = \frac{1}{2}\nabla(\mathbf{u} \cdot \mathbf{u}) + (curl\ \mathbf{u}) \times \mathbf{u}$
f) $\Delta\mathbf{u} = \nabla(div\ \mathbf{u}) - (curl\ (curl\ \mathbf{u}))$
g) $div\ (\mathbf{u}\mathbf{u}^T) = \mathbf{u}(div\ \mathbf{u}) + (\mathbf{u} \cdot \nabla)\mathbf{u}$. The divergence of the matrix $\mathbf{u}\mathbf{u}^T$ is a vector where component i is the divergence of row i in $\mathbf{u}\mathbf{u}^T$

5.3 SOME PDEs IN SCIENCE AND ENGINEERING

In this section, some well-known PDEs from different applications in science and engineering are presented. These equations exhibit a mathematical beauty in the sense that they can be written compactly on a few lines. However, they are complicated to treat both analytically and numerically and usually they have to be simplified both geometrically (from 3D to 2D or even 1D) and analytically (by introducing simplifying modeling assumptions) before numerical treatment is possible.

In this section, initial and boundary conditions are omitted as they can be of very many different types. Examples of initial and boundary conditions are shown in the chapters to follow.

5.3.1 Navier–Stokes Equations for Incompressible Flow

Navier–Stokes equations are fundamental in aerodynamics, hydrodynamics, meteorology, and other applications where the flow of a gas or a liquid is modeled. They are based on Newton's second law for deformable bodies together with a constitutive equation stating that the shear stresses are proportional to the velocity gradients. The equations are complicated in their most general form. Therefore, they are here presented under the condition that the density ρ (kg/m^3) and the dynamic viscosity μ (N $\cdot$ s/m^2) of the fluid are constant:

$$div\ \mathbf{u} = 0 \tag{5.26a}$$

$$\rho\left(\frac{\partial\mathbf{u}}{\partial t} + (\mathbf{u} \cdot \nabla)\mathbf{u}\right) = \rho\mathbf{g} - \nabla p + \mu\Delta\mathbf{u} \tag{5.26b}$$

In the PDEs (5.26), $\mathbf{u} = (u_1, u_2, u_3)^T$ (m/s) is the velocity field of the flow, p (N/m^2) the pressure in the flow, and $\mathbf{g}$ (m/s^2) the gravitational acceleration.

The first equation expresses conservation of mass and the second conservation of linear momentum (Newton's second law).

The equations (5.26) constitute four PDEs for the four unknowns u, v, w, and p in *incompressible* flow (ρ constant). The incompressible Navier–Stokes equations are nonlinear PDEs classified as mixed hyperbolic–parabolic. They were first derived by the French physicist Louis Navier and the Irish mathematician George Stokes in the beginning of the 19th century.

Associated with Navier–Stokes equation is Reynolds number Re, an important dimensionless parameter $Re = L \cdot U \cdot \rho/\mu$ introduced by the Irish engineer Osborne Reynolds in the second half of the 19th century. In the formula, L is a characteristic length of the region and U a characteristic velocity. Reynolds number gives a characterization of the flow. For small Re, say $Re < 1$, we have *laminar* flow while a large Re number corresponds to *turbulent* flow. The numerical treatment of Navier–Stokes equations for large Re-numbers is difficult as both *turbulence* and moving *transition layers* make it tricky to resolve the equations with a numerical method.

Depending on the physical assumptions, Navier–Stokes equations can be simplified to less complex PDEs. For a small Reynolds number corresponding to, e.g., slow velocity or large viscosity, the Navier–Stokes equations in steady (time-independent) flow are reduced to Stokes equations

$$\mu\Delta\mathbf{u} - \nabla p + \rho\mathbf{g} = 0, \qquad div\ \mathbf{u} = 0 \tag{5.27}$$

This is an elliptic problem consisting of four equations for the four unknowns $\mathbf{u} = (u, v, w)^T$ and p.

In case the flow can be regarded as irrotational, nonviscous, and steady, it is called *potential flow*. The reason is that if $curl\ \mathbf{u} = 0$, there is a potential function Φ such that $\mathbf{u} = \nabla\Phi$. This relation together with $div\ \mathbf{u} = 0$ gives Laplace equation for Φ

$$\Delta\Phi = 0 \tag{5.28}$$

When appropriate BCs has been added to (5.28), the flow field is obtained from the solution Φ from $\mathbf{u} = \nabla\Phi$. We see that the pressure p is not present in this flow model. In 2D, a lot of analysis has been done with conformal mapping, i.e., transformations in the complex plane. An example of potential flow is given in Chapter 7, Example 7.2.

Yet another simplification of Navier–Stokes equations is obtained for steady, nonviscous, and irrotational flow with Bernoulli's equation

$$\frac{1}{2}\rho U^2 + p = constant \tag{5.29}$$

This often used model for, e.g., large pipe systems where the wall-friction is neglected is not even a differential equation. It gives no information about the flow field, just the size of the velocity U. Newton's second law is not needed to derive this equation, just hydrostatical relations.

5.3.2 Euler's Equations for Compressible Flow

Often variants of the Navier–Stokes equations are used depending on the physical process to be modeled. For a compressible gas (ρ not constant) and inviscid (without inner friction, i.e., $v = 0$), we have the *compressible* Euler equations:

$$\frac{\partial\rho}{\partial t} + div(\rho\mathbf{u}) = 0 \tag{5.30a}$$

$$\frac{\partial(\rho\mathbf{u})}{\partial t} + div(\rho\mathbf{u}\mathbf{u}^T) + \nabla p = 0 \tag{5.30b}$$

$$\frac{\partial E}{\partial t} + div((E+p)\mathbf{u}) = 0 \tag{5.30c}$$

$$E = \frac{p}{\gamma - 1} + \frac{1}{2}\rho(u_1^2 + u_2^2 + u_3^2) \tag{5.30d}$$

where $\mathbf{u}\mathbf{u}^T$ is the outer product, $\mathbf{u} = (u_1, u_2, u_3)^T$, and $(\mathbf{u}\mathbf{u})_{ij} = u_i u_j$. The quantity E (J/m^3) is the total energy density and γ is a constant depending on the gas studied ($\gamma = 1.4$ for air).

The equations (5.26) constitute five PDEs and one algebraic equation for the six unknowns ρ, u_1, u_2, u_3, E, and p. The equations were first derived by Leonard Euler in 1755. They are hyperbolic and the solutions can contain propagating shocks, narrow regions separating continuous flow regions. Across the shock, there are rapid changes in pressure, velocity, and density. The resolution of shocks puts special demands on the numerical methods.

5.3.3 The Convection–Diffusion–Reaction Equations

The convection–diffusion–reaction equations (CDR equations) model processes where the transport of chemical substances and energy is governed by convection, diffusion, and reactions. Coupled to the mass conservation equations, expressed in units of concentrations, there is an energy conservation equation formulated for the temperature.

$$\frac{\partial \mathbf{c}}{\partial t} + div\ (\mathbf{c}\mathbf{u}^T) = div\ (D\nabla\mathbf{c}) + S\mathbf{r}(\mathbf{c}, T) \tag{5.31a}$$

$$\rho C\left(\frac{\partial T}{\partial t} + div\ (T\mathbf{u})\right) = div\ (\kappa\nabla T) + \Delta\mathbf{H}^T\ S\mathbf{r}(\mathbf{c}, T) \tag{5.31b}$$

In the CDR equations (5.31), $\mathbf{c}$ (mol/m^3) is a vector of concentrations of the chemical substances taking part in a set of reactions, $\mathbf{u}$ (m/s) the velocity field (usually assumed to be known, otherwise the CDR equations must be coupled to the Navier–Stokes equations), ρ (kg/m^3) the density of the mixture, D (m^2/s) the diffusion coefficient (or the diffusion matrix), $\mathbf{r}$[mol/(m^3 · s)] the reaction rate terms (see Chapter 2), S the stoichiometric matrix, T(K) the temperature, C[J/(kg · K)] the heat capacity, κ[J/(m · K · s)] the heat conduction coefficient and $\Delta\mathbf{H}$(J/mol) the enthalpies of the reactions, $\Delta H_i < 0$ if the reaction is endotermic (energy consuming), and $\Delta H_i > 0$ if the reaction is exotermic (energy producing).

The CDR equations model processes in chemical engineering, e.g., combustion, industrial reactors, and environmental processes such as atmospheric chemistry. From these equations, the variations in substance concentrations and temperature of a process can be computed. The PDEs are parabolic–hyperbolic and often stiff and nonlinear with a jacobian being large and sparse.

5.3.4 The Heat Equation

The heat equation is a special case of the CDR equations above.

$$\frac{\partial T}{\partial t} = div\ (\alpha \nabla T) + Q(x, y, z) \tag{5.32}$$

where α (m^2/s) is the thermal diffusivity and $Q(x, y, z)$ (K/s) a source or sink term that accounts for the heat production or consumption in the medium.

The heat equation was formulated in 1807 by the French physicist Joseph Fourier. He also invented a method for analytical solution of this equation in some cases, known as the *Fourier method*, published 1822.

5.3.5 The Diffusion Equation

The diffusion equation is mathematically similar to the heat equation, also being a special case of the CDR equations.

$$\frac{\partial c}{\partial t} = div\ (D \nabla c) + q(x, y, z) \tag{5.33}$$

where c (mol/m^3) is the concentration of the substance being diffused, D (m^2/s) the diffusion coefficient, and $q(x, y, z)$ [$mol/(m^3 \cdot s)$] a source or sink term that accounts for substance production or consumption.

The diffusion equation, also called Fick's second law, was formulated in 1855 by the German physicist Adolf Fick.

5.3.6 Maxwell's Equations for the Electromagnetic Field

Maxwell's equations model the relation between the electric and magnetic fields generated by electric charges.

$$curl\ \mathbf{H} = \mathbf{j} + \frac{\partial \mathbf{D}}{\partial t} \tag{5.34a}$$

$$curl\ \mathbf{E} = -\frac{\partial \mathbf{B}}{\partial t} \tag{5.34b}$$

In the PDEs (5.34), $\mathbf{H}$ is the magnetic field strength (A/m), $\mathbf{E}$ the electric field strength (V/m), $\mathbf{B}$ the magnetic flux density (Vs/m^2), $\mathbf{D}$ the electric flux density (As/m^2), and $\mathbf{j}$ the current density (A/m^2). Maxwell's equations consist of 6 equations and 12 unknowns, $\mathbf{H}$, $\mathbf{E}$, $\mathbf{B}$, and $\mathbf{D}$. The missing equations are 6 *constitutive* equations, i.e., the relation between the electric variables and the magnetic variables, e.g., the equations valid in vacuum $\mathbf{D} = \epsilon_0 \mathbf{E}$ and $\mathbf{B} = \mu_0 \mathbf{H}$, where ϵ_0 and μ_0 are the electric permittivity and magnetic permeability. If $\mathbf{j}$ is given, Maxwell's equations can be solved assuming suitable ICs and BCs are provided.

The equations (5.34) were first invented by the Scottish physicist James Maxwell in 1867. Maxwell's equations are hyperbolic.

The following equation, known as the *charge continuity equation*, is fundamental

$$\frac{\partial \rho}{\partial t} + div\, \mathbf{j} = 0 \tag{5.35}$$

where ρ is the electric charge density (As/m^3).

Two more equations are usually associated with Maxwell's equations

$$div\, \mathbf{B} = 0 \tag{5.36a}$$

$$div\, \mathbf{D} = \rho \tag{5.36b}$$

These equations are consequences of (5.34) and (5.35) under suitable assumptions.

The force acting on an electric particle with charge q in an electromagnetic field is given by the *Lorentz-force* expression

$$\mathbf{F} = q(\mathbf{E} + \mathbf{v} \times \mathbf{B}) \tag{5.37}$$

If $\mathbf{j} = 0$, Maxwell's equations in vacuum are

$$curl\, \mathbf{H} = \epsilon_0 \frac{\partial \mathbf{E}}{\partial t} \tag{5.38a}$$

$$curl\, \mathbf{E} = -\mu_0 \frac{\partial \mathbf{H}}{\partial t} \tag{5.38b}$$

From these two equations, the following PDE is derived

$$\Delta \mathbf{E} - \frac{1}{c^2} \frac{\partial^2 \mathbf{E}}{\partial t^2} = 0 \tag{5.39}$$

and a similar equation for $\mathbf{H}$. The parameter c (m/s) is the wave propagation speed $c = (\epsilon_0 \mu_0)^{-1/2}$, speed of light 299792458 m/s. The equation (5.39) is called the *vector wave equation* and is a system of hyperbolic PDEs. The solutions are usually smooth, but numerical methods are time consuming as very many periods have to be computed to follow the rays and each period must be accurately enough represented.

5.3.7 Acoustic Waves

The propagation equation for acoustic waves has the form

$$\Delta p - \frac{1}{c^2} \frac{\partial^2 p}{\partial t^2} = 0 \tag{5.40}$$

where p (N/m^2) is the pressure and c (m/s) the speed of sound.

Equation (5.40) is a *scalar wave equation*, a hyperbolic PDE. It is similar to the equation for a vibrating string, presented in Chapter 9.

5.3.8 Schrödinger's Equation in Quantum Mechanics

Schrödinger's equation models the wave nature of small particles, e.g., electrons and atoms.

$$-\frac{\hbar^2}{2m}\Delta\Psi + V(\mathbf{r},t)\Psi = i\hbar\frac{\partial\Psi}{\partial t} \tag{5.41}$$

where $\hbar = h/2\pi$, h is the Planck's constant $= 6.626068\,\mathrm{m}^2 \cdot \mathrm{kg/s}$, m the mass of the particle, and $V(\mathbf{r}, t)$ a potential function that affects the movement of the particle. This equation is hyperbolic and was first proposed by the Austrian physicist Erwin Schrödinger in 1926.

In (5.41), $\Psi(\mathbf{r}, t)$ is a complex wavefunction, representing a propagating wave for a particle. This function is used to compute the probability of finding the particle in a certain domain in space. The solutions of the Schrödinger equation are oscillating and usually very smooth. The computational complexity increases immensely with the number of particles studied.

Often the interest is not directed to the wave function Ψ itself but rather to the possible energy levels E of the particle, e.g., the electron states of an atom or molecule, leading to eigenvalue problems of the type

$$\left(-\frac{\hbar^2}{2m}\Delta + V(\mathbf{r})\right)\Psi = E\Psi \tag{5.42}$$

The computation of the E values must be done with high accuracy in order to be able to distinguish between the energy levels. This puts certain requirements on the grid used in the discretization of (5.42).

5.3.9 Navier's Equations in Structural Mechanics

Navier's equations are the fundamental PDEs in elasticity theory. They model the deformations of a linearly elastic body that is exposed to external forces

$$\rho\frac{\partial^2\mathbf{u}}{\partial t^2} = \frac{E}{2(1+\nu)}\left(\Delta\mathbf{u} + \frac{1}{1-2\nu}\nabla(div\ \mathbf{u})\right) + \mathbf{f} \tag{5.43}$$

These equations are hyperbolic and were formulated by Louis Navier in France 1821. In case the problem is time independent, the equations are elliptic.

In (5.43), $\mathbf{u}$ is the displacement vector, i.e., $\mathbf{u} = (u, v, w)^T$, where u (m) is the displacement in the x direction, v (m) in the y direction, and w (m) in the z direction. E (N/m^2) is the elasticity module, ν [] the contraction parameter (also called Poisson's number), $\mathbf{f}$ (N/m^3) the force field acting on the body responsible for the displacements, and ρ (kg/m^3) the density of the body.

5.3.10 Black–Scholes Equation in Financial Mathematics

The Black–Scholes equation presented in 1973 models the value u of a European stock option

$$\frac{\partial u}{\partial t} + \frac{1}{2}\sigma^2 x^2 \frac{\partial^2 u}{\partial x^2} + rx\frac{\partial u}{\partial x} - ru = 0 \tag{5.44}$$

where x is the underlying asset value, r the continually compounding interest rate, and σ the volatility (standard deviation of the rate of return of the asset).

***Exercise* 5.3.1.** Verify that Navier–Stokes equation (5.26) in 1D is

$$\rho\left(\frac{\partial u}{\partial t} + u\frac{\partial u}{\partial x}\right) = -\frac{\partial p}{\partial x} + \mu\frac{\partial^2 u}{\partial x^2} + \rho f$$

***Exercise* 5.3.2.** In case $p = 0$ and $f = 0$ in Exercise 5.3.1, we obtain Burger's equation

$$\frac{\partial u}{\partial t} + u\frac{\partial u}{\partial x} = \nu\frac{\partial^2 u}{\partial x^2}$$

where the parameter $\nu = \mu/\rho$ is the kinematic viscosity.

a) Is this PDE elliptic, parabolic, or hyperbolic?

b) If $\frac{\partial u}{\partial t} = 0$, the problem is time independent. Solve the ODE with the BCs

$$u(0) = 1, \quad \frac{du}{dx}(0) = 0$$

c) With the transformation

$$u = -2\nu\frac{1}{v}\frac{\partial v}{\partial x},$$

the nonlinear PDE is transformed into a linear PDE

$$\frac{\partial v}{\partial t} = \nu\frac{\partial^2 v}{\partial x^2} + C(t)v.$$

***Exercise* 5.3.3.** Formulate Euler's equations (5.30) in 1D.

***Exercise* 5.3.4.** Formulate the CDR equation in 1D assuming that there is only one chemical substance with concentration c involved, T is constant, and the rate function is given by $r(c) = kc$, where k is the rate constant. In chemical engineering, this model is known as the *tubular reactor model.*

Exercise 5.3.5. Formulate the heat equation in 2D, cylindrical coordinates.

Exercise 5.3.6. Show that Maxwell's equation (5.38a) and (5.38b) for plane waves propagating in the x direction with the only components $E_z \neq 0$ and $H_y \neq 0$ satisfy the PDE system

$$\frac{\partial H_y}{\partial t} = \frac{1}{\mu_0}\frac{\partial E_z}{\partial x}$$

$$\frac{\partial E_z}{\partial t} = \frac{1}{\epsilon_0}\frac{\partial H_y}{\partial x}$$

Also show that both E_z and H_y satisfy the wave equation.

5.4 INITIAL AND BOUNDARY CONDITIONS FOR PDEs

The PDEs presented in Section 5.3 have all been given without initial or boundary conditions. Such conditions are needed for a unique solution and hence necessary for numerical computation. Initial conditions define the state of a system at time $t = 0$. Boundary conditions specify the state of a problem against the environment. If the space domain is infinite, a *Cauchy problem*, it must be truncated to a finite domain before numerical treatment.

Boundary conditions can be of different types, e.g., Dirichlet, Neuman, mixed, periodic, numerical, absorbing, and reflecting. Examples of such conditions are presented in the chapters to follow.

For a given PDE, the initial and boundary conditions cannot be given arbitrarily; they should be given so that the problem is *well posed*. This means that the solution is continuous with respect to the given conditions.

5.5 NUMERICAL SOLUTION OF PDES, SOME GENERAL COMMENTS

When solving ODEs, there are some general methods, e.g., explicit and implicit Runge–Kutta methods, that could be applied to most problems, nonstiff and stiff ODE systems. The situation is different with PDEs. The classification (parabolic, elliptic, hyperbolic) determines the type of method that should be used. Some comments on numerical properties for *linear* PDEs are given below. For nonlinear PDEs, there will be additional difficulties of various types.

A parabolic PDE can be approximated by an ODE system, being large, sparse, and stiff, i.e., implicit methods are needed where the sparsity must be taken into account.

An elliptic PDE can be approximated by a large system of algebraic equations, being very sparse, i.e., numerical methods specially designed for such equations are needed.

A hyperbolic PDE needs special attention as a solution may contain a propagating discontinuity. Special methods are needed to capture the discontinuity and to properly conserve different quantities.

BIBLIOGRAPHY

Mathematical textbooks on PDEs

1. S.J. Farlow, "Partial Differential Equations for Scientists and Engineers", Wiley, 1982
2. F. John, "Partial Differential Equations", Springer, 1991
3. A.D. Snider, "Partial Differential Equations, Sources and Solutions", Prentice Hall, 1999
4. H.F. Davis, A.D. Snider, "Introduction to Vector Analysis", Quantum, 1995

Textbooks on mathematics and mathematical models in science and engineering

5. Courant, Hilbert, "Methods of Mathematical Physics", Interscience, 1955
6. K. Eriksson, D. Estep, C. Johnson, "Applied Mathematics: Body and Soul I-III, Springer, 2004
7. Lin, Segel, "Mathematics Applied to Deterministic Problems in the Natural Sciences", Macmilan, 1974

6

NUMERICAL METHODS FOR PARABOLIC PARTIAL DIFFERENTIAL EQUATIONS

Parabolic partial differential equations (PDEs) are frequently encountered in mathematical models for heat transfer and diffusion. Their solutions reflect the underlying physics, such as the dissipative nature of heat transfer.

A fundamental property of solutions to parabolic PDEs is that information travels with infinite speed in the sense that an initial value immediately effects the solution in other space points. Solutions to parabolic PDEs are stable and initial peaks are smoothed out when integrated in one of the t-directions (say $t > 0$) but *ill posed* in the other direction ($t < 0$).

Parabolic PDEs have a natural coupling to systems of ordinary differential equations (ODEs). Discretization of the space derivatives with the finite difference method (FDM) or with the finite element method (FEM) leads to *stiff* ODE systems. In Chapter 5, the generic parabolic PDE problem in 1D was introduced. In this chapter, this PDE is used as a *model problem* for various numerical methods, i.e.,

$$\frac{\partial u}{\partial t} = \kappa \frac{\partial^2 u}{\partial x^2}, \quad 0 < x < 1, \quad 0 < t \leq t_{end}, \quad \kappa > 0 \tag{6.1a}$$

with Dirichlet boundary conditions (BCs)

$$u(0,t) = \alpha(t), \qquad u(1,t) = \beta(t) \tag{6.1b}$$

and an initial condition (IC)

$$u(x,0) = u_0(x) \tag{6.1c}$$

Introduction to Computation and Modeling for Differential Equations, Second Edition. Lennart Edsberg.
© 2016 John Wiley & Sons, Inc. Published 2016 by John Wiley & Sons, Inc.

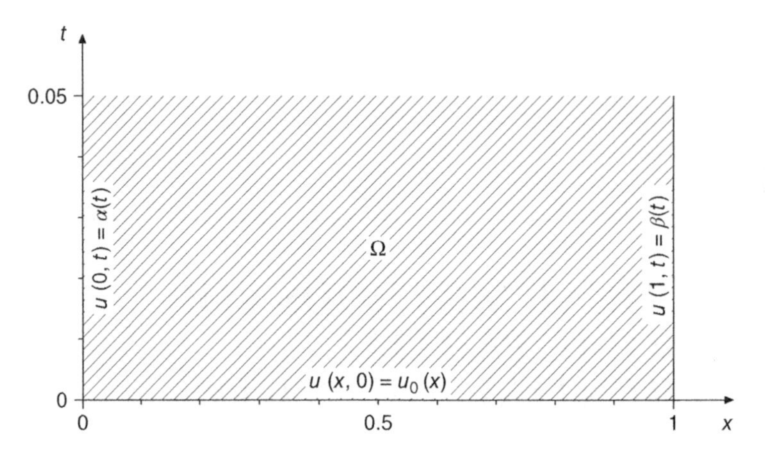

Figure 6.1 Region of definition, Ω, in the $x - t$ plane

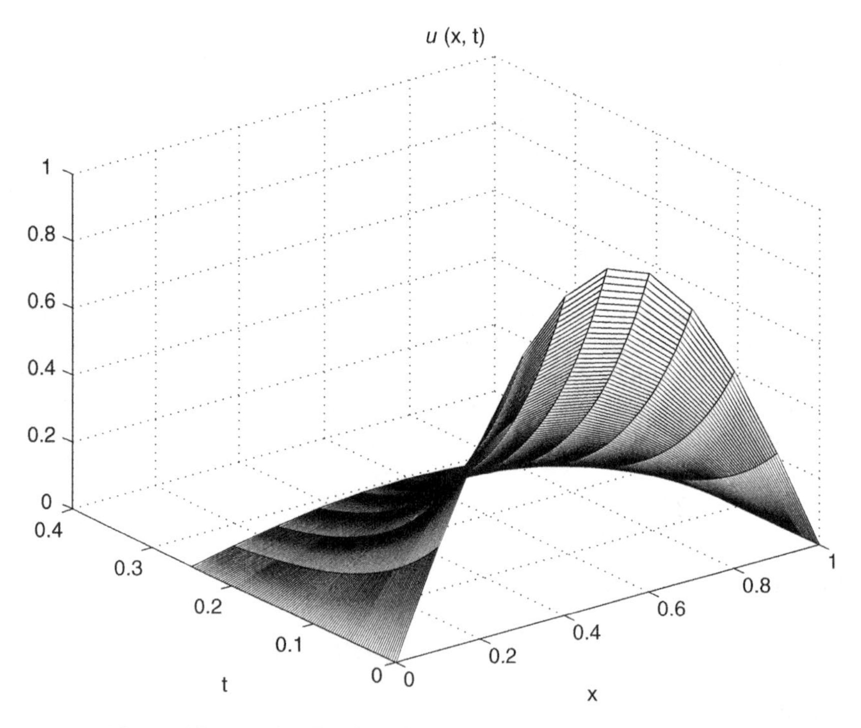

Figure 6.2 3D visualization of a solution $u(x, t)$ of the heat equation

Geometrically the solution $u(x, t)$ will be a surface above the region of definition $\Omega = \{(x, t), \quad 0 \leq x \leq 1, \quad 0 \leq t \leq t_{\text{end}}\}$ (see Figure 6.1). With the IC function $u_0(x) = \sin(\pi x)$ and the BC functions $\alpha(t) = 0$ and $\beta(t) = 0$, the graph of the solution (for $\kappa = 1$) is shown in Figure 6.2.

An important generalization of (6.1) is to include space dependence in the heat conduction coefficient $\kappa(x)$ and a driving function $f(x, t)$ (or $f(x)$) in the right hand side. We then have the *inhomogeneous* heat equation

$$\frac{\partial u}{\partial t} = \frac{\partial}{\partial x}\left(\kappa(x)\frac{\partial u}{\partial x}\right) + f(x,t), \qquad 0 \leq x \leq 1, \quad t \geq 0 \tag{6.2}$$

The steady-state solution of (6.2) is obtained when $t \to \infty$. In the steady-state case, there is no time variation in the solution. Therefore, the steady-state solution satisfies the boundary value problem (BVP)

$$-\frac{d}{dx}\left(\kappa(x)\frac{du}{dx}\right) = f(x), \qquad u(0) = \alpha_0, \quad u(1) = \beta_0 \tag{6.3}$$

where α_0 and β_0 are constants and $f(x)$ a function that is time independent. Compare with the model problem in (4.20).

6.1 APPLICATIONS

The occurrence of parabolic PDEs in application problems is now demonstrated on some examples.

Example 6.1. (Time-dependent 1D flow in a pipe). Consider the boundary value problems (BVP) in Example 4.1. Now assume that the temperature of the fluid has not come to a steady state (Figure 6.3) but is passing through a time-dependent transient from a given IC. The model for the temperature $T(z, t)$ (K) is changed into the PDE

$$\rho C\frac{\partial T}{\partial t} + \rho Cv\frac{\partial T}{\partial z} = \frac{\partial}{\partial z}\left(\kappa\frac{\partial T}{\partial z}\right) - \frac{2h}{R}(T - T_{out}), \quad 0 < z < L, \quad t > 0 \tag{6.4}$$

with IC

$$T(z, 0) = T_{init}(z)$$

and BCs

$$T(0, t) = T_0, \qquad -\kappa\frac{\partial T}{\partial z}(L) = k(T(L) - T_{out})$$

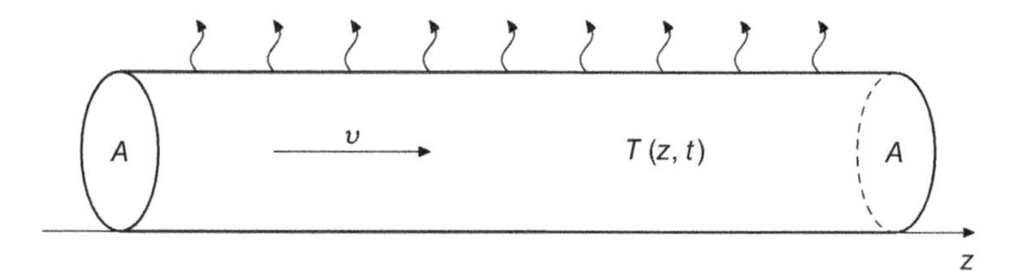

Figure 6.3 Hot flow in a cylindrical pipe

The only new variable and parameter introduced in this model compared to the model in Example 4.1 is time t (s) and the initial value $T_{\text{init}}(z)$ (K). We see that when steady state is reached ($t \to \infty$), the first term drops out, the IC is not necessary, and the PDE problem turns into the BVP (4.10).

Example 6.2. (Time-dependent 2D flow in a pipe). Consider again the flow in a pipe model but now assume that heat is transported by diffusion both in the z- and in the r directions. The model for the temperature $T(z, r, t)$ (K) is modified to

$$\rho C\frac{\partial T}{\partial t} + \rho C v\frac{\partial T}{\partial z} = \kappa\frac{\partial^2 T}{\partial z^2} + \frac{\kappa}{r}\frac{\partial}{\partial r}\left(r\frac{\partial T}{\partial r}\right), \quad 0 < z < L,\ 0 < r < R,\ t > 0 \quad (6.5)$$

with IC

$$T(z, r, 0) = T_{init}(z, r), \quad 0 \le z \le L, 0 \le r \le R$$

and BCs for $z = 0$ and $z = L$

$$T(0, r, t) = T_0, \quad T(L, r, t) = T_{out}, \quad 0 \le r \le R, t \ge 0$$

and BCs for $r = 0$ and $r = R$ (Figure 6.4)

$$\frac{\partial T}{\partial r}(z, 0, t) = 0, \quad -\kappa\frac{\partial T}{\partial r}(z, R, t) = k(T(z, R, t) - T_{out}), \quad 0 \le z \le L, t \ge 0$$

Note that the heat loss through the wall is now implemented as a BC at $r = R$.

Exercise 6.1.1. Formulate the steady-state problem of the PDE (6.5). The solution will depend on z and r, i.e., $T = T(z, r)$. Classify the PDE, i.e., is it parabolic, elliptic, or hyperbolic?

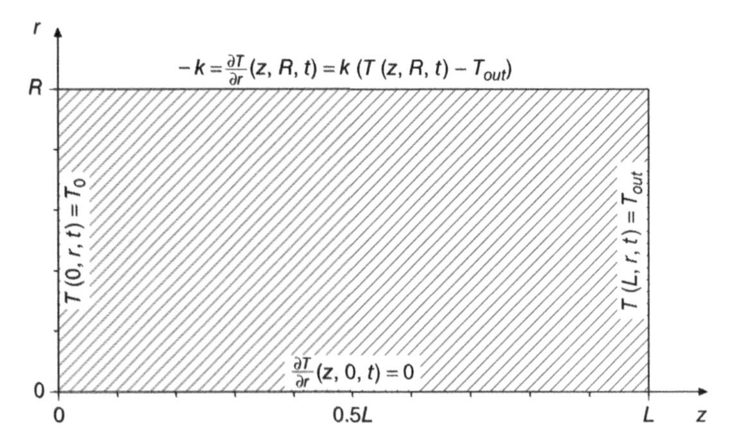

Figure 6.4 2D model of time-dependent flow in a pipe

Example 6.3. (Time-dependent 1D tubular reactor model). Consider again a cylindrical pipe but this time it is called a tube in accordance with the nomenclature in chemical engineering. A fluid is transported through the tube of length L (m) with an exothermal reaction going on in it. Simplifying the general convection-diffusion-reaction (CDR) equations in Section 5.3.3 to only one chemical species concentration c (mol/m^3) and a given constant flow speed v (m/s) gives the following system of two PDEs:

$$\frac{\partial c}{\partial t} + v\frac{\partial c}{\partial z} = \frac{\partial}{\partial z}\left(D\frac{\partial c}{\partial z}\right) - Ae^{-E/RT}c \tag{6.6a}$$

$$\rho C\frac{\partial T}{\partial t} + \rho Cv\frac{\partial T}{\partial z} = \frac{\partial}{\partial z}\left(\kappa\frac{\partial T}{\partial z}\right) + \Delta HAe^{-E/RT}c \tag{6.6b}$$

where units of the variables c, T, z, t and parameters v, ρ, C, D, κ, ΔH, A, E, and R are explained in Sections 2.4.4 and 5.3.3. The PDE system (6.6) is *nonlinear* due to the term $Ae^{-E/RT}c$.

In addition to the PDE (6.6), suitable ICs and BCs have to be given, e.g.,

$$c(z, 0) = c_{init}(z), \quad T(z, 0) = T_{init}(z)$$

and

$$c(0, t) = c_0, \; T(0, t) = T_0, \quad \frac{\partial c}{\partial z}(L, t) = 0, \; \frac{\partial T}{\partial z}(L, t) = 0$$

Exercise 6.1.2. Formulate the steady-state problem of (6.6) with suitable BCs. If the steady-0 state model is reduced further so that T is assumed to be constant, formulate the corresponding BVP with BCs.

6.2 AN INTRODUCTORY EXAMPLE OF DISCRETIZATION

When using the FDM to solve the parabolic problem (6.1) for $\kappa = 1$, we can follow the workscheme introduced in Chapter 4, i.e.,

DG—Discretize the region of definition Ω by using the stepsize h_x for the x-axis and the stepsize h_t for the t-axis:

$$x_i = ih_x, \quad i = 0, 1, 2, \ldots, N + 1, \qquad t_k = kh_t, \quad k = 0, 1, 2, \ldots$$

A rectangular grid is defined (see Figure 6.5).

DD—Discretize the differential equation (6.1). Denote by $u_{i,k}$ the numerical solution of the problem at the point (x_i, t_k), i.e., $u(x_i, t_k) \approx u_{i,k}$. Use the explicit

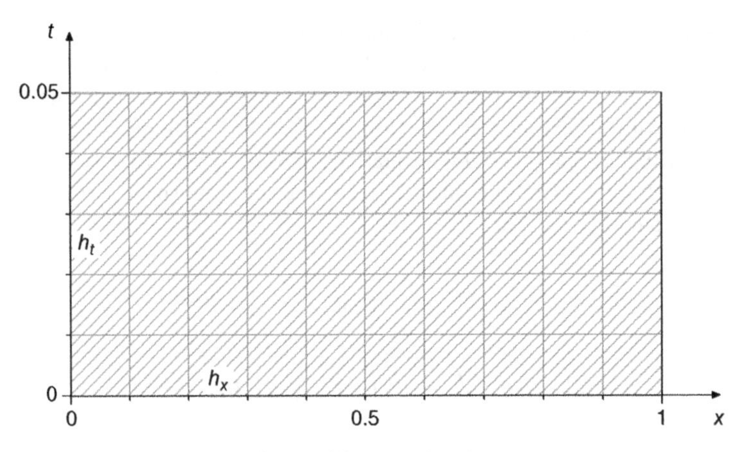

Figure 6.5 2D grid of Ω

Euler approximation for the time derivative

$$\frac{\partial u}{\partial t}(x_i, t_k) = \frac{u_{i,k+1} - u_{i,k}}{h_t} + \mathcal{O}h_t \tag{6.7}$$

and the central difference approximation for the space derivative

$$\frac{\partial^2 u}{\partial x^2}(x_i, t_k) = \frac{u_{i+1,k} - 2u_{i,k} + u_{i-1,k}}{h_x^2} + \mathcal{O}h_x^2 \tag{6.8}$$

and we obtain the following recursion formula for the heat equation:

$$\frac{u_{i,k+1} - u_{i,k}}{h_t} = \frac{u_{i+1,k} - 2u_{i,k} + u_{i-1,k}}{h_x^2} \tag{6.9}$$

This FDM can also be referred to as the *FTCS method* (forward-time-central-space).
Introduce the stepsize parameter $\sigma = h_t/h_x^2$ and (6.9) can be written as

$$u_{i,k+1} = \sigma u_{i-1,k} + (1 - 2\sigma)u_{i,k} + \sigma u_{i+1,k} \tag{6.10}$$

There is an appealing visualization of this formula as a *stencil* (also called a difference star or computational molecule) (see Figure 6.6). For FDMs, there is always such a stencil showing which $u_{i,k}$ values at the time level k (and/or earlier time levels) that are to be combined to give the $u_{i,k+1}$ values at the time level $k + 1$.

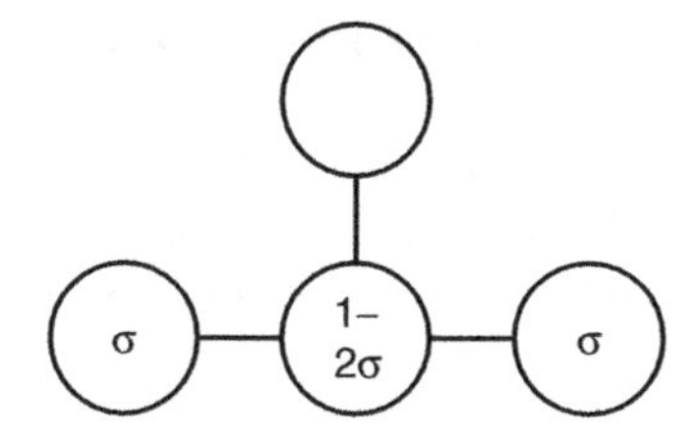

Figure 6.6 Stencil for the heat equation

DB—Discretize the IC and BCs. For problem (6.1), the IC and BCs are represented exactly

$$u_{i,0} = u_0(x_i), \qquad u_{0,k} = \alpha(t_k), \quad u_{N+1,k} = \beta(t_k) \tag{6.11}$$

In Figure 6.1, the corresponding grid points where these conditions are given are situated on the boundary $\partial\Omega$.
Observe, however, that if we have BCs containing derivatives, they have to be represented approximately (see Section 6.3).

After this discretization, the stencil is moved along the grid pattern from the lower left corner row by row from left to right up to the upper right corner according to Figure 6.7. After this computation, the $u_{i,k}$ values can be plotted above the grid, hence visualizing the solution just as in the solution graph (6.2). Observe that all values at the new time level can be computed simultaneously.

Exercise 6.2.1. Write a program giving the solution in Figure 6.2 using the method (6.10). Use the stepsizes $h_x = 0.1$, $h_t = 0.001$.

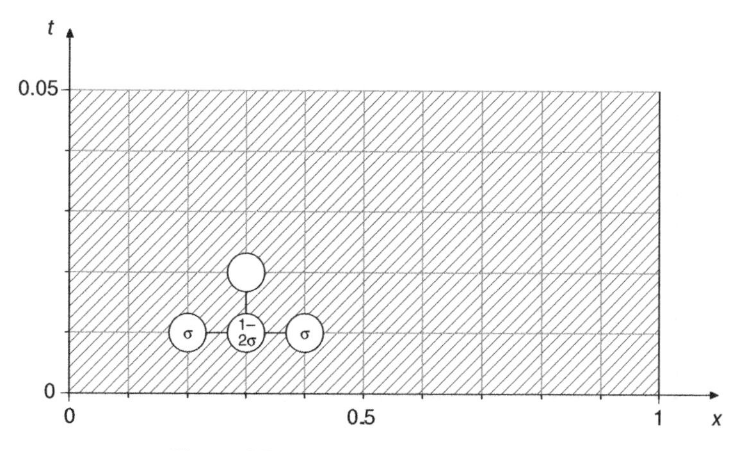

Figure 6.7 Stencil moving along the grid

6.3 THE METHOD OF LINES FOR PARABOLIC PDES

The Method of Lines (MoL) is based on the FDM and can be characterized as a general method for numerical solution of time-dependent PDEs. In the MoL, the space derivatives are discretized, but the time derivatives are kept. This principle of approximation is referred to as *semi-discretization.*

6.3.1 Solving the Test Problem with MoL

For the generic parabolic problem (6.1), semi-discretization means that $u(x_i, t) \approx u_i(t)$ where $u_i(t)$ is a time-dependent solution function associated to the space point x_i, where the $u_i(t)$ function satisfies the ODE ($\kappa = 1$)

$$\frac{du_i(t)}{dt} = \frac{u_{i+1}(t) - 2u_i(t) + u_{i-1}(t)}{h_x^2}, \quad u_i(0) = u_0(x_i), \quad i = 1, 2, \ldots, N \tag{6.12}$$

For $i = 1$ and $i = N$, the BCs enter as driving functions

$$\frac{du_1(t)}{dt} = \frac{u_2(t) - 2u_1(t) + \alpha(t)}{h_x^2}, \qquad u_1(0) = u_0(x_1) \tag{6.13}$$

$$\frac{du_N(t)}{dt} = \frac{\beta(t) - 2u_N(t) + u_{N-1}(t)}{h_x^2}, \qquad u_N(0) = u_0(x_N) \tag{6.14}$$

In Figure 6.8, the lines of definition for the solution curves are shown.
The ODEs (6.12)–(6.14) above can be written in matrix form

$$\frac{d\mathbf{u}}{dt} = T\mathbf{u} + \mathbf{b}(t), \qquad \mathbf{u}(0) = \mathbf{u}_0 \tag{6.15}$$

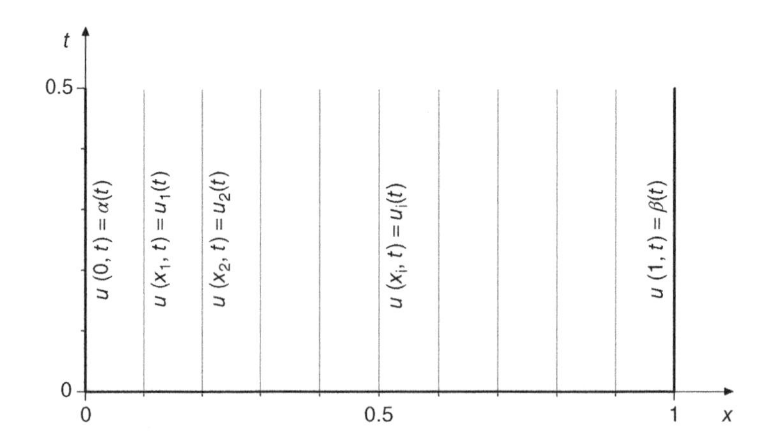

Figure 6.8 Grid for a MoL discretization

where

$$T = \frac{1}{h_x^2}\begin{pmatrix} -2 & 1 & 0 & \dots & 0 \\ 1 & -2 & 1 & \ddots & \vdots \\ 0 & \ddots & \ddots & \ddots & 0 \\ \vdots & \ddots & 1 & -2 & 1 \\ 0 & \dots & 0 & 1 & -2 \end{pmatrix} \tag{6.16}$$

and

$$\mathbf{b}(t) = \frac{1}{h_x^2}\begin{pmatrix} \alpha(t) \\ 0 \\ 0 \\ \vdots \\ \beta(t) \end{pmatrix}, \quad \mathbf{u}(0) = \begin{pmatrix} (u_0 x_1 \\ u_0(x_2) \\ u_0(x_3) \\ \vdots \\ u_0(x_N) \end{pmatrix} \tag{6.17}$$

Hence, the ODE system is linear with constant coefficients. The tridiagonal matrix T differs from the tridiagonal matrix A in (4.29) by a scalar factor

$$T = -\frac{1}{h_x^2}A = \frac{1}{h_x^2}\text{tridiag}(1, -2, 1) \tag{6.18}$$

where the notation *tridiag* is also used in Section 4.2.

In Section A.2, the eigenvalues of A are calculated exactly and the values are

$$\lambda_j(A) = 4\sin^2\left(\frac{j\pi}{2(N+1)}\right)$$

Hence, the eigenvalues of T are

$$\lambda_j(T) = -\frac{4}{h_x^2}\sin^2\left(\frac{j\pi}{2(N+1)}\right), \quad j = 1, 2, \dots, N \tag{6.19}$$

Since the x interval is discretized with constant stepsize h_x according to $h_x(N+1) = 1$, $\lambda_j(T)$ can be written as

$$\lambda_j(T) = -\frac{4}{h_x^2}\sin^2\left(\frac{j\pi h_x}{2}\right), \quad j = 1, 2, \dots, N \tag{6.20}$$

We see that all eigenvalues of T are real, negative, and of very different magnitude in the range from approximately $-\pi^2$ (for $j = 1$) to $-4/h_x^2$ (for $j = N$). Hence, (6.15) is a *stiff* system of ODEs.

If Euler's explicit method is used to solve (6.15), we get the recursion formula

$$\mathbf{u}_{k+1} = \mathbf{u}_k + h_t(T\mathbf{u}_k + \mathbf{b}(t_k)), \qquad \mathbf{u}_0 = \mathbf{u}(0) \tag{6.21}$$

Figure 6.2 is generated with Euler's explicit method, which is equivalent to the stencil method based on formula (6.10).

In Chapter 3, the stability areas of some methods for IVP are shown. For the explicit Euler method, the stepsize in time, here denoted by h_t, must fulfill the condition

$$h_t\,\lambda_j \in S_{\text{EE}}, \quad \rightarrow \quad h_t\frac{4}{h_x^2} \le 2, \quad \rightarrow \quad \frac{h_t}{h_x^2} \le \frac{1}{2} \tag{6.22}$$

Hence, the time step h_t must be quadratically smaller than the space step h_x, which puts a very severe restriction on h_t, making the explicit Euler method very inefficient for parabolic PDEs.

Violating this stepsize criterion will give a very dramatic numerical solution, which is wrong of course (see Figure 6.9). The stability result (6.22) is based on the eigenvalues of the matrix T and is therefore called *eigenvalue stability*. In Section 8.3, another method based on *Fourier analysis* is presented. This kind of stability analysis is called *von Neumann stability*, named after the Hungarian-American mathematician John von Neumann, active in the middle of the 20th century. The results from the von Neumann stability analysis are the same as the eigenvalue stability for parabolic PDEs (see Section A.6). However, in Chapter 8, hyperbolic PDEs, the von Neumann analysis is the only adequate stability analysis.

Using a stiff method gives stable numerical solutions for large time steps h_t. For the implicit Euler method (also called the BTCS method from

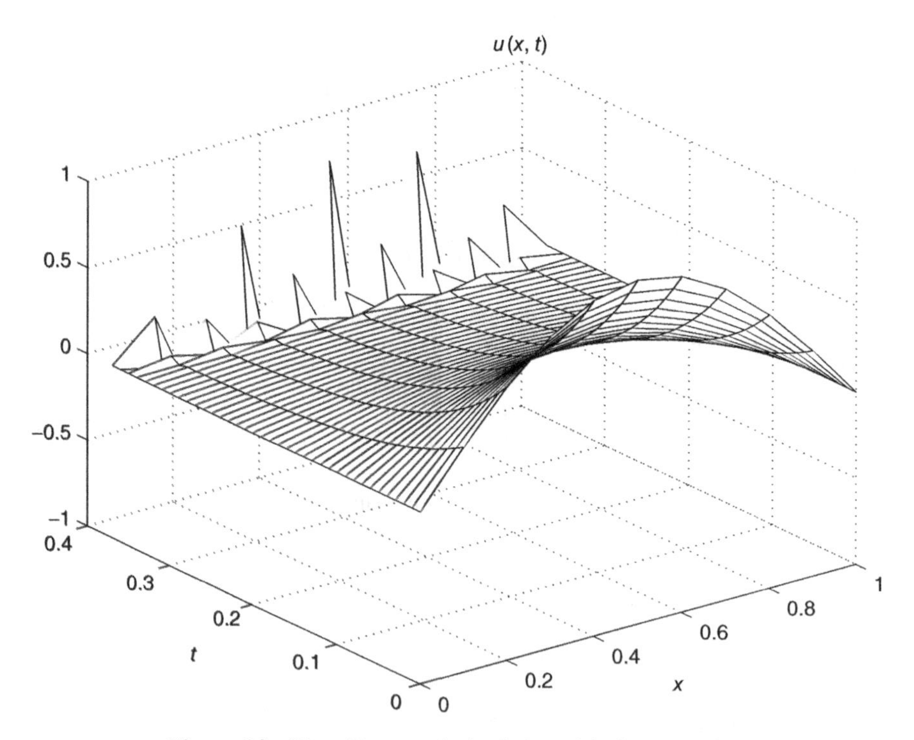

Figure 6.9 Unstable numerical solution of the heat equation

backward-time-central-space), we get the recursion formula

$$\mathbf{u}_{k+1} = \mathbf{u}_k + h_t(T\mathbf{u}_{k+1} + \mathbf{b}(t_{k+1})), \qquad \mathbf{u}_0 = \mathbf{u}(0) \tag{6.23}$$

After rearrangement, we get a linear system of equations to solve

$$(I - h_t T)\mathbf{u}_{k+1} = \mathbf{u}_k + h_t \mathbf{b}(t_{k+1}) \tag{6.24}$$

The matrix $(I - h_t T)$ is tridiagonal, symmetric, and positive definite

$$I - h_t T = \text{tridiag}(-\sigma, 1 + 2\sigma, -\sigma) \tag{6.25}$$

where $\sigma = h_t/h_x^2$. However, the accuracy in the $\mathbf{u}_k$ values is poor, since implicit Euler is only first order.

An often-used method for the heat/diffusion equation is Crank – Nicolson's method. John Crank and Phyllis Nicolson were British mathematicians and the method was introduced around 1950. The Crank – Nicolson method is of second order in both the x and the t directions. In fact, this method is based on the trapezoidal method for solving (6.15)

$$\mathbf{u}_{k+1} = \mathbf{u}_k + \frac{h_t}{2}(T\mathbf{u}_k + \mathbf{b}(t_k) + T\mathbf{u}_{k+1} + \mathbf{b}(t_{k+1})), \qquad \mathbf{u}_0 = \mathbf{u}(0) \tag{6.26}$$

On inspection, we see that the trapezoidal method can be regarded as a combination of the explicit and the implicit Euler methods. The trapezoidal method will also give a tridiagonal system of linear algebraic equations to be solved in each time step

$$\left(I - \frac{h_t}{2}T\right)\mathbf{u}_{k+1} = \left(I + \frac{h_t}{2}T\right)\mathbf{u}_k + \frac{h_t}{2}(\mathbf{b}(t_{k+1}) + \mathbf{b}(t_k)) \tag{6.27}$$

Crank – Nicolson's method is stable for all time steps h_t and spacesteps h_x, but if h_t is too large compared to h_x, there will be damped oscillations in the numerical solution as is seen in the Figure 6.10(a). If h_t is small enough, we get a smooth solution (b).

Exercise 6.3.1. Give the stencils corresponding to the implicit Euler method and the Crank – Nicolson method.

Exercise 6.3.2. Consider the following system of parabolic PDEs:

$$\frac{\partial c_1}{\partial t} = D_{11}\frac{\partial^2 c_1}{\partial x^2} + D_{12}\frac{\partial^2 c_2}{\partial x^2}$$

$$\frac{\partial c_2}{\partial t} = D_{21}\frac{\partial^2 c_1}{\partial x^2} + D_{22}\frac{\partial^2 c_2}{\partial x^2}$$

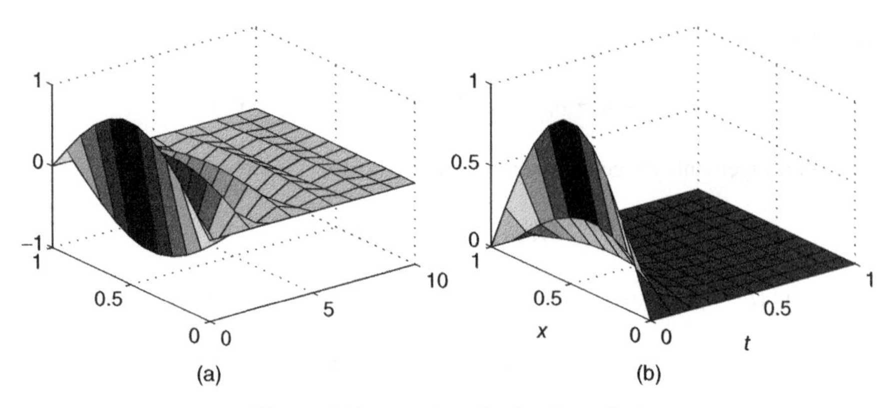

Figure 6.10 Crank – Nicolson's method

with ICs

$$c_1(x,0) = c_{10}, \quad c_2(x,0) = c_{20}$$

and BCs

$$c_1(0,t) = 1, \; c_2(0,t) = 1, \quad \frac{\partial c_1}{\partial x}(1,t) = 0, \; \frac{\partial c_2}{\partial x}(1,t) = 0$$

The parameters D_{ij}, $i = 1, 2, j = 1, 2$ are constants. Use the MoL to discretize the PDE system into a system of ODEs. Follow the DG, DD, and DB recipe and give the answer in the form

$$\frac{d\mathbf{c}}{dt} = A\mathbf{c} + \mathbf{b}, \quad \mathbf{c}(0) = \mathbf{c}_0$$

The structure of A depends on in which order the components of $\mathbf{c}_1$ and $\mathbf{c}_2$ are sorted. Try to find the way of sorting that gives the smallest bandwidth of A.

A more effective way to solve the ODE system (6.15) is to use some higher order method for stiff ODEs, such as the BDF method, described in Chapter 3. Parabolic PDEs are often solved in this way and the efficiency is taken into account by utilizing the sparse structure of the jacobian (often tridiagonal) of the ODE system obtained after space discretization.

6.3.2 Various Types of Boundary Conditions

Common BCs for parabolic PDEs are of the same type as the BCs for a second-order BVP (see Chapter 4). Assume that $x = a$ is a boundary point:

- If u is specified at $x = a$, we have a Dirichlet condition

$$u(a,t) = \alpha(t) \tag{6.28}$$

- If $\partial u/\partial x$ is specified at $x = a$, we have a Neumann condition

$$\frac{\partial u}{\partial x}(a,t) = \gamma(t) \tag{6.29}$$

- If a linear combination of u and $\partial u/\partial x$ is specified at $x = a$, we have a Robin condition (also called mixed or generalized Neumann condition)

$$\frac{\partial u}{\partial x}(a,t) = \mu(u(a,t) - u_{out}(t)) \tag{6.30}$$

  where μ is a constant and $u_{\text{out}}(t)$ is a given function.
- An example of a *nonlinear* BC is the heat radiation condition

$$\frac{\partial u}{\partial x}(a,t) = \lambda(u^4(a,t) - u^4_{out}(t)) \tag{6.31}$$

In connection with the heat equation, all the BCs given above have a physical meaning in terms of the temperature:

- The Dirichlet BC means that the temperature $\alpha(t)$ is known at the boundary point $x = a$.
- The Neumann condition means that the heat flux is known at the boundary point. The special case

$$\frac{\partial u}{\partial x}(a,t) = 0 \tag{6.32}$$

  means that the point $x = a$ is isolated from the environment.
- The mixed condition means that the heat flux is proportional to the temperature difference $u(a, t) - u_{\text{out}}(t)$.
- The nonlinear BC given above corresponds to heat radiation at the point $x = a$ according to Stefan – Boltzmann's law.

6.3.3 An Example of the Use of MoL for a Mixed Boundary Condition

The handling of BCs containing derivatives for parabolic PDE problems is similar to their treatment for two-point BVPs.

Given the model problem (6.1a), $\kappa = 1$, with one Dirichlet and one mixed BC

$$\frac{\partial u}{\partial t} = \frac{\partial^2 u}{\partial x^2}, \quad u(0,t) = \alpha(t), \quad \frac{\partial u}{\partial x}(1,t) = \mu(u(1,t) - u_{out}(t)) \tag{6.33}$$

we make a discretization with the MoL. Introduce an extra grid point after the right boundary point

0 h_x 1

x_0 x_1 x_{i-1} x_i x_{N-1} x_N x_{N+1} x

Hence, $x_i = ih_x$, $i = 0, 1, 2, \ldots, N + 1$, where $h_x = 1/N$.

The ODE system is as in (6.12)

$$\frac{du_i}{dt} = \frac{1}{h_x^2}(u_{i+1} - 2u_i + u_{i-1}), \quad i = 1, 2, \dots, N$$

At the point $x_1 = h$, we get after inserting the BC (6.13)

$$\frac{du_1}{dt} = \frac{1}{h_x^2}(u_2 - 2u_1 + \alpha(t))$$

At the last point $x_N = 1$, we have both an ODE and a discretized BC

$$\frac{du_N}{dt} = \frac{1}{h_x^2}(u_{N+1} - 2u_N + u_{N-1}) \tag{6.34}$$

$$\frac{u_{N+1} - u_{N-1}}{2h_x} = \mu(u_N - u_{out}(t)) \tag{6.35}$$

If u_{N+1} is eliminated, the last ODE is

$$\frac{du_N}{dt} = \frac{1}{h_x^2}(2u_{N-1} - (2 - 2h_x\mu)u_N - 2h_x\mu u_{out}(t)) \tag{6.36}$$

We now have a system of N ODEs for the N dependent variables $u_1, u_2, \dots, u_N$. ICs must be given, i.e., $u_i(0) = u_0(x_i)$

Exercise 6.3.3. Formulate the ODE system in Section 6.3 on the form $\dot{\mathbf{u}} = T\mathbf{u} + \mathrm{b}(t)$, i.e., find T and $\mathbf{b}$.

a) The matrix T is not symmetric. Find a diagonal transformation $\mathbf{u} = D\mathbf{v}$ so that the matrix $D^{-1}TD$ is symmetric.
b) Write a program for solving the ODE system in Section 6.3. Plot the solution.

6.4 GENERALIZATIONS OF THE HEAT EQUATION

6.4.1 The Heat Equation with Variable Conductivity

As indicated in (6.2), the parameter κ cannot always be assumed to be constant in applications. It can be, e.g., space dependent $\kappa = \kappa(x)$

$$\frac{\partial u}{\partial t} = \frac{\partial}{\partial x}\left(\kappa(x)\frac{\partial u}{\partial x}\right) \tag{6.37}$$

or temperature dependent $\kappa = \kappa(u)$

$$\frac{\partial u}{\partial t} = \frac{\partial}{\partial x}\left(\kappa(u)\frac{\partial u}{\partial x}\right) \tag{6.38}$$

In equation (6.37), the PDE is still linear, but in the (6.38), we have a nonlinear PDE to solve.

In the linear case, we discretize the space-dependent part of the PDE with the following second-order difference approximation:

$$\begin{aligned}\frac{\partial}{\partial x}\left(\kappa(x)\frac{\partial u}{\partial x}\right)(x_i) &= \frac{1}{h_x}\left(\kappa(x_{i+1/2})\frac{u_{i+1}-u_i}{h_x} - \kappa(x_{i-1/2})\frac{u_i - u_{i-1}}{h_x}\right) + \mathcal{O}h_x^2\\ &= \frac{1}{h_x^2}\left(\kappa(x_{i+1/2})u_{i+1} - (k(x_{i+1/2}) + \kappa(x_{i-1/2}))u_i + k(x_{i-1/2})u_{i-1}\right)\\ &\quad + \mathcal{O}h_x^2\end{aligned}$$

Hence, after applying MoL, we obtain the following system of ODEs:

$$\frac{d\mathbf{u}}{dt} = \frac{1}{h_x^2}\,\mathrm{tridiag}(\kappa(x_{i-1/2})\ , -(\kappa(x_{i-1/2}) + \kappa(x_{i+1/2}))\ ,\ \kappa(x_{i+1/2}))\mathbf{u} \tag{6.39}$$

i.e., a linear ODE system with a tridiagonal matrix.

In the nonlinear case, we use the same space discretization

$$\begin{aligned}\frac{\partial}{\partial x}\left(\kappa(u)\frac{\partial u}{\partial x}\right)(x_i) &= \frac{1}{h_x}\left(\kappa(u_{i+1/2})\frac{u_{i+1}-u_i}{h_x} - \kappa(u_{i-1/2})\frac{u_i - u_{i-1}}{h_x}\right) + \mathcal{O}h_x^2\\ &= \frac{1}{h_x^2}\left(\kappa(u_{i+1/2})u_{i+1} - (\kappa(u_{i+1/2}) + \kappa(u_{i-1/2}))u_i + \kappa(u_{i-1/2})u_{i-1}\right)\\ &\quad + \mathcal{O}h_x^2\end{aligned}$$

The u values $u_{i-1/2}$ and $u_{i+1/2}$ are, according to the notation, values that are not given in grid points. Therefore, we have to approximate them somehow. A simple symmetric approximation giving second-order accuracy is to set them to the mean values of the neighboring points, i.e.,

$$u_{i-1/2} = \frac{u_{i-1} + u_i}{2}, \qquad u_{i+1/2} = \frac{u_i + u_{i+1}}{2}$$

Applying the MoL gives the following nonlinear ODE system:

$$\begin{aligned}\frac{d\mathbf{u}}{dt} = \frac{1}{h_x^2}\,\mathrm{tridiag}\Bigg(&\kappa\left(\frac{u_{i-1}+u_i}{2}\right), -\left(\kappa\left(\frac{u_{i-1}+u_i}{2}\right) + \kappa\left(\frac{u_{i+1}+u_i}{2}\right)\right),\\ &\kappa\left(\frac{u_{i+1}+u_i}{2}\right)\Bigg)\mathbf{u}(3)\end{aligned} \tag{6.40}$$

When this system is discretized with a stiff method, e.g., implicit Euler, we obtain a nonlinear system of algebraic equations to be solved at each time step, which is accomplished with Newton's method. In each iteration, the jacobian is tridiagonal.

6.4.2 The Convection – Diffusion – Reaction PDE

The following equation occurs frequently in chemical technology and is a hyperbolic – parabolic PDE (see Chapter 5) the CDR equations:

$$\frac{\partial u}{\partial t} + v\frac{\partial u}{\partial x} = D\frac{\partial^2 u}{\partial x^2} + r(u) \tag{6.41}$$

If $v = 0$, we have a parabolic problem. If $D = 0$ the problem is hyperbolic.

If the MoL is used to discretize the PDE, we obtain (with central differences)

$$\frac{d\mathbf{u}}{dt} = T\mathbf{u} + \mathbf{b}(\mathbf{u}), \quad \mathbf{u}(0) \text{ given} \tag{6.42}$$

where T is an unsymmetric tridiagonal matrix

$$T = \text{tridiag}\left(\frac{D}{h_x^2} + \frac{v}{2h_x}, -\frac{2D}{h_x^2}, \frac{D}{h_x^2} - \frac{v}{2h_x}\right) \tag{6.43}$$

6.4.3 The General Nonlinear Parabolic PDE

A general formulation of a nonlinear scalar parabolic PDE is

$$\frac{\partial u}{\partial t} = f\left(x, t, u, \frac{\partial u}{\partial x}, \frac{\partial^2 u}{\partial x^2}\right), \quad t \geq 0, \quad 0 \leq x \leq 1 \tag{6.44}$$

Assuming Dirichlet conditions $u(0, t) = g_1(t), u(1, t) = g_2(t)$, the MoL can be used to discretize the problem to a system of nonlinear ODEs

$$\frac{du_i}{dt} = f\left(x_i, t, u_i, \frac{u_{i+1} - u_{i-1}}{2h}, \frac{u_{i+1} - 2u_i + u_{i-1}}{h^2}\right), \quad i = 1, 2, \dots, N \tag{6.45}$$

This system is of the form

$$\frac{d\mathbf{u}}{dt} = \mathbf{F}(t, \mathbf{u}), \quad \mathbf{u}(0) = \mathbf{u}_0 \tag{6.46}$$

and has a tridiagonal character, since the jacobian of the right hand side function is a tridiagonal $N \times N$ matrix.

6.5 ANSATZ METHODS FOR THE MODEL EQUATION

Ansatz methods can be applied to time-dependent PDEs. As for the heat equation, we illustrate the ansatz method on the model problem

$$\frac{\partial u}{\partial t} = \frac{\partial^2 u}{\partial x^2} + f(x), \qquad u(0,t) = u(1,t) = 0, \quad u(x,0) = u_0(x) \tag{6.47}$$

The ansatz, $u_h(x,t)$, is based on a time-dependent linear combination of basis functions

$$u(x,t) \approx u_h(x,t) = \sum_{j=1}^{N} c_j(t)\varphi_j(x) \tag{6.48}$$

where $\varphi_j(x)$ are given functions fulfilling the BCs $\varphi_j(0) = 0$, $\varphi_j(1) = 0$. The coefficients $c_j(t)$ are to be computed so that $u_h(x,t)$ is a "good" approximation of $u(x,t)$. Inserting $u_h(x,t)$ into the PDE gives a residual function

$$\begin{aligned} r(x,t) &= \frac{\partial u_h}{\partial t} - \frac{\partial^2 u_h}{\partial x^2} - f(x) \\ &= \sum_{j=1}^{N} \frac{dc_j}{dt}\varphi_j(x) - \sum_{j=1}^{N} c_j(t)\frac{d^2\varphi_j}{dx^2} - f(x) \neq 0 \end{aligned} \tag{6.49}$$

Now impose the condition that the residual function is orthogonal to the basis functions for all t

$$\int_0^1 r(x,t)\varphi_i(x)\,dx = 0, \quad i = 1, 2, \dots, N \tag{6.50}$$

This gives

$$\sum_{j=1}^{N} \frac{dc_j}{dt}\int_0^1 \varphi_i(x)\varphi_j(x)\,dx - \sum_{j=1}^{N} c_j(t)\int_0^1 \frac{d^2\varphi_j}{dx^2}\varphi_i(x)\,dx - \int_0^1 f(x)\varphi_i(x)\,dx = 0$$

By partial integration, we obtain

$$-\int_0^1 \frac{d^2\varphi_j}{dx^2}\varphi_i(x)\,dx = -\left[\varphi_i\frac{d\varphi_j}{dx}\right]_0^1 + \int_0^1 \frac{d\varphi_i}{dx}\frac{d\varphi_j}{dx}\,dx$$

Since $\varphi_i(0) = \varphi_i(1) = 0$, the outintegrated term disappears and we finally get the *Galerkin formulation*

$$M\frac{d\mathbf{c}}{dt} + A\mathbf{c} = \mathbf{f} \tag{6.51}$$

This is a system of ODEs where

$$M_{ij} = \int_0^1 \varphi_i(x)\varphi_j(x)\,dx, \qquad A_{ij} = \int_0^1 \frac{d\varphi_i}{dx}\frac{d\varphi_j}{dx}\,dx, \quad f_i = \int_0^1 f(x)\varphi_i(x)dx \tag{6.52}$$

The ICs are obtained from

$$u_0(x) \approx u_h(x,0) = \sum_{i=1}^{N} c_i(0)\varphi_i(x) \tag{6.53}$$

Multiply this relation with $\varphi_j(x)$, integrate over [0, 1] and we obtain the initial values $\mathbf{c}(0)$ from the linear system of equations

$$M\mathbf{c}(0) = \mathbf{u}_0 \tag{6.54}$$

where

$$u_{0,i} = \int_0^1 u_0(x)\varphi_i(x)\,dx.$$

Now, choose as basis functions the roof functions introduced in Chapter 4. Inserting the basis functions into (6.52), we obtain

$$M = \frac{h_x}{6}\begin{pmatrix} 4 & 1 & 0 & \dots & 0 \\ 1 & 4 & 1 & \ddots & \vdots \\ 0 & \ddots & \ddots & \ddots & 0 \\ \vdots & \ddots & 1 & 4 & 1 \\ 0 & \dots & 0 & 1 & 4 \end{pmatrix}, \quad A = \frac{1}{h_x}\begin{pmatrix} 2 & -1 & 0 & \dots & 0 \\ -1 & 2 & -1 & \ddots & \vdots \\ 0 & \ddots & \ddots & \ddots & 0 \\ \vdots & \ddots & -1 & 2 & -1 \\ 0 & \dots & 0 & -1 & 2 \end{pmatrix} \tag{6.55}$$

Observe that (6.51) is not presented in standard form $du/dt = f(t, u)$. We can achieve that by multiplying both sides with M^{-1}

$$\frac{d\mathbf{c}}{dt} + M^{-1}A\mathbf{c} = M^{-1}\mathbf{f} \tag{6.56}$$

If we do that, however, the tridiagonal structure of the matrices in (6.55) is destroyed, since M^{-1} is a *full* matrix. By keeping the tridiagonal form, a stiff method, e.g., the implicit Euler method applied to (6.51) gives a tridiagonal system of linear equations to be solved at each time step

$$M(\mathbf{c}_{k+1} - \mathbf{c}_k) + h_t A\mathbf{c}_{k+1} = h_t\,\mathbf{f}_{k+1} \Rightarrow (M + h_t A)\mathbf{c}_{k+1} = M\mathbf{c}_k + h_t\mathbf{f}_{k+1}$$

BIBLIOGRAPHY

The classical textbook on numerical solution of PDEs is:

1. G.D. Smith, Numerical Solution of Partial Differential Equations, 3rd ed, Oxford University Press, 1986

A modern book, more mathematical, on numerical analysis of ODEs and PDEs:

2. A. Iserles, A First Course in the Numerical Analysis of Differential Equations, Cambridge University Press, 1996

A book with several applications:

3. K.W. Morton, D.F. Myers, Numerical Solution of Partial Differential Equations, Cambridge University Press, 2005

7

NUMERICAL METHODS FOR ELLIPTIC PARTIAL DIFFERENTIAL EQUATIONS

Elliptic partial differential equations (PDEs) arise in equilibrium or steady-state problems, i.e., PDE problems that are time independent.

The solution of an elliptic PDE is often related to minimization of the total energy of a system formulated as an integral that depends on a state function and its derivatives. The minimization of this integral with respect to the state function is known as *variational calculus* and often leads to an elliptic PDE (or ordinary differential equation (ODE) if there is only one independent variable) and corresponding boundary conditions (BCs). Variational calculus is not taken up in this text but is referred to in some of the textbooks mentioned at the end of this chapter.

The BCs usually specify either the value of the solution function on the boundary or the value of its normal derivative or a combination of both.

The model problem for a 2D elliptic PDE problem is *Poisson's equation*,

$$-\frac{\partial^2 u}{\partial x^2} - \frac{\partial^2 u}{\partial y^2} = f(x, y), \quad (x, y) \in \Omega \tag{7.1a}$$

$$u = 0, \quad (x, y) \in \partial\Omega \tag{7.1b}$$

where Ω is assumed to be a bounded region in $\mathbf{R}^2$ with a smooth (or at least piecewise smooth) boundary $\partial\Omega$. When $f(x, y) \equiv 0$ in equation (7.1a), the PDE is called *Laplace's equation* and its solutions are called *harmonic functions*.

Different types of BCs can be given, just as in the case of 1D boundary value problems (see Chapter 4). The following common examples of BCs are *linear*

- Dirichlet's BC

$$u = g(x, y) \quad (x, y) \in \partial\Omega \tag{7.2}$$

Introduction to Computation and Modeling for Differential Equations, Second Edition. Lennart Edsberg.
© 2016 John Wiley & Sons, Inc. Published 2016 by John Wiley & Sons, Inc.

- Neumann's BC

$$\frac{\partial u}{\partial n} = h(x, y), \quad (x, y) \in \partial\Omega \tag{7.3}$$

Observe that Poisson's equation with a Neumann BD has a solution that is determined only up to an additive constant. To obtain a unique solution, a Dirichlet (or mixed) condition is needed in at least some part of $\partial\Omega$. Observe also that with Neumann's BC, the following relation must be fulfilled

$$\int_{\partial\Omega} \frac{\partial u}{\partial n} ds = \int \int_{\Omega} f(x, y) dx dy$$

- Mixed BC (also referred to as Robin's BC):

$$\frac{\partial u}{\partial n} = \alpha u + g(x, y) \quad (x, y) \in \partial\Omega \tag{7.4}$$

where α is a known constant and $g(x, y)$ a known function.
- The three BCs given above are *linear* but *nonlinear* BCs can also be formulated for elliptic PDEs.

At each point on the boundary $\partial\Omega$, only one type of BC can be given, but along the boundary the type of BC can shift, from, e.g., Dirichlet type to, e.g., Neumann type. However, at boundary points where either the BCs change or where the boundary is not smooth, singularities may occur in the solution's derivatives, which will imply difficulties in the numerical treatment when discretization methods are used.

The 1D correspondence to equation (7.1) is the following boundary value problem (BVP) with Dirichlet BCs, presented as the model equation in Chapter 4:

$$-\frac{d^2u}{dx^2} = f(x), \quad 0 < x < 1, \quad u(0) = u(1) = 0 \tag{7.5}$$

Many of the numerical properties of the discretized version of equation (7.5) will be preserved for PDE problems in 2D and 3D. In Section 4.2, it was stated that in the discretization of a BVP a *matching technique* leads to a coupled system of algebraic equations. This is true also for elliptic problems in 2D and 3D but the size of the systems will increase immensely. Examples of such numerical properties are

- For a linear elliptic PDE with linear BCs, discretization with finite difference method (FDM) or finite element method (FEM), leads to a linear algebraic system of equations to be solved: $A\mathbf{u} = \mathbf{b}$.
- The matrix A will be sparse.

For 1D and 2D problems, the solution of the linear system is usually performed with a sparse version of a *direct method*, e.g., sparse Gaussian elimination. For a 3D problem, the size of the linear system is usually so large that an *iterative method* is preferred (when it works), (see Appendix A.5).

Exercise 7.0.1. Given Poisson's equation with inhomogeneous BC:

$$-\frac{\partial^2 u}{\partial x^2} - \frac{\partial^2 u}{\partial y^2} = f(x, y), \quad (x, y) \in \Omega$$

$$u = g(x, y), \quad (x, y) \in \partial\Omega$$

Assume that $g(x, y)$ is twice differentiable in the region Ω. Let $v = u - g(x, y)$. Find the Poisson equation and the BCs that v satisfies.

Exercise 7.0.2. The forms of Laplace's equation for cartesian, cylindrical, and spherical coordinates are shown in equations (5.24a), (5.24b).

1. Assume that the solution to Laplace's equation in cylindrical coordinates depends only on the variable r (cylinder symmetry). Find the general solution.
2. The same as in (1) but for spherical coordinates (spherical symmetry).

7.1 APPLICATIONS

Poisson's and Laplace's equations occur in a great variety of applications in science and engineering. Some examples are given in Table 7.1.

To each one of the elliptic PDE problems given in Table 7.1, appropriate BCs have to be given in order to specify a unique solution.

Laplace's and Poisson's equations are not the only elliptic problems. The examples below show some other types of elliptic PDEs, e.g., a higher order PDE, an eigenvalue problem, a nonlinear PDE, and a system of two PDEs. For some of these examples, the numerical solution is demonstrated in Section 7.3.

Example 7.1. **(Stationary heat conduction).**

Consider a thin, metallic, homogeneous, flat, rectangular plate of length a (m) and width b (m). The plate has no heat sources. When the plate is placed in an

TABLE 7.1 Applications of Poisson's Equation

Application	Variable	PDE	Right Hand Side
Heat conduction	T = temperature	$-div(\kappa\nabla T) = Q$	Heat source/sink
Diffusion	c = concentration	$-div(D\nabla c) = q$	Mass source/sink
Electrostatics	V = el potential	$-div(\epsilon\nabla V) = \rho$	Charge density
Gravity	ψ = grav potential	$-\Delta\psi = \rho$	Mass density
Membrane deflection	u = deformation	$-div(\nabla u) = f/E$	Pressure
Torsion of a cylinder	ϕ = stress function	$-\Delta\phi = 2\theta G$	Torsion module
Fluid dynamics	Φ = velocity potential	$\Delta\Phi = 0$	

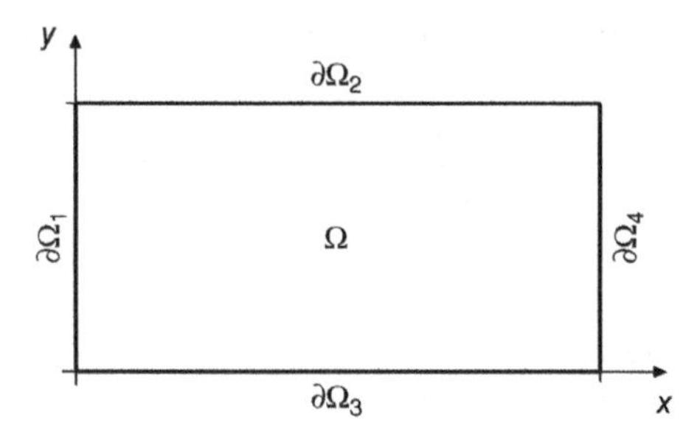

Figure 7.1 Region and boundaries for a rectangular heat conduction problem

xy-coordinate system, it covers the rectangular region Ω. Heat is conducted isotropically in the plate and the heat fluxes are subject to different BCs given on $\partial\Omega$. Let $T(x, y)$ (K) be the temperature in a point (x, y) in Ω or on $\partial\Omega$. $T(x, y)$ is determined by the heat equation, in this case Laplace's equation and appropriate BCs

$$-\nabla(\kappa \nabla T) = 0$$

where κ is the heat conduction coefficient [J/(K· m· s)] (Figure 7.1).

Along $\partial\Omega_1$, the temperature is given by a function $g(x, y)$, i.e.

$$T(x, y) = g_1(y), \qquad (x, y) \in \partial\Omega_1 \tag{7.6}$$

hence leading to a Dirichlet BC on this part of the boundary.

Along $\partial\Omega_2$, assume that the plate is warmer than the environment. Hence, heat is leaking out through this part of the boundary and a mixed BC is appropriate

$$-\kappa\frac{\partial T}{\partial n} = k(T - T_{out}), \qquad (x, y) \in \partial\Omega_2 \tag{7.7}$$

where k [J/(K· m^2· s)] is the convection heat transfer coefficient.

Along $\partial\Omega_3$, the boundary is assumed to be heat insulated against the environment, which leads to a Neumann BC

$$\kappa\frac{\partial T}{\partial n} = 0 \tag{7.8}$$

Finally along Ω_4, the temperature is given by a function $g_2(y)$, i.e., a Dirichlet BC just as along $\partial\Omega_1$. For a solution of this problem, see Section 7.3. Compare this example with example 6.2.

Example 7.2. **(Steady, incompressible, nonviscous, irrotational flow).**

If a flow is irrotational, i.e., the velocity field $\mathbf{u} = (u_x, u_y)$ satisfies $curl\mathbf{u} = 0$, there exists a potential function Φ such that $\mathbf{u} = -\nabla\Phi$. As the flow is incompressible we

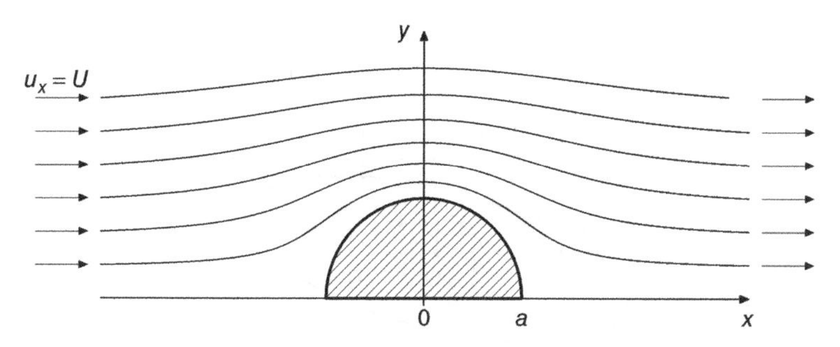

Figure 7.2 The flow around a circular obstacle

also have $div\mathbf{u} = 0$. Combining these two equations we find that Φ satisfies Laplace's equation

$$\Delta\Phi = 0$$

The physical assumptions needed to justify this model is that the effects of viscosity are neglected.

By specifying suitable BCs, the flow is determined. In this example, we consider a flow in the xy-plane that originally is parallel with the x-axis and then passing a cylinder with the symmetry axis perpendicular to the xy-plane. We want to compute the flowlines as the flow passes the cylinder.

Assume the 2D flow far away from the cylinder ($x << 0$) has the velocity components $u_x = U$ and $u_y = 0$. The radius of the cylinder is a and the symmetry axis goes through the origin (Figure 7.2).

For this problem, it is natural to use cylindrical coordinates. As the z-coordinate plays no part in the solution, the potential function Φ satisfies Laplace's equation in polar coordinates

$$\frac{\partial^2\Phi}{\partial r^2} + \frac{1}{r}\frac{\partial\Phi}{\partial r} + \frac{1}{r^2}\frac{\partial^2\Phi}{\partial\varphi} = 0 \tag{7.9}$$

From Φ, the cylindrical velocity components u_r and u_φ are computed from

$$u_r = -\frac{\partial\Phi}{\partial r}, \qquad u_\varphi = -\frac{1}{r}\frac{\partial\Phi}{\partial\varphi} \tag{7.10}$$

To solve Laplace's equation, BCs are needed in the region Ω yet to be defined. Owing to symmetry, it is enough to formulate BCs in the half plane $0 \le \varphi \le \pi$. In a numerical formulation based on the FDM, the region must be bounded. This can be achieved by restricting the r-variable to be bounded by $a \le r \le R$, where $R >> a$. Hence, Ω is a large half disc from which a small half disc has been cut out: $\Omega = (r, \varphi),\ a \le r \le R,\ 0 \le \varphi \le \pi$.

1. On the boundary $\partial\Omega_1$, the small half circle we have $u_r = 0$, i.e., no radial velocity component.

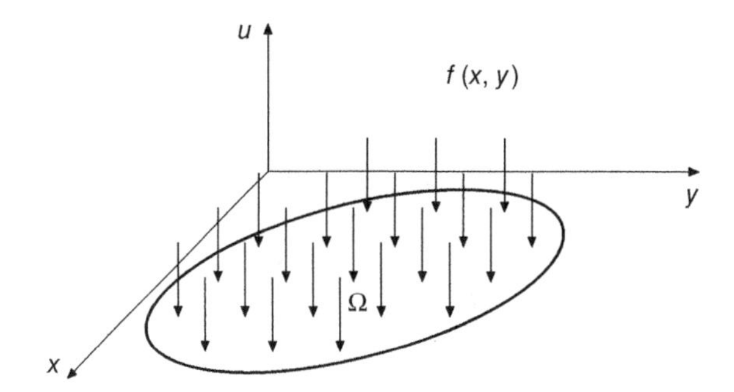

Figure 7.3 Deformation of an elastic plate by a transversal load

2. Along the boundary $\partial\Omega_2$ and $\partial\Omega_4$, the two parts on the x-axis between the small and the big half circle, we have $u_\varphi = 0$ from the symmetry of the problem.
3. Far away from the cylinder, the flow is unaffected, i.e. $u_x = U$ and $u_y = 0$. Hence, on the boundary $\partial\Omega_3$ the big half circle, the following Dirichlet condition satisfies the BC: $\Phi = -Ux = -Ur\cos(\varphi)$.

Example 7.3. **(Deformation of an elastic plate by a transversal load).**

Consider a thin, metallic, homogeneous flat plate (see Figure 7.3). When it is placed in an xy-coordinate system, it covers the region Ω. We want to study small deformations imposed by a transversal pressure $f(x, y)$ (N/m^2). The deformation u (m) satisfies the biharmonic equation

$$\Delta^2 u = \frac{\partial^4 u}{\partial x^4} + 2\frac{\partial^4 u}{\partial x^2 \partial y^2} + \frac{\partial^4 u}{\partial y^4} = \frac{f(x,y)}{D}, \quad (x,y) \in \Omega \tag{7.11a}$$

where D (N$\cdot$ m) is the flexural rigidity. For a clamped plate, the following BCs are valid

$$u = 0, \quad \frac{\partial u}{\partial n} = 0, \quad (x,y) \in \partial\Omega \tag{7.11b}$$

Equation (7.11a) is a fourth-order elliptic problem. The corresponding 1D problem is the ODE modeling the deformation of a *beam* clamped between two walls and transversally loaded by $f(x)$ (N/m)

$$EI\frac{d^4 u}{dx^4} = f(x), \quad u(0) = \frac{du}{dx}(0) = 0, \quad u(L) = \frac{du}{dx}(L) = 0 \tag{7.12}$$

Example 7.4. **(Eigenvalue problem).**

In different engineering applications, one is often interested in the frequencies allowing a system to vibrate by itself without influence of external forces. Such problems occur in, e.g., mechanical and electrical problems and such frequencies are called *eigenfrequencies*. They would vibrate for ever only in idealized situations, when damping forces are neglected. Nevertheless, these eigenfrequencies are important in engineering. The reason is that we want to avoid them, as external forces vibrating with the same frequency as an eigenfrequency introduces *resonance* effects that will increase the amplitude of the vibrations of the system and may cause damage.

Consider an elastic membrane that is first deformed and then left to itself to perform vibrations with amplitude $u(x, y)$. The vibrations $v(x, y, t)$ satisfy the *wave equation* (see Chapter 5)

$$\frac{\partial^2 v}{\partial t^2} = c^2 \Delta v, \quad (x, y) \in \Omega \tag{7.13a}$$

$$v = 0, \quad (x, y) \in \partial\Omega \tag{7.13b}$$

For vibrations with the angular frequency ω, we make the following ansatz for the time-dependent solution

$$v = u(x, y)e^{i\omega t} \tag{7.14}$$

Inserting this ansatz into the wave equation gives the following elliptic PDE problem for the amplitude (Helmholtz' equation), also called the eigenvalue problem for the vibrating membrane

$$-\Delta u = \lambda u, \quad (x, y) \in \Omega \tag{7.15a}$$

$$u = 0, \quad (x, y) \in \partial\Omega \tag{7.15b}$$

where

$$\lambda = \left(\frac{\omega}{c}\right)^2 \tag{7.16}$$

Observe that $u(x, y)$ is determined only up to a multiplicative constant, see also example 4.6.

***Example 7.5.* (Minimal surface problem, a nonlinear elliptic PDE).**

Consider a closed curve $\partial\Omega_S$ in 3D. Such a curve can be defined by $u = g(x, y)$ evaluated along a closed curve $\partial\Omega$ enclosing the region Ω in the xy-plane. The problem is to find the minimal surface S and the corresponding function $u(x, y)$ with $\partial\Omega_S$ as boundary (Figure 7.4).

This can be stated as a minimization problem

$$S = \min_{u(x,y)} \int_\Omega \sqrt{1 + (\nabla u)^2}\, dxdy \tag{7.17a}$$

subject to the constraint

$$u = g(x, y), \quad (x, y) \in \partial\Omega \tag{7.17b}$$

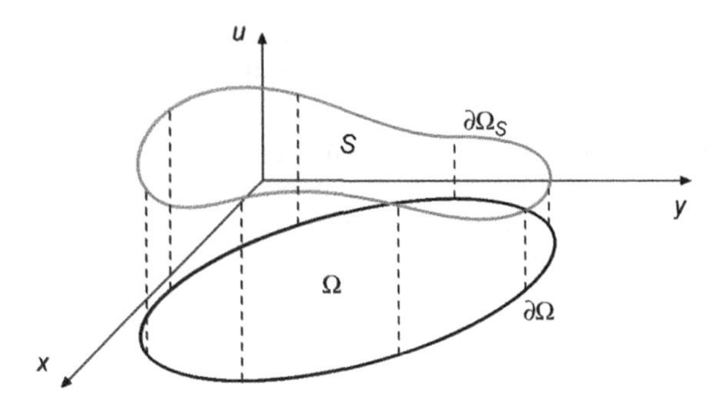

Figure 7.4 Minimal surface problem

With variational calculus, it can be shown that equation (7.17) is equivalent to solving the following *nonlinear* elliptic PDE

$$div\left\{\frac{\nabla u}{\sqrt{1+(\nabla u)^2}}\right\} = 0, \quad (x, y) \in \Omega \tag{7.18a}$$

with Dirichlet BC

$$u = g(x, y), \quad (x, y) \in \partial\Omega \tag{7.18b}$$

This problem is also known as the soap film problem, as the PDE (7.18) describes how the shape of a soap film will adjust itself when it is fastened to a metallic wire with a given shape, defined by $\partial\Omega_S$.

Example 7.6. **(Deformation of an elastic plate in a plane).**
Plane strain of an elastic plate in 2D is modeled by two coupled elliptic PDEs

$$\frac{\partial^2 u}{\partial x^2} + \frac{1-\nu}{2}\frac{\partial^2 u}{\partial y^2} + \frac{1+\nu}{2}\frac{\partial^2 v}{\partial x\partial y} = -\frac{1-\nu^2}{E^2}f(x, y) \tag{7.19a}$$

$$\frac{\partial^2 v}{\partial y^2} + \frac{1-\nu}{2}\frac{\partial^2 v}{\partial x^2} + \frac{1+\nu}{2}\frac{\partial^2 u}{\partial x\partial y} = -\frac{1-\nu^2}{E^2}g(x, y) \tag{7.19b}$$

where u and v are the displacements in the x and y directions. E is the elasticity module and ν is Poisson's number. The deformation forces in the x and y directions are, respectively, $f(x, y)$ and $g(x, y)$. With appropriate BCs, this is a system of two elliptic PDEs.

7.2 THE FINITE DIFFERENCE METHOD

When the FDM is used, the algorithm for solving an elliptic PDE problem can be divided into three steps, just as for BVPs in Chapter 4. Hence, the steps are

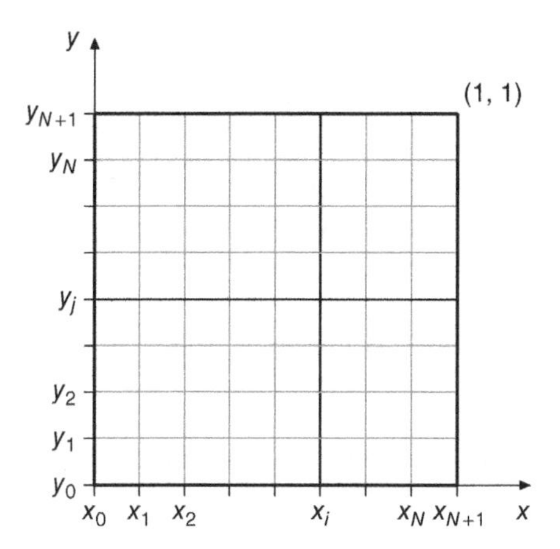

Figure 7.5 2D grid

1. discretize the region to a grid
2. discretize the PDE
3. discretize the BCs

To illustrate the algorithm, we apply it to the model problem (7.1) in the case Ω is the region $\{(x, y),\ 0 \leq x \leq 1,\ 0 \leq y \leq 1\}$.

DG Discretize the region to a grid

Discretize the x- and the y-axis with the same stepsize h corresponding to $N \times N$ inner points equidistantly distributed over the two axis, $h(N + 1) = 1$ giving $x_i, i = 1, 2, \ldots N$ and $y_j = jh, j = 1, 2, \ldots N$ (Figure 7.5).

DD Discretize the PDE

Use the well-known difference formula for the second derivative:

$$\frac{\partial^2 u}{\partial x^2}(x_i, y_j) = \frac{u_{i+1,j} - 2u_{i,j} + u_{i-1,j}}{h^2} + O(h^2) \tag{7.20a}$$

$$\frac{\partial^2 u}{\partial y^2}(x_i, y_j) = \frac{u_{i,j+1} - 2u_{i,j} + u_{i,j-1}}{h^2} + O(h^2) \tag{7.20b}$$

We get the following difference approximation of Poisson's equation:

$$4u_{i,j} - u_{i-1,j} - u_{i+1,j} - u_{i,j-1} - u_{i,j+1} = h^2 f(x_i, y_j) \tag{7.21}$$

DB The BCs are represented exactly for the given Dirichlet conditions:

$$u_{0,j} = 0, \quad u_{N+1,j} = 0, \quad u_{i,0} = 0, \quad u_{i,N+1} = 0 \tag{7.22}$$

We can illustrate the left hand side of equation (7.21) by the five-point stencil $h^2 \Delta_5^h$ (Figure 7.6).

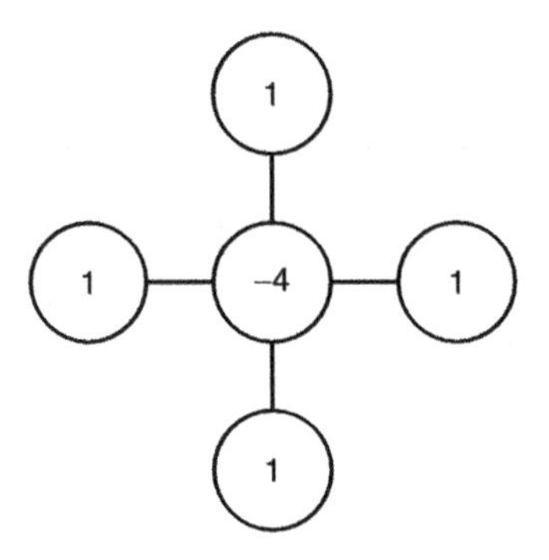

Figure 7.6 Stencil for discretized laplacian

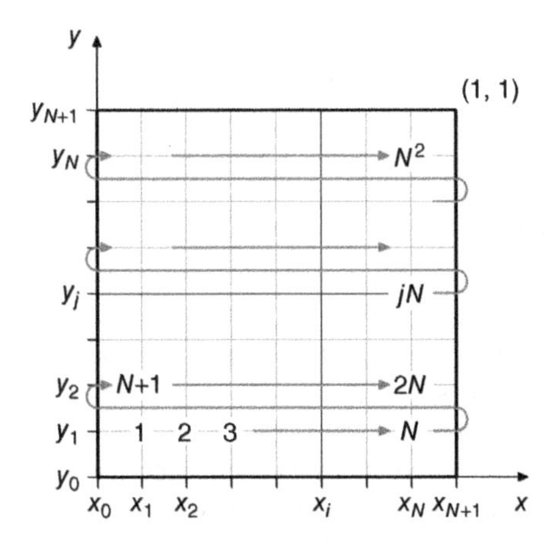

Figure 7.7 Ordering of unknowns

The difference equations (7.21) can be set up in all N^2 inner points (x_i, y_j). giving N^2 equations for N^2 unknowns $u_{i,j}$. By introducing the matrix U for the unknowns, it *might* seem reasonable that equations (7.21) and (7.22) can be formulated as a linear matrix equation:

$$AU = B$$

with appropriate elements in the A and B matrices. This, however, is NOT possible!

Instead we have to represent the matrix of unknowns U as a long vector u. One way to do this is to enumerate the unknowns and the gridpoints *row-wise* (also known as natural ordering) according to the Figure 7.7: corresponding to the following ordering of the $u_{i,j}$ values:

$$\mathbf{u} = (u_{1,1}, u_{2,1}, \ \dots \ , u_{N,1}, u_{1,2}, u_{2,2}, \ \dots \ , u_{N,2}, \ \dots\dots \ u_{1,N}, u_{2,N}, \ \dots \ u_{N,N})^T$$

With this enumeration, the linear system of equations will be

$$A\mathbf{u} = \mathbf{b}$$

where A is a sparse symmetric positive definite $N^2 \times N^2$ matrix and $\mathbf{b}$ is an $N^2 \times 1$ column vector.

In the case $N = 3$, A and $\mathbf{b}$ are

$$A = \begin{pmatrix} 4 & -1 & 0 & -1 & 0 & 0 & 0 & 0 & 0 \\ -1 & 4 & -1 & 0 & -1 & 0 & 0 & 0 & 0 \\ 0 & -1 & 4 & 0 & 0 & -1 & 0 & 0 & 0 \\ -1 & 0 & 0 & 4 & -1 & 0 & -1 & 0 & 0 \\ 0 & -1 & 0 & -1 & 4 & -1 & 0 & -1 & 0 \\ 0 & 0 & -1 & 0 & -1 & 4 & 0 & 0 & -1 \\ 0 & 0 & 0 & -1 & 0 & 0 & 4 & -1 & 0 \\ 0 & 0 & 0 & 0 & -1 & 0 & -1 & 4 & -1 \\ 0 & 0 & 0 & 0 & 0 & -1 & 0 & -1 & 4 \end{pmatrix}, \quad \mathbf{b} = h^2\mathbf{f} = h^2 \begin{pmatrix} f_{1,1} \\ f_{2,1} \\ f_{3,1} \\ f_{1,2} \\ f_{2,2} \\ f_{3,2} \\ f_{1,3} \\ f_{2,3} \\ f_{3,3} \end{pmatrix}$$

The matrix A is symmetric and positive definite (compare with the matrix A in (4.29)) and has the structure of a block tridiagonal matrix

$$A = trid_N(-I, T, -I), \qquad \text{where} \quad T = trid_N(-1, 4, -1) \tag{7.23}$$

The matrix A is very sparse, only five diagonals have nonzero elements. On the other hand, the bandwidth is $2N - 1$ and (see Section A.5) in the Cholesky factorization, $A = LL^T$, there will be “fill-in” in L, i.e., the zeros inside the band between the outer and inner diagonals will be filled with nonzero elements denoted by x:

$$A_L = \begin{pmatrix} x & 0 & 0 & 0 & 0 & 0 & 0 & 0 & 0 \\ x & x & 0 & 0 & 0 & 0 & 0 & 0 & 0 \\ 0 & x & x & 0 & 0 & 0 & 0 & 0 & 0 \\ x & 0 & 0 & x & 0 & 0 & 0 & 0 & 0 \\ 0 & x & 0 & x & x & 0 & 0 & 0 & 0 \\ 0 & 0 & x & 0 & x & x & 0 & 0 & 0 \\ 0 & 0 & 0 & x & 0 & 0 & x & 0 & 0 \\ 0 & 0 & 0 & 0 & x & 0 & x & x & 0 \\ 0 & 0 & 0 & 0 & 0 & x & 0 & x & x \end{pmatrix} -> L = \begin{pmatrix} x & 0 & 0 & 0 & 0 & 0 & 0 & 0 & 0 \\ x & x & 0 & 0 & 0 & 0 & 0 & 0 & 0 \\ x & x & x & 0 & 0 & 0 & 0 & 0 & 0 \\ x & x & x & x & 0 & 0 & 0 & 0 & 0 \\ 0 & x & x & x & x & 0 & 0 & 0 & 0 \\ 0 & 0 & x & x & x & x & 0 & 0 & 0 \\ 0 & 0 & 0 & x & x & x & x & 0 & 0 \\ 0 & 0 & 0 & 0 & x & x & x & x & 0 \\ 0 & 0 & 0 & 0 & 0 & x & x & x & x \end{pmatrix}$$

where A_L is the lower triangular part of A. For more comments on the solution of large sparse systems with different methods, see Section A.5. Observe the increasing complexity as the problem is changed from 1D to 2D and 3D. If the x-axis is discretized into N inner points for 1D the problem (7.5), the matrix A will be tridiagonal, $N \times N$ and the solution time $O(N)$.

With a similar discretization for the 2D problem (7.1) with Ω being a quadrangle (N inner points in both the x and the y directions), A is $N^2 \times N^2$ with bandwidth $O(N)$ and solution time $O(N^4)$

In the 3D case, A is $N^3 \times N^3$ with bandwidth $O(N^2)$ and solution time $O(N^7)$.

Exercise 7.2.1. Use the five-point stencil to compute an approximation of $u(0, 0)$, where $u(x, y)$ satisfies the PDE

$$\Delta u = u, \qquad (x, y) \in \Omega$$

$$u = |x| + |y|, \quad (x, y) \in \partial\Omega$$

where Ω is the square $(x, y), -1 \leq x \leq 1, -1 \leq y \leq 1$. Use the stepsizes, $h = 1$ and $h = 1/2$. Use extrapolation to improve the result.

Exercise 7.2.2. Find the stencil obtained with central difference approximations for the following PDE expressions. Assume that the stepsize h in the x and y directions is the same.

$$\frac{\partial^2 u}{dx^2} + \frac{\partial^2 u}{\partial y^2} + \frac{\partial^2 u}{\partial x \partial y}$$

$$\frac{\partial^2 u}{dx^2} + \frac{\partial^2 u}{\partial y^2} + \frac{\partial u}{\partial x} + \frac{\partial u}{\partial y}$$

Exercise 7.2.3. When the boundary $\partial\Omega$ is curved and intersects the quadratic grid at points that are not grid points, the five-point stencil $h^2\Delta_5^h$ must be modified when applied to Dirichlet BC.

Assume the stencil is positioned with the midpoint in (x_i, y_j), which is situated inside the region Ω. The point on the curve above the midpoint is situated αh, $\alpha < 1$ from the midpoint. The point to the right of the midpoint has the distance βh, $\beta < 1$ from the midpoint. The points below and to the left of the midpoint are both situated inside the region and therefore at distance h from the midpoint.

The parameters α and β in the unsymmetric stencil are chosen so that it fits the boundary points correctly. Derive the coefficients a, b, c, d, e (they depend on α and β) in the difference approximation

$$\Delta u(x_i, y_j) = \frac{au_{ij-1} + bu_{i-1j} + cu_{ij} + du_{i+1j} + eu_{ij+1}}{h^2} + O(h^p)$$

so that the order of approximation is as high as possible. Show the result as a stencil.

7.3 DISCRETIZATION OF A PROBLEM WITH DIFFERENT BCs

Based on the FDM, the problems in Examples 7.1–7.5 can be solved, essentially using the same three step algorithm as outlined in Section 7.2. We present here a discretized solution technique for Example 7.1.

Assume that the region Ω is a rectangle placed in an xy-coordinate system with the corners on $(0, 0), (a, 0), (0, b)$ and (a, b).

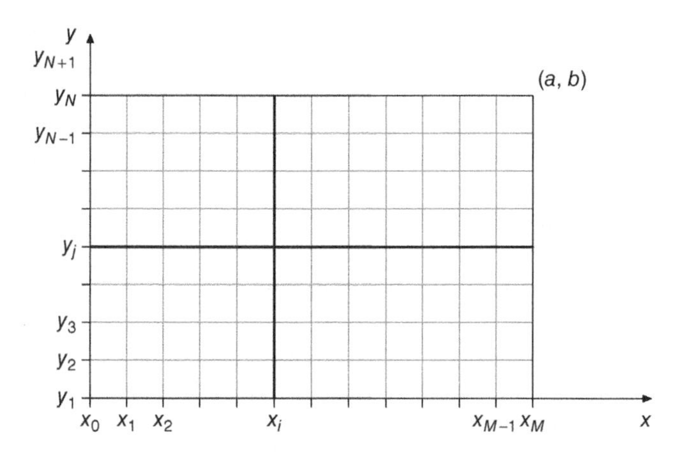

Figure 7.8 The region discretized with the FDM

1. **DG**—Discretize Ω using a stepsize h, which is assumed to be the same in both the x and the y directions, according to $Mh = a$, $(N-1)h = b$ (Figure 7.8). Having the type of BCs in mind, the points are numbered according to $x_i = ih, i = 0, 1, \ldots M$, $x_0 = 0, x_M = a$ and $y_j = (j-1)h, j = 0, 1, \ldots, N+1$, $y_0 = -h, y_{N+1} = b + h$. The four corner points of the grid, however, are not included. This gives $(M+1)(N+2) - 4 = MN + 2M + N - 2$ grid points. This is also the number of unknowns $T_{i,j}$ of the discretized problem.
2. **DD** Discretize the PDE. The following equations are valid at the inner points:

$$4T_{i,j} - T_{i-1,j} - T_{i+1,j} - T_{i,j-1} - T_{i,j+1} = 0,$$

   where $i = 1, 2, \ldots, M-1, j = 1, 2, \ldots, N$ giving a total of $(M-1)N$ equations.
3. **DB** Discretize the BCs along the four boundaries.

   On $\partial\Omega_1$, there is a Dirichlet BC: $T_{0,j} = g(0, y_j), j = 1, 2, \ldots, N$ giving N equations.

   On $\partial\Omega_2$, there is a mixed BC: $-\kappa(T_{i,N+1} - T_{i,N-1})/2h = k(T_{i,N} - T_{out}), i = 1, 2, \ldots, M-1$ giving $M-1$ equations.

   On $\partial\Omega_3$, there is a Neumann BC: $(T_{i,0} - T_{i,2})/2h = 0, i = 1, 2, \ldots, M-1$ giving $M-1$ equations.

   On $\partial\Omega_4$, there is a Dirichlet BC: $T_{M,j} = g(a, y_j), j = 1, 2, \ldots, N$ giving N equations.

The total number of equations in (2) + (3) is $(M-1)N + N + M - 1 + M - 1 + N = MN + 2M + N - 2$. Hence, there are as many equations as unknowns and a linear system of equations can be set up when the unknown $T_{i,j}$ elements have been renumbered into a vector $T_i, i = 1, 2, \ldots, MN + 2M + N - 2$.

Exercise 7.3.1. Discretize the Laplace equation in polar coordinates presented in Example 7.2. Follow the recipe DG,DD,DB. Observe that the region Ω is a rectangle in the (r, φ)-space. Find second-order approximations to the PDE and the BCs.

Exercise 7.3.2. Discretize the eigenvalue problem in Example 7.4 assuming the region Ω is a square with side length = 1. Formulate the corresponding algebraic eigenvalue problem $A\mathbf{u} = \lambda\mathbf{u}$.

7.4 ANSATZ METHODS FOR ELLIPTIC PDEs

So far we have demonstrated the FDM on problems defined on rectangular domains. However, real problems often involve domains with irregular shapes. To demonstrate the FEM, we apply the method to Poisson's problem (7.1), but assume that the region Ω has a more complicated shape (not a square) (Figure 7.9).

7.4.1 Starting with the PDE Formulation

Instead of the FDM we now choose an ansatz for the approximate solution of equation (7.1), i.e.,

$$-\frac{\partial^2 u}{\partial x^2} - \frac{\partial^2 u}{\partial y^2} = f(x, y), \quad (x, y) \in \Omega$$

$$u = 0, \quad (x, y) \in \partial\Omega$$

and use Galerkin's method. The generalization of the 1D problem to a 2D problem is straightforward: change $\int$ to $\iint$ and d/dx to *grad* and repeat the formulas in Section 4.3 to the 2D case. Denote by $u_h(x, y)$ the ansatz function

$$u_h(x, y) = \sum_{j=1}^{N} c_j \varphi_j(x, y) \tag{7.24}$$

where the basis functions $\varphi_j(x, y)$ are chosen to satisfy the BCs, i.e.

$$\varphi_j(x, y)_{|\partial\Omega} = 0, \quad j = 1, 2, \ldots, N \tag{7.25}$$

When $u_h(x, y)$ is inserted into the PDE, we obtain a residual function $r(x, y)$:

$$r(x, y) = \sum_{j=1}^{N} c_j \Delta\varphi_j(x, y) + f(x, y) \neq 0 \tag{7.26}$$

In order to make $r(x, y)$ "small", we use Galerkin's method, which means that $r(x, y)$ is orthogonal to $\varphi_i(x, y)$, $i = 1, 2, \ldots, N$:

$$\iint_{\Omega} r(x, y)\varphi_i(x, y)dxdy = 0 \tag{7.27}$$

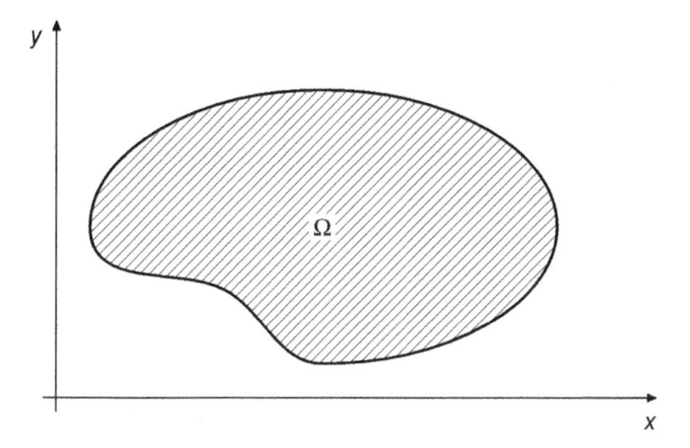

Figure 7.9 A region to be discretized with the FEM

$$\int\int_{\Omega}(\sum_{j=1}^{N} c_j \Delta\varphi_j(x,y) + f(x,y))\varphi_i(x,y)dxdy = 0$$

$$\sum_{j=1}^{N} c_j \int\int_{\Omega} \varphi_i \Delta\varphi_j dxdy + \int\int_{\Omega} f(x,y)\varphi_i dxdy = 0$$

Now use integration "by parts" in 2D

$$\int\int_{\Omega} \varphi_i \Delta\varphi_j dxdy = \oint_{\partial\Omega} \varphi_i \frac{\partial\varphi_j}{\partial n} ds - \int\int_{\Omega} \nabla\varphi_i \cdot \nabla\varphi_j dxdy \tag{7.28}$$

Compare this formula with the corresponding integration by parts in 1D

$$\int_0^1 \varphi_i \varphi_j'' dx = [\varphi_i \varphi_j']_0^1 - \int_0^1 \varphi_i' \varphi_j' dx$$

As $\varphi_i = 0$ on $\partial\Omega$ this integral $= 0$, and we finally obtain

$$\sum_{j=1}^{N} c_j \int\int_{\Omega} \nabla\varphi_i \cdot \nabla\varphi_j dxdy = \int\int_{\Omega} f(x,y)\varphi_i dxdy, i = 1, 2, \dots, N \tag{7.29}$$

which is an $N \times N$ linear system of equations for the coefficients c_j:

$$A\mathbf{c} = \mathbf{f}$$

where

$$A_{i,j} = \int\int_{\Omega} \nabla\varphi_i \cdot \nabla\varphi_j dxdy, \quad f_i = \int\int_{\Omega} f(x,y)\varphi_i dxdy \tag{7.30}$$

A is called the *stiffness matrix*, where stiffness is related to a material property and not the concept "stiff" in connection with certain ODE problems (see Chapter 3). The right hand side vector **f** is called the *load vector*.

7.4.2 Starting with the Weak Formulation

The analytical solution of equation (7.1) is referred to as the *strong (or classical) solution*, but for PDEs, in general, there is also a *weak (or variational)* formulation just as for the 1D problems in Section 4.3. The weak formulation is demonstrated on a slightly more general PDE than (7.1).

$$-div(\kappa\nabla u) + \mathbf{p}\cdot\nabla u + qu = f(x,y) \quad \text{in} \quad \Omega \tag{7.31a}$$

$$\kappa\frac{\partial u}{\partial n} = \alpha u + g(x,y) \quad \text{on} \quad \partial\Omega \tag{7.31b}$$

where κ, $\mathbf{p}$, and q can be constants or depend on (x,y). This PDE with the BC is linear and corresponds to a generalization of the BVP problem (4.57) to 2D.

For the weak formulation, multiply equation (7.31a) with a test function $v = v(x,y) \in U$ and integrate over the region Ω

$$\int\int_{\Omega}(-div(\kappa\nabla u) + \mathbf{p}\cdot\nabla u + qu)vdxdy = \int\int_{\Omega} fvdxdy \tag{7.32}$$

Use integration "by parts" in 2D on the first term

$$-\int\int_{\Omega} div(\kappa\nabla u)vdxdy = -\oint_{\partial\Omega}\kappa\frac{\partial u}{\partial n}vds + \int\int_{\Omega}\kappa\nabla u\cdot\nabla vdxdy \tag{7.33}$$

Using the BC (7.31b) in equation (7.33), we obtain the weak formulation: Find $u \in U$ so that

$$\int\int_{\Omega}(\kappa(\nabla u\cdot\nabla v) + (\mathbf{p}\cdot\nabla u)v + quv)dxdy - \oint_{\partial\Omega}\alpha uvds = \int\int_{\Omega} fvdxdy + \oint_{\partial\Omega} gvds \tag{7.34}$$

for all $v \in U$, where $U = C^1(\Omega)$. In equation (7.34), the quadratic terms are collected in the left hand side and the linear terms in the right hand side.

When there are different types of BCs on $\partial\Omega$, the weak formulation must be changed accordingly, e.g., if we have both mixed and Dirichlet conditions

$$\kappa\frac{\partial u}{\partial n} = \alpha_1 u + g_1(x,y), \quad (x,y) \in \partial\Omega_1$$

$$u = 0, \quad (x,y) \in \partial\Omega_2$$

Then equation (7.34) is changed to: Find $u \in U$ such that

$$\int\int_{\Omega}(\kappa(\nabla u\cdot\nabla v) + \mathbf{p}\cdot\nabla uv + quv)dxdy - \oint_{\partial\Omega_1}\alpha uvds = \int\int_{\Omega} fvdxdy + \oint_{\partial\Omega_1} gvds \tag{7.35}$$

for all $v \in U$, where $U = C^1_{[0]}(\Omega)$ and [0] means $v = 0$, *on* $\partial\Omega_2$.

Now, use Galerkin's method on the weak formulation (7.34), i.e., let

$$u_h(x,y) = \sum_{j=1}^{N} c_j\varphi_j(x,y), \quad v_h(x,y) = \varphi_i(x,y),\ i = 1,2,\ \dots\ ,N$$

For the coefficients c_j, we obtain the linear system of equations

$$(A\kappa + P + Q + B)\mathbf{c} = \mathbf{f}$$

where

$$A\kappa_{ij} = \int\int_\Omega \kappa\nabla\varphi_i\cdot\nabla\varphi_j dxdy, \quad P_{ij} = \int\int_\Omega \varphi_i\mathbf{p}\cdot\nabla\varphi_j dxdy, \quad Q_{ij} = \int\int_\Omega q\varphi_i\varphi_j dxdy$$

$$B_{ij} = \int\int_{\partial\Omega} \alpha\varphi_i\varphi_j ds \quad f_i = \int\int_\Omega f(x,y)\varphi_i dxdy + \oint_{\partial\Omega} g\varphi_i ds$$

Exercise 7.4.1. Modify the weak formulation to Poisson's equation with a nonhomogeneous BC

$$-\frac{\partial^2 u}{\partial x^2} - \frac{\partial^2 u}{\partial y^2} = f(x,y), \quad (x,y) \in \Omega$$
$$u = g(x,y), \quad (x,y) \in \partial\Omega$$

Hint: Make a transformation similar to the one in Exercise 7.0.1 to obtain a homogeneous BC.

7.4.3 The Finite Element Method

Up to now nothing has been said about the basis functions φ_j. We want the matrix A to be sparse, in order to be competitive with the FDM. In the 1D case, $\varphi_j(x)$ were chosen as "roof functions" (see Section 4.3) defined on a grid of subintervals.

In the 2D case, the grid is defined by subdivision of Ω into *triangles* T_k, $k = 1,2,\ \dots\ ,M$ (Figure 7.10).

It is seen from the example above that the boundary $\partial\Omega$ will not be exactly represented, unless the boundary consists of straight lines coinciding with the triangle sides. Hence

$$\Omega \approx \Omega_T = \cup_1^M T_k$$

However, by making the triangles sufficiently small, the boundary can be accurately enough represented. In connection with this modification of Ω, we also change the BCs of $\varphi_j(x,y)$ to

$$\varphi_j(x,y)_{|\partial\Omega_T} = 0, \quad j = 1,2,\ \dots\ ,N$$

For the definition of the triangular grid, we must define a numbering of both the inner nodes $P_j, j = 1,2,\ \dots\ ,N$ and the triangles $T_k, k = 1,2,\ \dots\ ,M$. Hence, for an arbitrary triangle T_k, there are three nodes (corners).

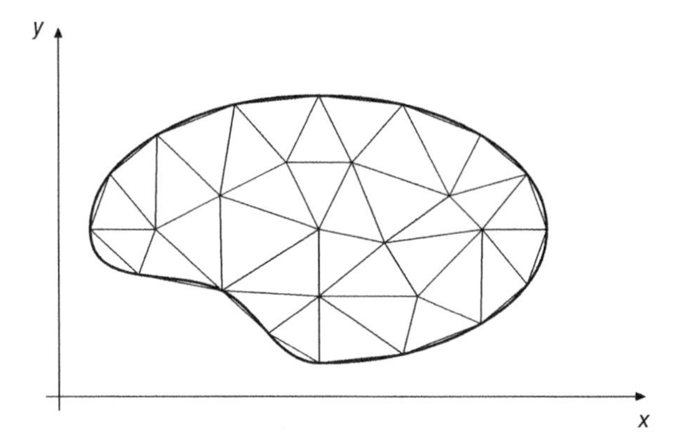

Figure 7.10 Triangular discretization of a region

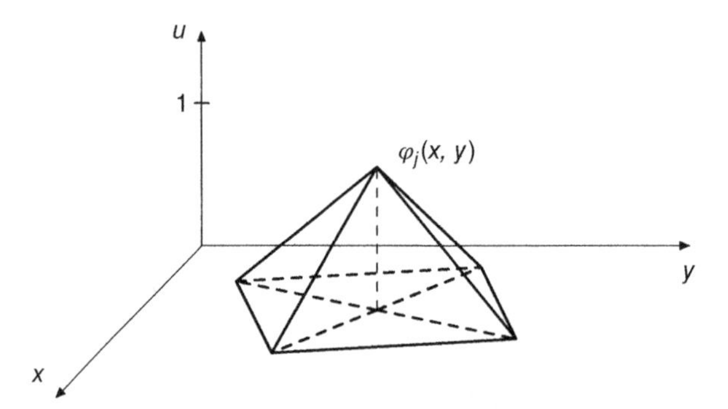

Figure 7.11 A pyramid function

When the grid has been defined for the region Ω_T, the basis functions can be defined by associating to each node j a "pyramid" function $\varphi_j(x, y)$ (Figure 7.11).

The side of a pyramid is a triangle, hence every basis function $\varphi_i(x, y)$ consists of a number of triangular parts of planes and can, therefore, be given an analytical description, which, however, is rather extensive, so we refrain from this. Instead, a splitting technique of the integrals is used.

$$A_{ij} = \int\int_{\Omega_T} \nabla\varphi_i \cdot \nabla\varphi_j dxdy, \quad f_i = \int\int_{\Omega_T} f(x, y)\varphi_i dxdy$$

The matrix elements consist of double integrals evaluated over the whole region Ω_T. A trick simplifying this computation is to split this integral into a sum of integrals where each integral is computed over each triangle:

$$A_{ij} = \sum_{k=1}^{M} \int\int_{T_k} \nabla\varphi_i \nabla\varphi_j dxdy, \quad f_i = \sum_{k=1}^{N} \int\int_{T_k} f(x, y)\varphi_i dxdy \tag{7.36}$$

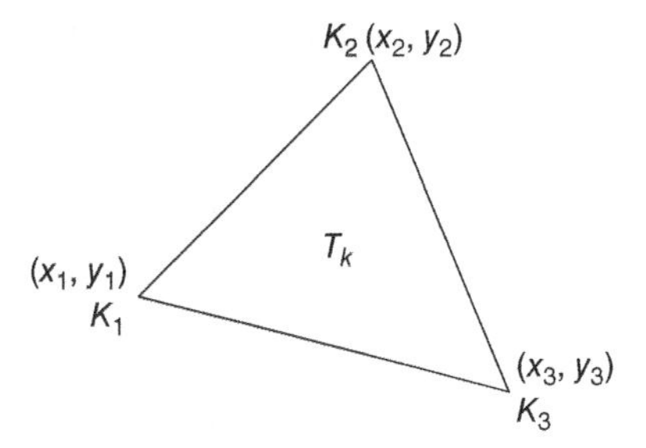

Figure 7.12 Local numbering of the nodes of a linear element

Assume that triangle T_k has the nodes K_1, K_2, and K_3. To simplify the notation, introduce a *local* numbering of the nodes for the triangle T_k: node 1, 2, and 3 with coordinates (x_1, y_1), (x_2, y_2), and (x_3, y_3) (Figure 7.12).

In the triangle T_k, there are three nonzero basis functions, namely those associated with node $i = 1, 2$, and 3. These basis functions have the form

$$\varphi_i(x, y) = a_i + b_i x + c_i y, \quad (x, y) \in T_k, \quad i = 1, 2, 3$$

Furthermore, as all the basis functions have the property

$$\varphi_i(x_j, y_j) = \begin{cases} 1, & \text{if } i = j \\ 0, & \text{if } i \neq j \end{cases}$$

the coefficients a_i, b_i, c_i satisfy the linear system of equations

$$\begin{pmatrix} 1 & x_1 & y_1 \\ 1 & x_2 & y_2 \\ 1 & x_3 & y_3 \end{pmatrix} \begin{pmatrix} a_1 & a_2 & a_3 \\ b_1 & b_2 & b_3 \\ c_1 & c_2 & c_3 \end{pmatrix} = \begin{pmatrix} 1 & 0 & 0 \\ 0 & 1 & 0 \\ 0 & 0 & 1 \end{pmatrix}$$

and therefore

$$\begin{pmatrix} a_1 & a_2 & a_3 \\ b_1 & b_2 & b_3 \\ c_1 & c_2 & c_3 \end{pmatrix} = \begin{pmatrix} 1 & x_1 & y_1 \\ 1 & x_2 & x_3 \\ 1 & x_3 & y_3 \end{pmatrix}^{-1}$$

In addition, as

$$\nabla \varphi_i = \begin{pmatrix} b_i \\ c_i \end{pmatrix}$$

we obtain

$$\int\int_{T_k} \nabla \varphi_i \cdot \nabla \varphi_j dxdy = (b_i b_j + c_i c_j) Y_{T_k} \tag{7.37}$$

where Y_{T_k} is the area of T_k. Note that the elements of A are here computed *analytically* with an explicit formula.

Hence, each triangle T_k contributes with a 3×3 matrix A_k^{loc}, the local stiffness matrix. However, as the true node numbers are K_1, K_2, and K_3, the elements of A_k^{loc}

are spread out to the corresponding positions in a sparse $N \times N$ matrix A_k:

$$A_k^{loc} = \begin{pmatrix} x & x & x \\ x & x & x \\ x & x & x \end{pmatrix} \quad \rightarrow \quad A_k = \begin{pmatrix} \ldots\ldots & \ldots\ldots & \ldots\ldots & \ldots\ldots \\ \ldots.x & \ldots.x & \ldots.x & \ldots\ldots \\ \ldots\ldots & \ldots\ldots & \ldots\ldots & \ldots\ldots \\ \ldots.x & \ldots.x & \ldots.x & \ldots\ldots \\ \ldots\ldots & \ldots\ldots & \ldots\ldots & \ldots\ldots \\ \ldots.x & \ldots.x & \ldots.x & \ldots\ldots \\ \ldots\ldots & \ldots\ldots & \ldots\ldots & \ldots\ldots \end{pmatrix}$$

As the BC in equation (7.1) is homogeneous, triangles with nodes on the boundary will be treated separately. For a triangle having two nodes on the boundary, there will be only one element $\neq 0$ in A_k^{loc} and if there is one node on the boundary, there will be four elements $\neq 0$ in the local stiffness matrix.

When A_k^{loc} for triangle T_k has been computed it is added, or rather *assembled* to the stiffness matrix

$$A = 0, \qquad \text{for} \quad k = 1 : M \quad A = A + A_k \quad \text{end}$$

The load vector $\mathbf{f}$ is computed in the same way. Each triangle T_k contributes with a local load vector $\mathbf{f}_k^{loc}$ being 3×1 and the components of which are spread out to the true positions in a sparse $N \times 1$ vector $\mathbf{f}_k$:

$$\mathbf{f}_k^{loc} = \begin{pmatrix} x \\ x \\ x \end{pmatrix} \quad \rightarrow \quad \mathbf{f}_k = \begin{pmatrix} . \\ x \\ . \\ x \\ . \\ x \\ . \end{pmatrix}$$

These vectors are assembled to the load vector $\mathbf{f}$:

$$\mathbf{f} = 0, \qquad \text{for} \quad k = 1 : M \quad \mathbf{f} = \mathbf{f} + \mathbf{f}_k \quad \text{end}$$

The components of $\mathbf{f}_k$ should be computed *numerically* as $f(x, y)$ may be complicated and therefore not possible to compute the integral analytically.

When the linear system

$$A\mathbf{c} = \mathbf{f}$$

has been solved with some sparse direct (or possibly iterative) method, the coefficients c_j are known and hence the ansatz function

$$u_h(x, y) = \sum_{j=1}^{N} c_j \varphi_j(x, y)$$

can be computed at any point $(x, y) \in \Omega_T$. The ansatz solution is a piecewise linear function in x and y with the property

$$u_h(x_j, y_j) = c_j \tag{7.38}$$

When $(x, y) \in T_k$, $u_h(x, y)$ is computed as a linear function obtained by linear interpolation through the node points of the triangle.

Observe that this discussion of the FEM is based on *homogeneous* BCs, i.e., $u = 0$ on the boundary $\partial\Omega$. If the BCs are nonhomogeneous, i.e., $u = g(x, y)$ on the boundary, the ansatz must be modified to

$$u_h(x, y) = \sum_{IP} c_j^{IP} \varphi_j^{IP}(x, y) + \sum_{BP} g(x_j^{BP}, y_j^{BP}) \varphi_j^{BP}(x, y) \tag{7.39}$$

where IP are the inner points and BP are the boundary points.

The FEM can be generalized to

- other PDEs
- other BCs
- other basis functions giving higher accuracy
- 3D problems
- time-dependent problems

Hence, FEM is a very flexible, accurate, and stable method for solving PDE problems of different types. FEM was first applied to problems in solid mechanics but is nowadays used to solve all kinds of PDE models in science and engineering. The Comsol Multiphysics program is designed for scientific computing of such models (see Section B.2).

Exercise 7.4.2. In equation (7.37), the elements of the *local stiffness matrix* A_k^{loc} for triangle T_k

$$(A_k^{loc})_{ij} = \int\int_{T_k} \nabla\varphi_i \cdot \nabla\varphi_j dxdy$$

were computed. Derive a corresponding formula or suggest a numerical integration method for the elements of a *local mass matrix* Q_k^{loc} for the triangle T_k.

$$(Q_k^{loc})_{ij} = \int\int_{T_k} \varphi_i\varphi_j dxdy$$

Exercise 7.4.3. Derive the linear system of equations $A\mathbf{c} = \mathbf{f}$ in the case the BCs are nonhomogeneous (7.2). The ansatz function is given in equation (7.39).

BIBLIOGRAPHY

1. G.D. Smith, "Numerical solution of partial differential equations", 3rd ed, 1986, Oxford University Press.
2. A. Iserles, "A first course in the numerical analysis of differential equations", 1996, Cambridge University Press
3. K. Eriksson, D. Estep, P. Hansbo, C. Johnson: Computational differential equations", 2004, Studentlitteratur

8

NUMERICAL METHODS FOR HYPERBOLIC PDEs

Hyperbolic partial differential equations (PDEs) occur as mathematical models of conservation laws and are found in, e.g., transport process and wave propagation problems. Information travels with finite speed in contrast to parabolic PDEs where an initial value immediately effects the solution in all other space points.

A fundamental property of hyperbolic PDEs is that geometric features such as steep fronts and discontinuities in the solution (in the weak sense, see below), are propagated with time. For linear problems, jumps come from the initial data, while in nonlinear cases, discontinuities can develop in the solution even from smooth initial data. Therefore, for hyperbolic PDEs, we are not only interested in the order of accuracy of a method but also in the treatment of discontinuities. Important concepts in the description of the conservation properties are *dissipation* (loss of energy) and *dispersion* (distorted phase relations owing to variable wave speed).

The classical hyperbolic PDE is the wave equation (see Chapter 5):

$$\frac{\partial^2 u}{\partial t^2} - c^2\frac{\partial^2 u}{\partial x^2} = 0, \quad -\infty < x < \infty, \quad t > 0 \tag{8.1}$$

The general solution of this equation, the d'Alembert solution, is

$$u(x,t) = f(x - ct) + g(x + ct) \tag{8.2}$$

where $f(x)$ and $g(x)$ are arbitrary twice differentiable functions. The solution $u(x,t)$ corresponds to two waves, traveling along the x-axis, one to the left with speed $-c$ and one to the right with speed c (see Figure 8.1)

Introduction to Computation and Modeling for Differential Equations, Second Edition. Lennart Edsberg.
© 2016 John Wiley & Sons, Inc. Published 2016 by John Wiley & Sons, Inc.

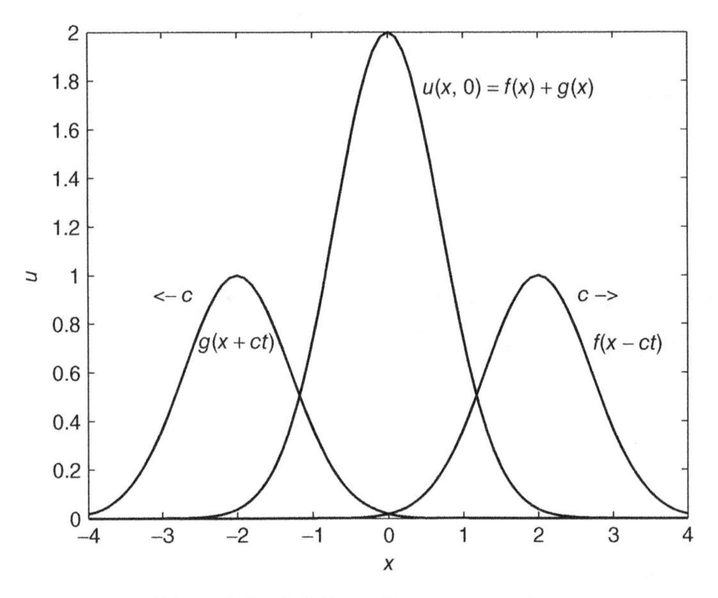

Figure 8.1 Solution of the wave equation

The wave equation (8.1) is a second-order PDE and can be written as a system of two first-order PDEs. Let $\mathbf{y}$ be a vector and let

$$y_1 = \frac{\partial u}{\partial t}, \quad y_2 = c\frac{\partial u}{\partial x}$$

Inserting these new variables into (8.1) and using the fact that $u_{tx} = u_{xt}$, we get the first-order system

$$\frac{\partial \mathbf{y}}{\partial t} + A\frac{\partial \mathbf{y}}{\partial x} = 0 \tag{8.3}$$

where A is the matrix

$$A = -c\begin{pmatrix} 0 & 1 \\ 1 & 0 \end{pmatrix} \tag{8.4}$$

Note that the eigenvalues of A, $-c$, and c, are the velocities of the waves in the general solution.

Definition 1: A system of first-order PDEs

$$\frac{\partial \mathbf{u}}{\partial t} + A\frac{\partial \mathbf{u}}{\partial x} = 0 \tag{8.5}$$

is hyperbolic if the real matrix A is diagonalizable with real eigenvalues.

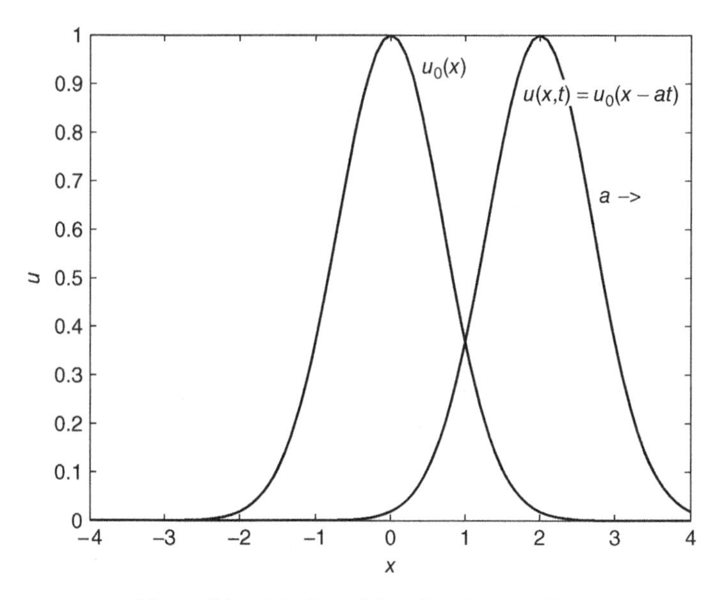

Figure 8.2 Solution of the advection equation

If the matrix A consists of only one element a, we get the following scalar hyperbolic homogeneous PDE, used as *model problem*:

$$\frac{\partial u}{\partial t} + a\frac{\partial u}{\partial x} = 0, \quad -\infty < x < \infty, \quad t > 0 \tag{8.6}$$

This equation is known as the *advection equation* (see Chapter 1) and is also called the transport equation or the one-way wave equation. With an initial condition given on the whole x-axis (Cauchy's problem)

$$u(x, 0) = u_0(x), \quad -\infty < x < \infty \tag{8.7}$$

the solution is a traveling wave

$$u(x, t) = u_0(x - at) \tag{8.8}$$

Hence, the initial data is simply advected with constant velocity a to the right if $a > 0$ (see Figure 8.2). The equation (8.6) is the simplest hyperbolic PDE, being linear and with a constant coefficient.

A scalar *conservation law* is a nonlinear hyperbolic PDE

$$\frac{\partial u}{\partial t} + \frac{\partial}{\partial x}f(u) = 0, \tag{8.9a}$$

defined by the function $f(u)$, called the *flux function*. The form (8.9a) is called the *conservative formulation* of the PDE.

If $f(u)$ is differentiable, the PDE can be written

$$\frac{\partial u}{\partial t} + a(u)\frac{\partial u}{\partial x} = 0 \tag{8.9b}$$

where $a(u) = f'(u)$. The form (8.9b) is called the *nonconservative formulation*. From numerical point of view, it is important if (8.9a) or (8.9b) is used as starting point for numerical solution, see Section 8.3.

A special case is

$$\frac{\partial u}{\partial t} + u\frac{\partial u}{\partial x} = 0 \tag{8.10a}$$

known as the *nonconservative formulation* of the *inviscid Burgers' equation* often used as a model problem for a nonlinear hyperbolic PDE. This equation will produce solutions with shocks even for smooth initial data. The conservative form of (8.10a) is

$$\frac{\partial u}{\partial t} + \frac{1}{2}\frac{\partial u^2}{\partial x} = 0 \tag{8.10b}$$

Compare (8.10a) with Burgers' viscous PDE where a diffusive term has been added

$$\frac{\partial u}{\partial t} + u\frac{\partial u}{\partial x} = \nu\frac{\partial^2 u}{\partial x^2} \tag{8.11}$$

This equation is much easier to solve numerically than (8.10), see also Exercise 5.3.2. Jan Burgers was a Dutch physicist active in the first half of the 20th century.

An important concept for hyperbolic PDEs is the *characteristic*.

Definition 2: The characteristics of (8.9) are the curves in the (x, t)-plane defined by the ordinary differential equation (ODE)

$$\frac{dx}{dt} = a(u(x(t), t)) \tag{8.12}$$

Along a characteristic the solution $u(x, t)$ is constant, which is seen from

$$\frac{du(x(t), t)}{dt} = \frac{\partial u}{\partial t} + \frac{dx(t)}{dt}\frac{\partial u}{\partial x} = \frac{\partial u}{\partial t} + a(u(x(t), t))\frac{\partial u}{\partial x} = 0$$

As $u(x, t)$ has a constant value u_C along a characteristic, we see that (8.12) fulfills

$$\frac{dx(t)}{dt} = a(u_C) = a_C \rightarrow x(t) = a_C t + C$$

Hence, the characteristics of (8.9) are *straight lines* (see Figure 8.3):

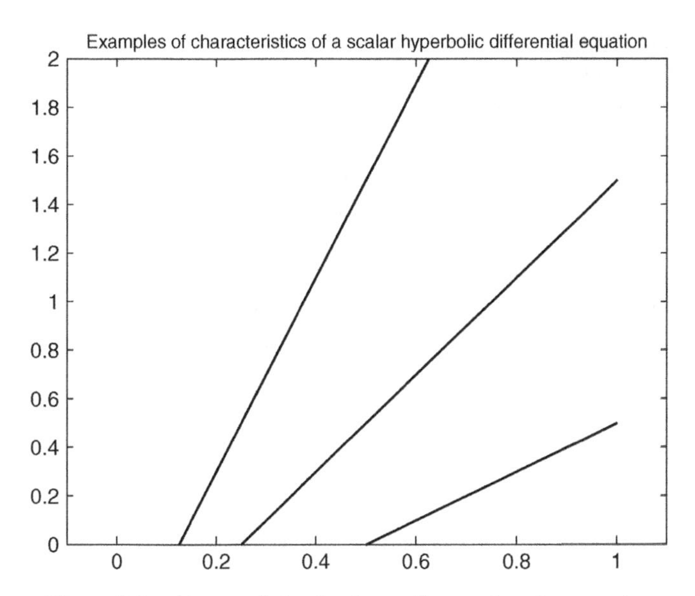

Figure 8.3 Characteristics for the nonlinear advection equation

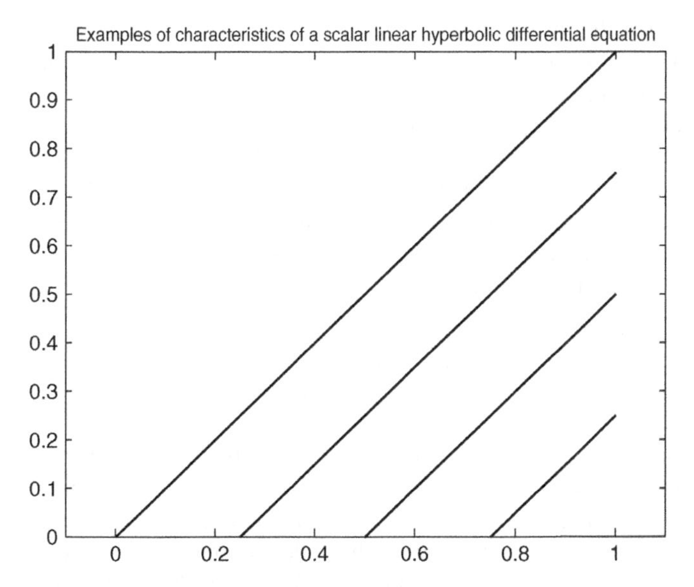

Figure 8.4 Characteristics for the linear advection equation

For the advection equation (8.6), $a(u) = a = constant$ and all characteristics have the same slope (see Figure 8.4).

The characteristics can also converge to a point in the (x, t)-plane thereby creating a shock in the solution, see Section 8.2.1 for numerical solution of Burger's equation (8.10).

In case the initial function is defined for $x \geq 0$ only, a semi-infinite problem, a boundary function is needed for $x = 0$, i.e., $u(0, t) = \alpha(t)$. The solution of the model problem then takes the form

$$u(x,t) = \begin{cases} \alpha(t - x/a) & \text{if } x - at < 0 \\ u_0(x - at) & \text{if } x - at \geq 0 \end{cases}$$

Hence, the initial and boundary values are advected along the characteristics.

For a bounded region $0 \leq x \leq 1$, BCs are needed together with an IC. For the hyperbolic model problem (8.6), the formulation is

$$\frac{\partial u}{\partial t} + a\frac{\partial u}{\partial x} = 0, \quad 0 < x < 1, \quad t > 0 \tag{8.13}$$

$$IC : u(x, 0) = u_0(x), \quad 0 \leq x \leq 1$$

$$BC : u(0, t) = \alpha(t), \quad t > 0 \quad \textit{if} \quad a > 0$$

$$BC : u(1, t) = \beta(t), \quad t > 0 \quad \textit{if} \quad a < 0$$

Observe that BC for problem (8.13) can be imposed only on the boundary $x = 0$ (not $x = 1$) if $a > 0$ as the characteristics are emanating from that boundary. If $a < 0$, however, the BC has to be given for $x = 1$.

Consider the wave equation (8.1) (which can also be written as the first-order system (8.3)) defined on the finite interval $[0, 1]$. For this problem, there are characteristics in both directions. Hence, BCs are needed both for $x = 0$ and $x = 1$.

An interesting property of the advection equation is that discontinuities in the initial function $u_0(x)$ are propagated along the x-axis when t increases.

Assume that the initial condition for (8.7) is a stepfunction (also called a Riemann initial condition):

$$u_0(x) = \begin{cases} 1, & x < 0 \\ 0, & x \geq 0 \end{cases} \tag{8.14}$$

The solution $u_0(x - at)$ is propagated without distortion by the advection equation. Hence at a later time $t = T > 0$, the solution has moved to $x = aT$ and the shape of the solution curve remains the same as the initial function $u_0(x)$. Bernhard Riemann was a German mathematician, active in the middle of the 19th century.

Now, discontinuous functions cannot be solutions of differential equations. The function $u_0(x)$ must be continuously differentiable on $-\infty < x < \infty$. However, if $u_0(x)$ is a discontinuous function, we can still talk about $u(x, t) = u_0(x - at)$ as a solution in the *weak* sense, i.e., an integrated version of the PDE is satisfied for all smooth functions $\varphi(x, t)$

$$\int_0^\infty \int_{-\infty}^\infty u\varphi_t + au\varphi_x dxdt + \int_{-\infty}^\infty u_0(x, 0)\varphi(x, 0)dx = 0 \tag{8.15}$$

This relation is obtained by multiplying (8.6) by a smooth function $\varphi(x, t)$, integrate over the (x, t) region and then integrate by parts. Compare this technique with the Galerkin method in Chapter 4, where weak solutions for ODEs are treated.

In applications, e.g., gas dynamics, the moving jump is called a *shock wave* or *contact discontinuity*, depending on which quantities jump. The following relation, known as *Rankine–Hugoniot's theorem*, gives the speed with which the shock propagates: Assume that a discontinuity is moving with speed s and that the value of u to the left of the jump is u_L and to the right u_R. Then the following relation holds for the scalar conservation law (8.9a):

$$s(u_L - u_R) = f(u_L) - f(u_R) \tag{8.16}$$

Macquorn Rankine was a Scotish physicist and Pierre-Henri Hugoniot was a French mathematician both active in the 19th century.

The generalization of the scalar conservation law (8.9) to a system of nonlinear hyperbolic PDEs is

$$\frac{\partial \mathbf{u}}{\partial t} + \frac{\partial}{\partial x}\mathbf{f}(\mathbf{u}) = 0, \quad -\infty < x < \infty, \quad t > 0 \tag{8.17}$$

If the spatial derivation is performed, we get the nonconservative form

$$\frac{\partial \mathbf{u}}{\partial t} + A(\mathbf{u})\frac{\partial \mathbf{u}}{\partial x} = 0 \tag{8.18}$$

where $A(\mathbf{u})$ is the jacobian of $\mathbf{f}(\mathbf{u})$. The system is hyperbolic if $A(\mathbf{u})$ is diagonalizable and has real eigenvalues.

8.1 APPLICATIONS

Example 8.1. **(Time-dependent hot flow in a pipe revisited)**

As a first example of a hyperbolic first-order PDE, consider Example 6.1. Assume that transport by diffusion is negligible, i.e., $\kappa = 0$. The PDE then takes the form

$$\frac{\partial T}{\partial t} + v\frac{\partial T}{\partial x} + \frac{2h}{\rho CR}(T - T_{cool}) = 0, \quad 0 < x < L, \quad t > 0 \tag{8.19}$$

We need an initial condition $T(x, 0) = T_{init}(x)$ and a boundary condition $T(0, t) = T_0(t)$, where $T_{init}(x)$ and $T_0(t)$ are known functions.

Example 8.2. **(Time-dependent counterflow heat exchanger)**

In Example 4.4, a heat exchanger is presented. If the temperature profiles have not come to a steady state, the following system of hyperbolic PDEs model the corresponding time-dependent problem on the x interval $0 < x < L$:

$$\frac{\partial H}{\partial t} + u_H \frac{\partial H}{\partial x} = -a(H - C), \qquad H(0,t) = H_0, \quad H(x,0) = H_0(x) \tag{8.20}$$

$$\frac{\partial C}{\partial t} - u_C \frac{\partial C}{\partial x} = a(H - C), \qquad C(L,t) = C_0, \quad C(x,0) = C_0(x) \tag{8.21}$$

Example 8.3. **(Wave equation)**

Another example of a hyperbolic PDE is a model of the movements $u(x,t)$ of a vibrating string

$$\frac{\partial^2 u}{\partial t^2} - c^2 \frac{\partial^2 u}{\partial x^2} = 0, \quad 0 < x < L, \quad t > 0 \tag{8.21a}$$

where the parameter c is the velocity of the wave. For this equation, we need two ICs,

$$u(x,0) = u_0(x) \quad \frac{\partial u}{\partial t}(x,0) = v_0(x), \quad 0 \le x \le L$$

and two BCs

$$u(0,t) = 0 \quad u(L,t) = 0, \quad t \ge 0$$

In Example 9.2, the wave equation is derived for a vibrating string of length L (m), density ρ ($\mathrm{kg/m^3}$), cross-section area A ($\mathrm{m^2}$), and string tension F (N). These parameters are related to c through $c^2 = F/\rho A$. The total energy of the string is the sum of the kinetic and the potential energy

$$E_{tot} = \int_0^L \left(\frac{1}{2}\rho A \left(\frac{\partial u}{\partial t}\right)^2 + \frac{1}{2}F\left(\frac{\partial u}{\partial x}\right)^2 \right) dx \tag{8.21b}$$

By differentiating E_{tot} with respect to t we obtain $dE_{tot}/dt = 0$, hence the total energy is conserved.

The PDE (8.21a) is also valid for plane electromagnetic waves traveling in the x direction. The variable u then represents the components of the orthogonal electric and magnetic fields and c is the speed of light. Other types of initial and boundary conditions are usually imposed.

Example 8.4. **(Shock propagation in Euler's equations)**

As was mentioned in Chapter 5, hyperbolic PDEs occur in fluid dynamics modeled by the Euler equations. Euler's equations model *shocks* and contact discontinuities, i.e., thin transition layers where the pressure, density, and speed of the fluid changes significantly.

An example of a hyperbolic model sometimes used in gas dynamics is the compressible isothermal Euler equations for a perfect gas in 1D, formulated as

$$\frac{\partial \rho}{\partial t} + \frac{\partial (\rho u)}{\partial x} = 0 \tag{8.22a}$$

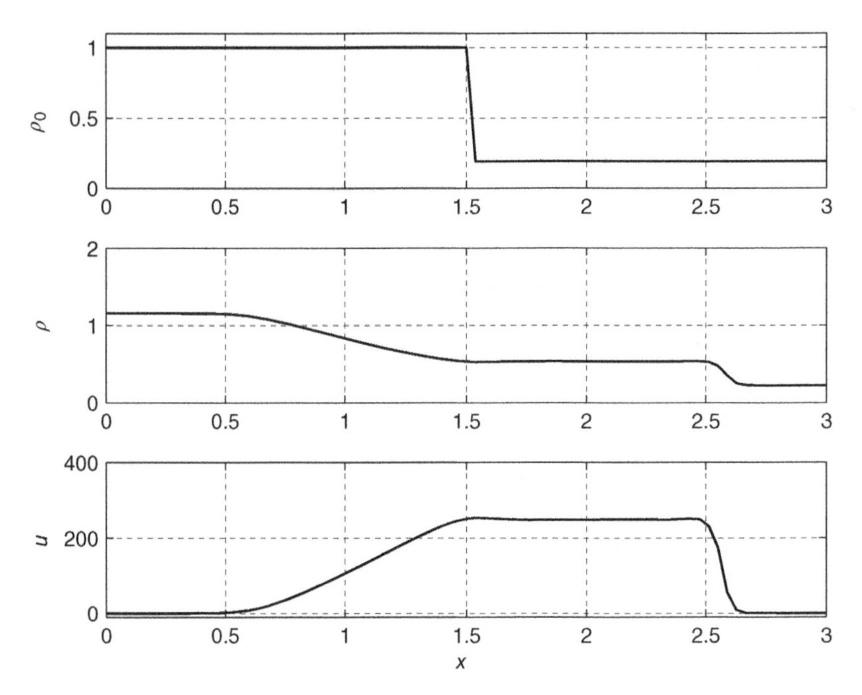

Figure 8.5 Evolution of density and velocity for Euler's 1D model

$$\frac{\partial(\rho u)}{\partial t} + \frac{\partial(\rho u^2 + p)}{\partial x} = 0 \tag{8.22b}$$

$$p = K\rho^{\gamma} \tag{8.22c}$$

There are three nonlinear equations (one is algebraic) for the three unknowns ρ (kg/m^3) (density), u (m/s) (velocity), and p (N/m^2) (pressure).

In the so-called shock-tube problem, both ends of the tube are closed and a diaphragm is separating two regions, one with high pressure and the other with low pressure.

The ICs for density is

$$\rho(x,0) = \begin{cases} \rho_0, & x \leq L/2 \\ \rho_1, & x > L/2 \end{cases}, \quad u(x,0) = 0 \tag{8.23a}$$

and for velocity, it is $u(x, 0) = 0$

As BCs we have

$$u(0,t) = u(L,t) = 0 \tag{8.23b}$$

For $\rho(0,t)$, numerical BCs are used (see Section 8.2.6), i.e., the value of ρ at the boundaries is extrapolated from the values of ρ inside the tube. The evolution of ρ and u is sketched in Figure 8.5.

For $\gamma = 2$, the equations also model the flow of a thin layer of fluid under gravity, the "shallow water equations." The problem above then becomes a model of a "dam-break" simulation if ρ is exchanged to h (m), the height of the water.

Exercise 8.1.1. Verify that E_{tot} in (8.21b) is constant in time.

Exercise 8.1.2. Verify that the Euler equations in (8.22) can be written in the form (8.18). Find the jacobian $A(\mathbf{u})$ and calculate the eigenvalues of the jacobian. Verify that the system is hyperbolic.

8.2 NUMERICAL SOLUTION OF HYPERBOLIC PDEs

Hyperbolic PDEs can be solved numerically with finite difference, finite volume, or finite element methods. In this chapter, the first two discretization methods are presented. We use (8.13) as model problem.

When the method of lines (MoL), see Section 6.3.1, is applied to (8.13), we start by discretizing the x interval $0 \leq x \leq 1$ into a grid $x_0, x_1, \ldots, x_N$, where $x_0 = 0, x_N = 1$, and $x_i = ih_x$, $h_x = 1/N$. At the point $x = x_i$, we discretize the PDE to an ODE with Euler's implicit method:

$$\frac{du_i}{dt} = -a\frac{u_i - u_{i-1}}{h_x}, \quad a > 0, \quad i = 1, 2, \ldots N$$

We get the following ODE system:

$$\frac{d\mathbf{u}}{dt} = \frac{a}{h_x}A\mathbf{u} + \frac{a}{h_x}\mathbf{f}(t), \quad \mathbf{u}(0) = \mathbf{u}_0 \tag{8.24}$$

where

$$A = \begin{pmatrix} -1 & 0 & 0 & \ldots & 0 \\ 1 & -1 & 0 & \ldots & 0 \\ 0 & 1 & -1 & \ldots & 0 \\ . & . & . & \ldots & 0 \\ . & . & . & \ldots & 0 \\ 0 & 0 & 0 & \ldots & -1 \end{pmatrix}, \quad \mathbf{f}(t) = \begin{pmatrix} \alpha(t) \\ 0 \\ 0 \\ . \\ . \\ 0 \end{pmatrix}, \quad \mathbf{u}_0 = \begin{pmatrix} u_0(x_1) \\ u_0(x_2) \\ u_0(x_3) \\ . \\ . \\ u_0(x_N) \end{pmatrix} \tag{8.25}$$

The eigenvalues of the matrix $(a/h_x)A$ are all $\lambda_i = -a/h_x$. Hence, they are all negative and the system is stable.

If Euler's explicit method is used to solve the time-dependent ODE system (8.24), we know from Chapter 3 that the time step h_t must fulfill the following condition to obtain a numerically stable solution

$$h_t\lambda_i \in S_{EE} \quad \rightarrow \quad a\frac{h_t}{h_x} \leq 2 \tag{8.26}$$

With a numerical test on (8.24) with Euler's explicit method, we notice, however, that when $ah_t/h_x = 1.1$, the numerical solution is *unstable*.

Hence, the theoretical stability result based on *eigenvalue analysis* as presented for ODE systems in Chapter 3 is not correct for this semidiscretized hyperbolic PDE model problem! The reason is that the matrix A in (8.25) is not diagonalizable! This implies that the solution of the ODE system will have a strong initial polynomial growth of size $t^{N-1}e^{-t}$. The difference equation system obtained with Euler's method will have a similar behavior. We therefore need another type of stability concept for hyperbolic problems. In Section 8.4, the *von Neumann analysis* based on Fourier analysis is presented.

Unless anything else is stated, the methods presented in this section are applied to the model problem (8.13) only.

8.2.1 The Upwind Method (FTBS)

The method formulated above can be written as a difference equation, the FTBS method (Forward-Time-Backward-Space):

$$\frac{u_{i,k+1} - u_{i,k}}{h_t} = -a\frac{u_{i,k} - u_{i-1,k}}{h_x} \tag{8.27}$$

or

$$u_{i,k+1} = (1 - \sigma)u_{i,k} + \sigma u_{i-1,k}$$

where σ is defined as

$$\sigma = a\frac{h_t}{h_x} \tag{8.28}$$

The FTBS method (8.27) can also be illustrated by a stencil (see Figure 8.6).

The FTBS method, also called the upwind (or upstream) method, has first-order accuracy in both space and time and the stability condition (see Section 8.4)

$$0 < \sigma \leq 1 \tag{8.29}$$

This inequality is known as the *Courant–Friedrich–Lewy condition* or simply the *CFL condition*, from 1928 and σ is called the *Courant number*. Richard Courant,

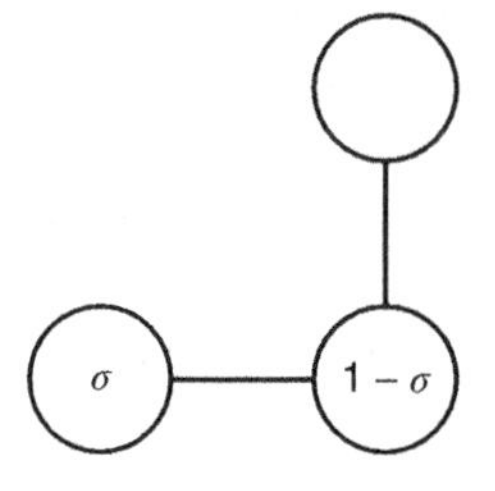

Figure 8.6 Stencil for the FTBS method

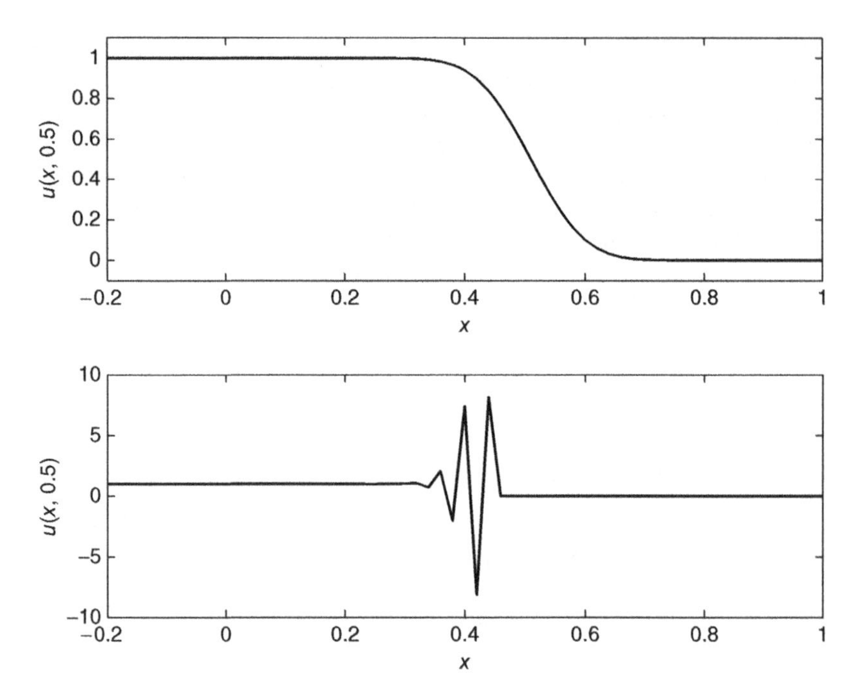

Figure 8.7 Stable and unstable numerical solution of the advection equation

Kurt Friedrichs, and Hans Lewy were German American mathematicians, active in the 20th century. Courant was the founder of the famous Courant Institute in New York.

Applying the upwind method (8.27) to the model problem (8.13) with the IC $u(x,0) = 0,\ x > 0$ and the BC $u(0,t) = 1,\ t > 0$, i.e., a stepfunction, gives a numerical solution shown as a graph in Figure 8.7. In the figure, the stepsize combinations $h_t = 0.8h_x/a$ and $h_t = 1.1h_x/a$ have been used. For the larger stepsize h_t, the solution will be unstable (the amplitude of the oscillations increases as time t increases).

With the "magic stepsize combination" $h_t = h_x/a$, i.e., $\sigma = 1$, however, it turns out that the stepfunction will be perfectly preserved, moving along the positive x-axis with speed a! See the modified equation (8.30) for an explanation.

For the smaller stepsize, we see that the numerical solution is smoothed out compared to the exact solution. To understand this dissipation phenomenon, we can use the *modified equation*, the PDE that is exactly satisfied by the numerical solution $u_{i,k}$. It can be derived using Taylor's expansion. However, the modified PDE will be a series having an infinite number of terms. If the series is truncated after a few terms, we obtain a PDE from which information about the nature of its analytical solution can be found.

To find the modified equation, replace $u_{i,k}$ by the function $v = v(x_i, t_k)$ and Taylor expand

$$v(x_i, t_{k+1}) = v(x_i, t_k) + h_t \frac{\partial v}{\partial t}(x_i, t_k) + \frac{h_t^2}{2} \frac{\partial^2 v}{\partial t^2}(x_i, t_k) + \dots$$

$$v(x_{i-1}, t_k) = v(x_i, t_k) - h_x \frac{\partial v}{\partial x}(x_i, t_k) + \frac{h_x^2}{2} \frac{\partial^2 v}{\partial x^2}(x_i, t_k) + \ldots$$

Insert these expansions into (8.27) and the following PDE at the point (x_i, t_k) is found:

$$\frac{\partial v}{\partial t} + a\frac{\partial v}{\partial x} + \frac{h_t}{2}\frac{\partial^2 v}{\partial t^2} - \frac{ah_x}{2}\frac{\partial^2 v}{\partial x^2} + \cdots = 0$$

Use the PDE (8.13) to obtain the following relations:

$$\frac{\partial^2 v}{\partial t^2} = -a\frac{\partial^2 v}{\partial x \partial t} = a^2 \frac{\partial^2 v}{\partial x^2}$$

Hence, replace v_{tt} by $a^2 v_{xx}$ and we obtain the following modified equation after having neglected higher order terms:

$$\frac{\partial v}{\partial t} + a\frac{\partial v}{\partial x} = \frac{ah_x}{2}(1 - \sigma)\frac{\partial^2 v}{\partial x^2} \tag{8.30}$$

The right hand side term is a diffusion term and this causes the solution to be smoothed out. Hence, the upwind method introduces dissipation into the numerical solution. We see also that when $\sigma = 1$, there is no diffusion term and no dissipation.

In the beginning of this chapter, Burgers' inviscid equation (8.10) was presented. The following example shows how a shock can develop from smooth initial data.

Example 8.5. **(Burgers' equation with the upwind method)**

We choose the conservative form of Burgers' equation, see also Example 8.5.

$$\frac{\partial u}{\partial t} + \frac{\partial u^2}{\partial x} = 0, \quad 0 < x < 1, \quad t > 0$$

With IC $u(x, 0) = cos(2\pi x) + 1$ and periodic BC $u(0, t) = u(1, t)$, the upwind method gives the following graph after a number of steps (Figure 8.8).

Exercise 8.2.1. Write a program or convince yourself by some hand calculations that the upstream method with $\sigma = 1$, the "magic stepsize combination" applied to the model problem (8.13) with a step function as IC will preserve the step function for $t > 0$.

8.2.2 The FTFS Method

When the parameter a in the advection equation (8.13) is negative, the upwind method gives unstable solutions. Instead the FTFS method (Forward-Time-Forward-Space) can be used

$$\frac{u_{i,k+1} - u_{i,k}}{h_t} = -a\frac{u_{i+1,k} - u_{i,k}}{h_x}$$

i.e.,

$$u_{i,k+1} - u_{i,k} = -\sigma(u_{i+1,k} - u_{i,k}) \tag{8.31}$$

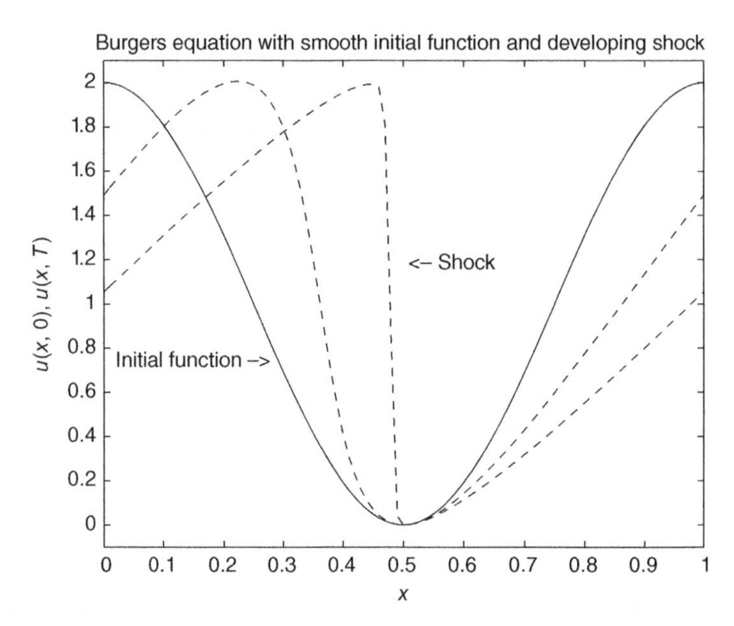

Figure 8.8 Burger's equation with smooth initial function and developing shock

8.2.3 The FTCS Method

To increase the accuracy, we try central differences in space, giving the FTCS method (Forward-Time-Central-Space):

$$\frac{u_{i,k+1} - u_{i,k}}{h_t} = -a\frac{u_{i+1,k} - u_{i-1,k}}{2h_x} \tag{8.32}$$

i.e.,

$$u_{i,k+1} = u_{i,k} - \frac{\sigma}{2}(u_{i+1,k} - u_{i-1,k}) \tag{8.33}$$

This method, however, turns out to be *unstable* for all h_t, h_x, see Section 8.3.

8.2.4 The Lax–Friedrich Method

Another method used in practice is the *Lax–Friedrich method*, which is first order in time, second order in space, and symmetric, hence insensitive to the sign of a

$$u_{i,k+1} = \frac{u_{i-1,k} + u_{i+1,k}}{2} - \frac{\sigma}{2}(u_{i+1,k} - u_{i-1,k}), \tag{8.34}$$

The Lax–Friedrich method is stable for $-1 \leq \sigma \leq 1$. The only difference compared with the FTCS method is that $u_{i,k}$ in the right hand side of (8.33) is replaced by the mean value of $u_{i-1,k}$ and $u_{i+1,k}$ in (8.34)

Exercise 8.2.2. Write a program or convince yourself by some hand calculations that the Lax–Friedrich method with $\sigma = 1$ applied to the model problem (8.13) with a stepfunction as IC will preserve the stepfunction for $t > 0$.

Just like the upwind method, the Lax–Friedrich method also introduces smoothing in the numerical solution. This can be explained by the modified equation.

Peter Lax is a Hungarian mathematician, now Prof.Em. at the Courant Institute. He received the prestigious Abel price in 2005.

Exercise 8.2.3. Show that the modified equation for Lax–Friedrichs' method applied to (8.13) is

$$\frac{\partial v}{\partial t} + a\frac{\partial v}{\partial x} = \frac{a^2 h_t}{2}\left(\frac{1}{\sigma^2} - 1\right)\frac{\partial^2 v}{\partial x^2}$$

Hence when $\sigma = \pm 1$, there is no diffusion term.

8.2.5 The Leap-Frog Method

The *leap-frog* method (see Section 3.4)

$$\frac{u_{i,k+1} - u_{i,k-1}}{2h_t} + a\frac{u_{i+1,k} - u_{i-1,k}}{2h_x} = 0 \tag{8.35}$$

is second order in both time and space. It is stable for all values of σ and does not involve any diffusivity. For nonlinear problems, however, this method becomes unstable for the model problem (8.13).

8.2.6 The Lax–Wendroff Method

Yet another method second order in both time and space is the *Lax-Wendroff method*. This method is based on a trick, where derivatives in t are expressed as derivatives in x. Start with Taylor expansion at the point (x_i, t_k):

$$u(x_i, t_{k+1}) = u(x_i, t_k) + h_t\frac{\partial u}{\partial t}(x_i, t_k) + \frac{h_t^2}{2}\frac{\partial^2 u}{\partial t^2}(x_i, t_k) + O(h_t^3) \tag{8.36}$$

Use the PDE (8.13) to obtain the relations:

$$\frac{\partial u}{\partial t} = -a\frac{\partial u}{\partial x}, \quad \frac{\partial^2 u}{\partial t^2} = -a\frac{\partial^2 u}{\partial x \partial t} = a^2\frac{\partial^2 u}{\partial x^2}$$

Insert space derivatives instead of time derivatives into the Taylor expansion, discretize and we obtain:

$$u_{i,k+1} = u_{i,k} - \frac{\sigma}{2}(u_{i+1,k} - u_{i-1,k}) + \frac{\sigma^2}{2}(u_{i+1,k} - 2u_{i,k} + u_{i-1,k}) \tag{8.37}$$

The method is stable for $-1 \le \sigma \le 1$. When the Lax–Wendroff method is used there is a problem at the right boundary. As there are no BCs at $x = 1$, we have to impose

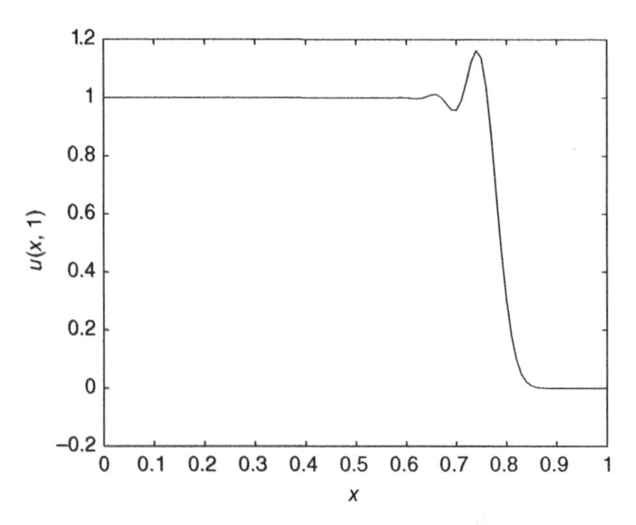

Figure 8.9 The model problem with Lax-Wendroff's method, dispersion effects

artificial *numerical boundary conditions* in order to be able to use the stencil. A simple way to construct values at $x_N = 1$ is to extrapolate the numerical solution by linear extrapolation

$$u_{Nk} = u_{N-1,k} + u_{N-1,k} - u_{N-2,k} = 2u_{N-1,k} - u_{N-2,k} \tag{8.38}$$

or, even better, quadratic extrapolation

When the Lax–Wendroff method is applied to the model problem with a step function as initial function, we obtain the result seen in Figure 8.9.

In Figure 8.9, we see that the Lax–Wendroff method generates spurious oscillations. This is typical of a method being second order in time and is related to a dispersion error introduced by the numerical method.

Burton Wendroff is an American mathematician still active.

If we let $h_x, h_t \to 0$ in the Lax–Wendroff discretization, we get the PDE expression (8.13) in the left hand side and a diffusion term in the right hand side

$$\frac{\partial u}{\partial t} + a\frac{\partial u}{\partial x} = \frac{a^2 h_t^2}{2}\frac{\partial^2 u}{\partial x^2} \tag{8.39}$$

Hence, the original hyperbolic PDE is approximated by a parabolic PDE, and the solution can no longer contain discontinuities. Just as for the previous methods, this is an example of *artificial dissipation*. Other methods with similar numerical behavior are obtained if the test problem (8.13) is replaced by the PDE

$$\frac{\partial u}{\partial t} + a\frac{\partial u}{\partial x} = \epsilon\frac{\partial^2 u}{\partial x^2}$$

where ϵ is a small positive quantity.

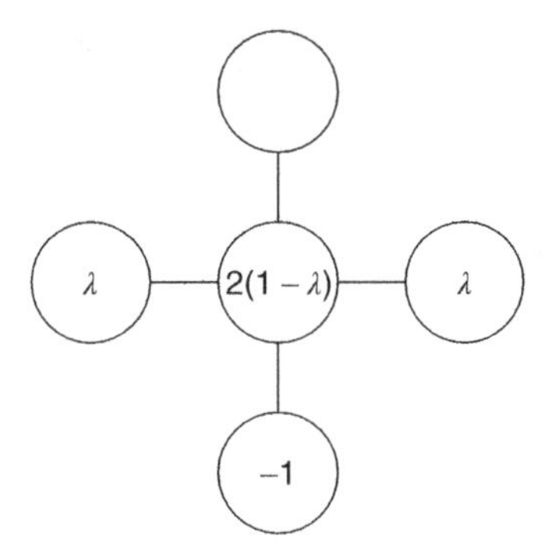

Figure 8.10 Stencil for central differences applied to the wave equation

8.2.7 Numerical Method for the Wave Equation

A special case of hyperbolic PDE is the wave equation in Example 8.3. It is reasonable to believe that there are special methods, efficient and accurate for this PDE. The numerical treatment can be based on central difference approximations in both x and t giving the following difference equation

$$\frac{u_{i,k+1} - 2u_{i,k} + u_{i,k-1}}{h_t^2} = c^2 \frac{u_{i+1,k} - 2u_{i,k} + u_{i-1,k}}{h_x^2}$$

This can be written as

$$u_{i,k+1} = \sigma^2 u_{i+1,k} + 2(1-\sigma^2)u_{i,k} + \sigma^2 u_{i-1,k} - u_{i,k-1} \tag{8.40a}$$

where $\sigma = ch_t/h_x$. The corresponding stencil is shown in Figure 8.10, where $\lambda = \sigma^2$. This method gives stable solutions for $-1 \leq \sigma \leq 1$.

For the magic stepsize $h_t = h_x/c$, a traveling wave $u(x,t) = f(x - ct)$ with exact initial values $u_{i,0} = f(ih_x)$, $u_{i,1} = f(ih_x - ch_t)$, $i = 0, 1, 2, \ldots$ is exactly preserved (in exact arithmetic) when $\sigma = 1$ in (8.40a)

$$u_{i,k+1} = u_{i+1,k} + u_{i-1,k} - u_{i,k-1} \tag{8.40b}$$

Example 8.6. **(Wave equation with the magic stepsize)**

Consider the wave equation in Example 8.3 (see also Example 5.3) with the parameter values $c = 1, L = 3$, and initial conditions $u_0(x) = -\sin(kx)$, $v_0(x) = \omega\cos(kx)$ where $\omega = 2\pi$ and $k = 2\pi$. Discretization according to (8.40b) with $h_x = L/50$, $h_t = h_x/c$ and taking $N = 10$ time steps gives the numerical solution (Figure 8.11).

The wave equation is closely connected to Maxwell's equations described in Section 5.3.6. A simple form of these equations is (see Exercise 5.3.6)

$$\frac{\partial H_y}{\partial t} = \frac{1}{\mu_0}\frac{\partial E_z}{\partial x}$$

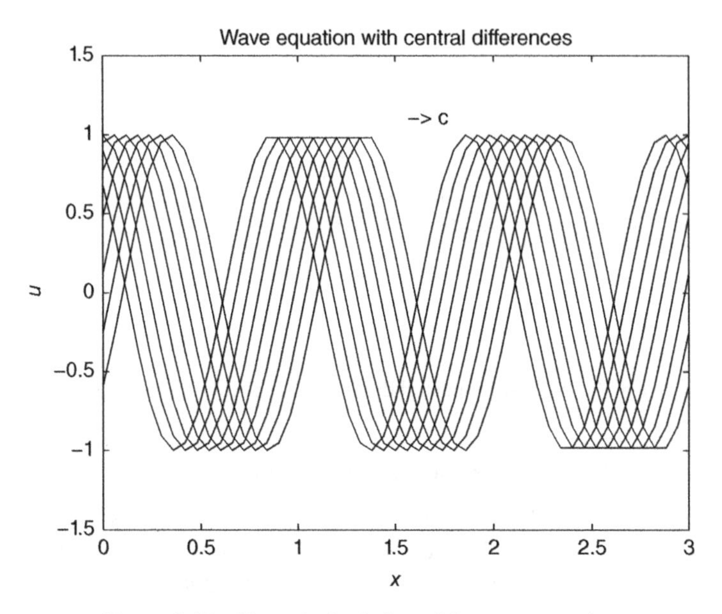

Figure 8.11 Numerical solution of the wave equation

$$\frac{\partial E_z}{\partial t} = \frac{1}{\epsilon_0}\frac{\partial H_y}{\partial x}$$

One possible way to compute E_z and H_y would be to solve the two wave equations for the two components. A better and more efficient way, however, is to use Yee's method on the first-order PDE system. That method is based on staggered grids (see Section 3.5.3) in both the x and the t direction and is demonstrated on the system (8.3), which is equivalent to the Maxwell equation system above

$$y_{2_{i+1/2}^{k+1/2}} = y_{2_{i+1/2}^{k-1/2}} + c\frac{h_t}{h_x}(y_{1_{i+1}^{k}} - y_{1_i^k})$$

$$y_{1_i^{k+1}} = y_{1_i^k} + c\frac{h_t}{h_x}(y_{2_{i+1/2}^{k+1/2}} - y_{2_{i-1/2}^{k+1/2}})$$

Hence, the two components y_1 and y_2 are computed on different grids. A comparison with (3.57) shows that the method is based on the leap-frog method and has second-order accuracy. It can also be shown to preserve the energy of the wave (8.21b) if the energy is defined from a trapezoidal expression instead of the integral itself.

Exercise 8.2.4. Formulate the upwind method for the inhomogeneous advection equation, i.e.,

$$\frac{\partial u}{\partial t} + a\frac{\partial u}{\partial x} = g(x, t)$$

Exercise 8.2.5. Give the stencil corresponding to the FTFS method in (8.31).

Exercise 8.2.6. Formulate a suitable difference approximation based on FTBS and FTFS to solve numerically the counterflow heat problem in Example 8.2.

Exercise 8.2.7. Give the stencils corresponding to Lax–Friedrich's method and Lax–Wendroff's method.

Exercise 8.2.8. As the method (8.40) is a two-step method, we need two ICs to start up the difference equation. Use the ICs given in Example 8.3. Work out the *discretized* ICs needed.

Exercise 8.2.9. Consider the wave equation in Example 8.3.

a) Discretize the x interval according to $x_i = ih_x, i = 0, 1, 2, .., N + 1$, where $h_x = L/(N + 1)$. With the MoL and central differences, a system of second-order ODEs is obtained.

$$\frac{d^2\mathbf{u}}{dt^2} = A\mathbf{u}, \quad \mathbf{u}(0) = \mathbf{u}_0, \quad \frac{d\mathbf{u}}{dt}(0) = \mathbf{v}_0$$

Find the values of the matrix A and the vectors $\mathbf{u}_0$ and $\mathbf{v}_0$.

b) Write the system in (a) as a system of first-order ODEs

$$\frac{d\mathbf{w}}{dt} = B\mathbf{w}, \quad \mathbf{w}(0) = \mathbf{w}_0$$

Give the values of B and $\mathbf{w}_0$.

c) Calculate the eigenvalues of B.

d) If the explicit midpoint method (3.42) is used on the first-order system in (b), for which time steps h_t is the numerical solution stable?

8.3 THE FINITE VOLUME METHOD

In this section, the finite volume method (FVM) is treated. Just like the FDM and the FEM, solution values are calculated at discrete grid points. In the FVM, each grid point is surrounded by a *control volume* or *cell*. What the FVM has in advantage compared to the other two methods is that the FVM gives approximate u values that satisfy a discretized version of the integrated form of the conservation law (8.9a), i.e.,

$$\int_{x_i}^{x_{i+1}} (u(x, t_{k+1}) - u(x, t_k))dx = \int_{t_k}^{t_{k+1}} (f(u(x_i, t) - f(u(x_{i+1}, t)))dt \tag{8.41}$$

where $f(u(x, t))$ is the flux function. The relation (8.41) simply means that the change of a quantity $u(x, t)$ accumulated in the "volume" $[x_i, x_{i+1}]$ balances the net influx, i.e., the difference between influx to and outflux from the volume during the time interval $[t_k, t_{k+1}]$.

To explain the FVM in 1D, introduce a grid with grid points x_i and the stepsize h_x. Also introduce control "volumes" V_i defined by the intervals $[x_{i-1/2}, x_{i+1/2}]$, where $x_{i\pm1/2} = (x_{i\pm1} + x_i)/2$.

Now consider the conservation law (8.9)

$$\frac{\partial u}{\partial t} + \frac{\partial}{\partial x} f(u) = 0$$

The volume average of $u(x, t)$ at time t_k in the volume V_i is

$$\overline{u}_i(t_k) = \frac{1}{x_{i+1/2} - x_{i-1/2}} \int_{x_{i-1/2}}^{x_{i+1/2}} u(x, t_k)dx \tag{8.42}$$

and similar for $t = t_{k+1}$. Now integrate (8.9) over the volume V_i

$$\int_{x_{i-1/2}}^{x_{i+1/2}} \frac{\partial u}{\partial t} dx = f(u_{i-1/2}) - f(u_{i+1/2})$$

where $f(u_{i\pm1/2}) = f(u(x_{i\pm1/2}, t))$ are the fluxes at the cell interfaces. Changing the order of integration and differentiation gives

$$\frac{d\overline{u}_i}{dt} = \frac{1}{h_x}(f(u_{i-1/2}) - f(u_{i+1/2})) \tag{8.43}$$

which is an exact relation for the volume averages, i.e., no approximations have been made so far. The ODE system (8.43) resembles the MoL, see Section 6.3.1, but the right hand side contains expressions of u that are not known. If the explicit Euler method with time step h_t is used to approximate the time derivate and $\overline{u}_i$ is approximated by u_i, we obtain the following conservative method, where $u_{i\pm1/2}$ can be computed with, e.g., interpolation of the u_i values.

$$u_{i,k+1} = u_{i,k} + \frac{h_t}{h_x}(f(u_{i-1/2}) - f(u_{i+1/2})) \tag{8.44}$$

By rearranging terms in (8.44) and summing over i, we obtain

$$h_x \sum_{i=1}^{N} (u_{i,k+1} - u_{i,k}) = h_t(f(u_{1/2}) - f(u_{N+1/2}))$$

which is the discrete version of the integral balance (8.41), equivalent to the basic conservation law (8.9a).

Example 8.7. **(Numerical solution of Burgers' equation, revisited)**
Consider Burgers' equation on conservative form

$$\frac{\partial u}{\partial t} + \frac{1}{2}\frac{\partial u^2}{\partial x} = 0, \ -\infty < x < \infty, \quad t > 0$$

with IC

$$u(x, 0) = \begin{cases} 2, & x < 0 \\ 1, & x \geq 0 \end{cases}$$

and BC

$$u(0, t) = \begin{cases} 1, & t = 0 \\ 2, & t > 0 \end{cases}$$

Hence, the problem describes a shock coming in at $x = 0$ and moving in the positive x direction.

Here the flux function is $f(u) = u^2/2$ and the upwind method gives

$$u_{i,k+1} = u_{i,k} - \frac{h_t}{2h_x}(u_{i,k}^2 - u_{i-1,k}^2)$$

which is conservative in this form. Make a grid of the x interval $[0, 2.5]$ using the stepsize $h_x = 0.025$ and of the t interval $[0, 1]$ using the time step $h_t = 0.0125$. Owing to Rankine–Hugoniot's relation (8.16), the speed s of the shock is $s = 1.5$. Hence, at time $t = 1$, the shock has reached the x value $x = s \cdot 1 = 1.5$, which is in agreement with Figure 8.10.

However, if the nonconservative form of Burgers' equation (8.10a) is used together with the upwind scheme

$$u_{i,k+1} = u_{i,k} - \frac{h_t}{h_x}u_{i,k}(u_{i,k} - u_{i-1,k})$$

and the same values of the numerical parameters h_x and h_t, the shock speed is lower than the speed obtained with the conservative form of the method, see Figure 8.12.

8.4 SOME EXAMPLES OF STABILITY ANALYSIS FOR HYPERBOLIC PDEs

As was pointed out earlier and shown in an example, eigenvalue analysis is not adequate for investigation of the stability of numerical solutions of hyperbolic PDEs. As the wave equation generates periodic solutions, the appropriate tool for stability analysis should be based on Fourier analysis.

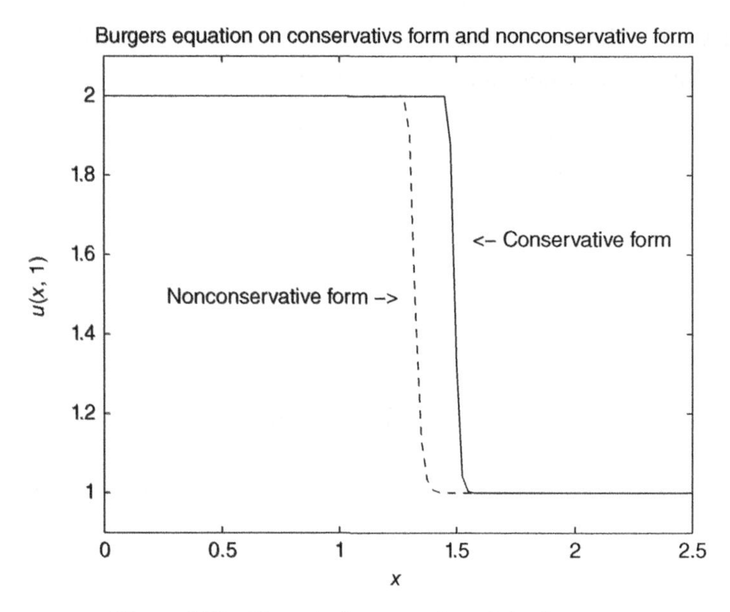

Figure 8.12 Numerical computation of shock waves

As hyperbolic equations preserve waveforms, it seems natural to make the following ansatz for the analytical solution:

$$u(x,t) = q(t)e^{j\omega x} \tag{8.45}$$

where $j = \sqrt{-1}$, ω is a spatial frequency (wave number) and $q(t)$ is a complex amplitude function.

Hence, for the numerical solution $u_{i,k}$, we make the ansatz:

$$u_{i,k} = q_k e^{j\omega x_i} \tag{8.46}$$

Insert this ansatz into the FTBS method in Section 8.2.1:

$$q_{k+1}e^{j\omega x_i} = (1-\sigma)q_k e^{j\omega x_i} + \sigma q_k e^{j\omega(x_i - h_x)}, \quad \sigma = a\frac{h_t}{h_x} \tag{8.47}$$

$$q_{k+1} = (1 - \sigma + \sigma e^{-j\omega h_x})q_k = G(\sigma)q_k \tag{8.48}$$

The absolute value of the complex factor $G(\sigma)$ must be ≤ 1 for stability:

$$|G(\sigma)|^2 = (1-\sigma+\sigma cos(\omega h_x))^2 + (\sigma sin(\omega h_x))^2 = 1 - 4\sigma(1-\sigma)sin^2(\omega h_x) \tag{8.49}$$

We see that

$$|G(\sigma)| \leq 1, \quad if \quad 0 < \sigma \leq 1$$

i.e.,

$$0 < a\frac{h_t}{h_x} \leq 1$$

Hence, we get a smaller stability interval by this analysis called *von Neumann analysis* than with the eigenvalues analysis! For hyperbolic problems, the von Neumann analysis is the adequate one to use for stability investigations.

For the FTCS method in Section 8.2.3

$$u_{i,k+1} = u_{i,k} - \frac{\sigma}{2}(u_{i+1,k} - u_{i-1,k})$$

the von Neumann analysis gives the formulas:

$$q_{k+1} = G(\sigma)q_k$$

where

$$|G(\sigma)|^2 = 1 + \sigma^2 sin^2(\omega h_x) > 1$$

Hence, the FCTS method is always unstable.

BIBLIOGRAPHY

1. A. Iserles, "A First Course in the Numerical Analysis of Differential Equations", Cambridge University Press, 1996
2. R. LeVeque, "Numerical Methods for Conservation Laws", 1992
3. R.M.M. Mattheij, S.W. Rienstra, J.H.M. ten Thije Boonkkamp, "Partial Differential Equations Modeling, Analysis, Computation", Chapter 13, SIAM, 2005
4. G.D. Smith, "Numerical Solution of Partial Differential Equations", Chapter 4, 3rd ed, Oxford University Press, 1986
5. J.C. Strikwerda, "Finite Difference Schemes and Partial Differential Equations", Chapman and Hall, 1989
6. R. Vichnevetsky, J.B. Bowles, "Fourier Analysis of Numerical Approximations of Hyperbolic Equations", SIAM, 1982

9

MATHEMATICAL MODELING WITH DIFFERENTIAL EQUATIONS

When a differential equation is presented in a textbook, you may sometimes ask yourself how it is derived. Are there any basic principles to follow when a differential equation is set up? Can you perform the modeling of an engineering problem yourself? There is no general answer to that question. Modeling is much of an art and the result obtained depends on what type of problem you want to model and which simplifying assumptions you make. In an area such as chemical engineering, it is customary to set up differential equation models from conservation principles, i.e., you derive the model yourself. In electromagnetic field theory, on the other hand, it is common to start from Maxwell's equations in general form, impose simplifying assumptions for your special problem to obtain a simpler differential equation model. In control theory, it is customary to describe a model with a block diagram, which can be translated into an ordinary differential equation (ODE) system.

To give an overview discussion of the art of modeling, it may be helpful to divide differential equations occurring in applications into the following classes:

1. differential equations considered as nature laws
2. constitutive differential equations
3. conservative differential equations

With these mathematical tools, it is possible to set up models for many scientific and engineering problems. In this chapter, we give an overview of deriving differential equation models in one space dimension, 1D. Some generalizations to 2D problems and simplifications of general differential equations to less complex problems are also shown.

Introduction to Computation and Modeling for Differential Equations, Second Edition. Lennart Edsberg.
© 2016 John Wiley & Sons, Inc. Published 2016 by John Wiley & Sons, Inc.

As before, in mathematical modeling, we stress upon using the *units* of the variables, e.g., density (kg/m^3), flow (kg/s), flux [kg/(m$^2 \cdot$ s)], and parameters, e.g., g (m/s^2).

9.1 NATURE LAWS

A nature law expressed as a differential equation is a mathematical relation that cannot be derived from physical facts. Its validity depends on observations and the fact that no experiment has been designed that contradicts the law. Examples of such laws are Newton's law for the motion of a particle, Maxwell's laws for the electromagnetic field, and Schrödinger's equation in quantum mechanics.

To show how a mathematical model can be built on a nature law, we present here Newton's law for particle dynamics

$$m\frac{d^2\mathbf{r}}{dt^2} = \mathbf{F} \tag{9.1}$$

where $\mathbf{F}$ (N) is the force acting on the particle, m (kg) its mass, $\mathbf{r}$ (m) the position, t (s) the time, and $\vec{a} = d^2\mathbf{r}/dt^2$, the acceleration (m/s^2) of the particle.

Example 9.1. (Particle dynamics in 1D). A particle with the mass m (kg) is part of a system consisting of a spring, a viscous damper, and the particle (see Figure 9.1). The spring has the stiffness constant k (N/m) and the damper has the damping coefficient c (N $\cdot$ s/m).

The system is at rest, when the particle is in the position $y = y_0$ relative to a y-axis pointing vertically downwards. The particle is then forced to the position $y_1 > y_0$ and left to move freely along the y-axis. If the position of the particle at time t is $y(t)$, the differential equation for the particle motion can be set up with the help of Newton's second law of motion

$$m\ddot{y} = F_1 + F_2 + F$$

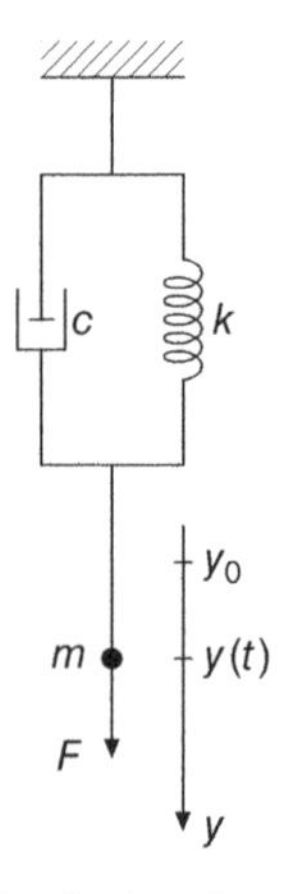

Figure 9.1 Spring-damper system

In the right hand side, we have the sum of

- the viscous force $F_1 = -cy$
- the spring force $F_2 = -\, ky$ and
- the external force F

giving the following ODE model, known as the *vibration equation*:

$$m\frac{d^2y}{dt^2} + c\frac{dy}{dt} + ky = F \tag{9.2}$$

To get a unique solution, we need to specify the position and the velocity of the particle at time $t = 0$, e.g., $y(0) = y_1, \dot{y}(0) = 0$.

The external force can be of different types, e.g.,

a constant force, e.g., gravity: $F = mg$
an oscillating force: $F = F_0 \sin\ (\omega t)$

Example 9.2. (The vibrating string, a distributed model). An elastic string (e.g., a rubber band) with cross-section area A (m^2) and density ρ (kg/m^3) is tensely fastened between two fixed points at the distance L (m) from each other. The tension of the string is F (N). At time $t = 0$, the string is moved vertically from the equilibrium position along the x-axis to a form given by the initial value function $u\ (x,0) = f(x), 0 \leq x \leq L$ and then released to move freely. Let the vertical distance from the x-axis at a point P on the string be $u(x, t)$. Let Q be another point where the vertical distance to the x-axis is $u(x + \Delta x, t)$ (Figure 9.2).

In this simplified model, the points P and Q can only move vertically. Hence we have no motion in the x-direction and Newton's second law for horizontal and vertical motion can be written

$$-F_1 \cos\alpha + F_2 \cos\beta = 0 \quad \Rightarrow \quad F_1 \cos\alpha = F_2 \cos\beta = F$$

$$\rho A \Delta x \frac{d^2u}{dt^2} = -F_1 \sin\alpha + F_2 \sin\beta$$

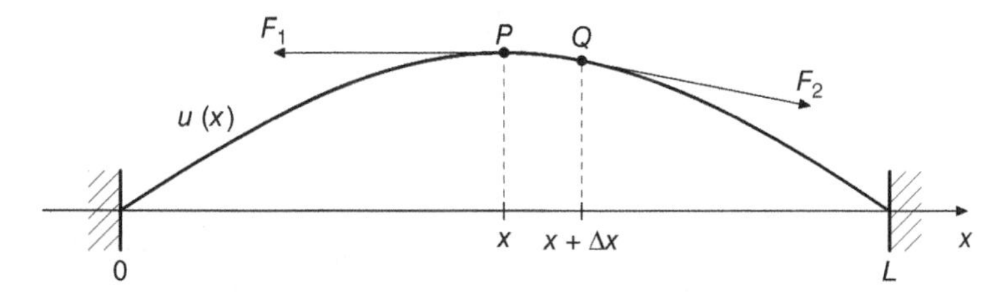

Figure 9.2 The vibrating string

Elimination of F_1 and F_2 gives

$$\rho A \Delta x \frac{d^2 u}{dt^2} = F(\tan \beta - \tan \alpha)$$

As $\tan \alpha = \frac{\partial u}{\partial x}(x, t)$ and $\tan \beta = \frac{\partial u}{\partial x}(x + \Delta x, t)$, we get

$$\frac{\rho A}{F} \frac{\partial^2 u}{\partial t^2} = \frac{\frac{\partial u}{\partial x}(x + \Delta x, t) - \frac{\partial u}{\partial x}(x, t)}{\Delta x}$$

Let $\Delta x \to 0$ and we obtain the partial differential equation, the wave equation

$$\frac{\partial^2 u}{\partial t^2} = c^2 \frac{\partial^2 u}{\partial x^2} \tag{9.3}$$

where $c^2 = F/\rho A$. The parameter c (m/s) is the velocity of the wave.

To get a unique solution, we need boundary conditions (BCs) and initial conditions (ICs). The endpoints of the string are fixed, hence the BCs are

$$u(0, t) = 0, \quad u(L, t) = 0, \qquad t \geq 0$$

The ICs originate from the state of the string at $t = 0$, i.e., the string has a given form and initial velocity zero, hence

$$u(x, 0) = f(x) \quad \frac{\partial u}{\partial t}(x, 0) = 0, \quad 0 \leq x \leq L$$

9.2 CONSTITUTIVE EQUATIONS

A constitutive equation is a mathematical model of the physical properties of a gas, fluid, or solid and can be either a differential equation or an algebraic equation. A constitutive equation is usually the mathematical result of very many observations of a phenomenon or measurements of an experiment and is therefore of empirical nature.

We give here some examples of constitutive equations formulated in 1D.

9.2.1 Equations in Heat Transfer Problems

Fourier's law of heat diffusion

$$q = -\kappa \frac{dT}{dx} \tag{9.4}$$

This is a relation between heat flux q [J/(m$^2 \cdot$ s)] and temperature gradient. The parameter κ [J/(K $\cdot$ m $\cdot$ s)] is the thermal conductivity of the material. Observe the minus sign (the heat flow travels from higher to lower temperatures)!

This can also be written as a law for diffusion of thermal energy $E = \rho C_p T$

$$q = -\alpha \frac{dE}{dx} \tag{9.5}$$

where q is the heat flux [J/(m$^2 \cdot$ s)], α thermal diffusivity (m^2/s), ρ the density (kg/m^3), and C_p the heat capacity at constant pressure [J/(K $\cdot$ kg)].

Newton's law of cooling

$$Q = -kA(T_0 - T) \tag{9.6}$$

This is a relation between the heat flow Q (J/s) from a region with temperature T to an environment with temperature T_0. The regions are separated by a wall of area A (m^2) with convection heat transfer coefficient k [J/(m$^2 \cdot$ s $\cdot$ K)].

Stefan–Boltzmann's law for temperature radiation loss

$$Q = -Ae\sigma(T_0^4 - T^4) \tag{9.7}$$

Radiation comes into account for high temperatures. Q (J/s) is the heat flow, A (m^2) the area of radiation, e a material constant, and $\sigma = 5.67 \cdot 10^{-8}$ [J/(m$^2 \cdot$ s $\cdot$ K^4)] the Boltzmann's constant

9.2.2 Equations in Mass Diffusion Problems

Fick's law of mass diffusion

$$f = -D\frac{d\rho}{dx} \tag{9.8}$$

This is a relation between mass flux f [kg/(m$^2 \cdot$ s)] and the density gradient. The parameter D is the mass diffusivity (m^2/s). There is also a derivative-free version of Fick's law (cf Newton's cooling law)

$$F = -\frac{\mu A}{l}(\rho_0 - \rho) \tag{9.9}$$

This is a relation between the mass flow F (kg/s) from a region with density ρ to an environment with density ρ_0. The regions are separated by a wall through which particles can diffuse. The diffusivity is μ(m^2/s), the area is A (m^2), and the thickness is l (m).

9.2.3 Equations in Mechanical Moment Diffusion Problems

Newton's law of viscosity

$$\tau_{xy} = -\mu\frac{dv_x}{dy} \tag{9.10}$$

This is a relation between the shear stress τ_{xy} (N/m^2) in a plane parallel to the direction of the flow with velocity v_x (m/s) along the x-axis and orthogonal to the y-axis and the velocity gradient along the y-axis. The parameter μ ($N \cdot s/m^2$) is called the *dynamic viscosity*.

If we write this law in the form

$$\tau_{xy} = -\nu \frac{d(\rho v_x)}{dy} \tag{9.11}$$

the parameter ν (m^2/s) is called the *kinematic viscosity* (can also be regarded as the diffusion coefficient of the mechanical moment).

9.2.4 Equations in Elastic Solid Mechanics Problems

Hooke's law for an elastic bar

$$\sigma = E\epsilon, \qquad \epsilon = \frac{du}{dx} \tag{9.12}$$

This is a relation between stress σ (N/m^2) and strain ϵ [ ] (no dimension), where the strain is the displacement u (m) per unit length of the bar. The parameter E is the elasticity module (Young's module) (N/m^2).

9.2.5 Equations in Chemical Reaction Engineering Problems

The mass action law in chemical kinetics

$$r = k \prod_{i=1}^{m} c_i^{\alpha_i} \tag{9.13}$$

This algebraic relation gives the rate r [$mol/(m^3 \cdot s)$] of the reaction

$$\alpha_1 \ A_1 + \ \dots \ + \alpha_m \ A_m \rightarrow \beta_1 \ B_1 + \ \dots \ + \beta_n \ B_n$$

The reactants A_i, $i = 1, \dots, m$ and the products B_i, $i = 1, \dots, n$ are measured in concentrations c_i (mol/m^3) and the unit of the rate constant k depends on $\alpha_1, \dots, \alpha_m$. The rate constant k is temperature dependent according to Arrhenius' law

$$k = Ae^{-\frac{E}{RT}} \tag{9.14}$$

where A [ ] is the preexponential factor, E (J/mol) the activation energy, T (K) the temperature, and $R = 8.314$ J/(K $\cdot$ mol) the gas constant.

The general gas law

$$pV = nRT \tag{9.15}$$

This is a relation between the pressure p (N/m^2), the volume V (m^3), the number of moles n (mol), the temperature T (K), and the gas constant R.

9.2.6 Equations in Electrical Engineering Problems

Ohm's law in electromagnetics

$$\mathbf{j} = \sigma \mathbf{E} \tag{9.16}$$

where $\mathbf{j}$ (A/m^2) is the electrical current density, σ [A/(V · m)] the conductivity, and $\mathbf{E}$ (V/m) the electric field strength. *Lorentz' law*

$$\mathbf{F} = q(\mathbf{E} + \mathbf{v} \times \mathbf{B}) \tag{9.17}$$

where $\mathbf{F}$ is the force (N) exerted on a charged particle with electric charge q (C) in an electric field (V/m) and a magnetic field $\mathbf{B}$ (V · s/m^2). $\mathbf{v}$ (m/s) the velocity of the particle.

9.3 CONSERVATIVE EQUATIONS

Conservation equations are based on the principle that physical quantities, such as mass, moment, and energy, are conserved in a system. Assume we study the changes of some quantity in a control volume that is fixed in space. From a purely logical point of view, the following balance principle must hold for the quantity during a given time interval:

$$\Delta Acc = In - Out + Prod - Cons \tag{9.18}$$

where

- ΔAcc = the change of amount of the accumulated quantity
- *In* = the amount of the quantity that has flowed in
- *Out* = the amount of the quantity that has flowed out
- *Prod* = the amount of the quantity produced and
- *Cons* = the amount of the quantity consumed

In scientific and engineering contexts, quantities usually change smoothly with time. Let $M(t)$ be a differentiable function denoting the amount of a quantity at time t. We then define the *flow* $Q(t)$ as

$$Q(t) = \frac{dM}{dt} \tag{9.19}$$

The flow is measured in amount per unit time.

Define *in*, *out*, *prod*, and *cons* as the flows of *In*, *Out*, *Prod*, and *Cons*. Using (9.18) during the time interval $[t, t + \Delta t]$ gives the following relation:

$$\Delta Acc = \int_t^{t+\Delta t} (in(t) - out(t) + prod(t) - cons(t))\, dt \tag{9.20}$$

Divide by Δt, let $\Delta t \to 0$ and we obtain

$$\frac{dAcc}{dt} = in - out + prod - cons \tag{9.21}$$

This relation is often referred to as the *continuity equation*. Here it is expressed as an ODE, but if the amount of the quantity depends on both time and space, the continuity equation will be a partial differential equation (PDE).

A conservation principle leading to an ODE with time as independent variable is called a *lumped* model. Such a model does not take into account any space variations of the quantities. A conservation law leading to a PDE in time and space is called a *distributed* model.

We show in a number of examples how this principle can be used to derive differential equations for the conservation of a given quantity.

9.3.1 Some Examples of Lumped Models

Example 9.3. (Model for the water contents in a tank). Assume we have a tank which is filled with water from a pipe delivering the flow Q_1 (m^3/s) (see Figure 9.3). It is drained at the bottom through another pipe with the flow Q_2 (m^3/s).

Let V (m^3) be the amount of water accumulated in the tank. Application of the continuity principle (9.18) during the time interval $[t, t + \Delta t]$ gives the following equality:

$$\Delta V = Q_1 \cdot \Delta t - Q_2 \cdot \Delta t$$

corresponding to

$$\Delta Acc = In - Out$$

Divide this relation by Δt, let $\Delta t \to 0$ and we obtain the following ODE:

$$\frac{dV}{dt} = Q_1 - Q_2$$

corresponding to

$$\frac{dAcc}{dt} = in - out$$

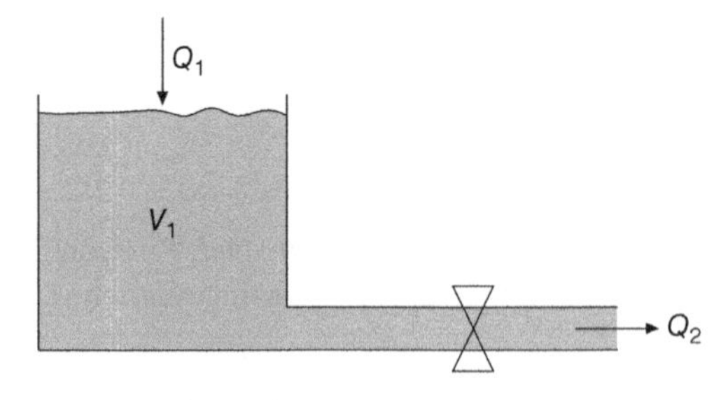

Figure 9.3 Tank filled with water

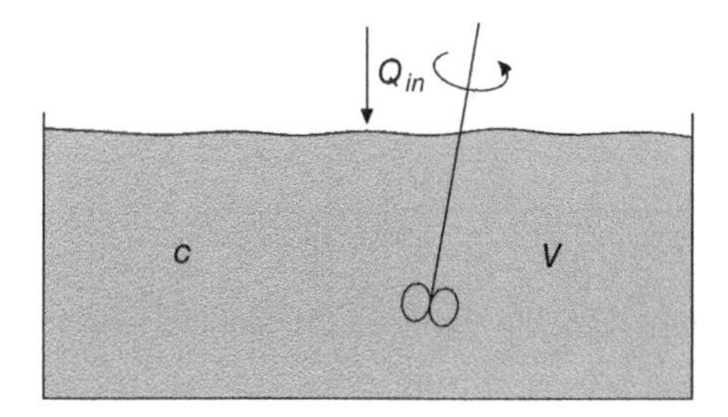

Figure 9.4 Continuously stirred tank reactor

If the volume is known at $t = 0$, we also have an IC. Denote the initial volume by V_0 (m^3). We now have the initial value problem (IVP)

$$\frac{dV}{dt} = Q_1 - Q_2, \quad V(0) = V_0 \tag{9.22}$$

Example 9.4. (A continuously stirred tank reactor). A continuously stirred tank reactor is a chemical reactor where chemical species are stirred together from inflow of reactants to a homogeneous mix (see Figure 9.4). In this mix, the reactions take place and products are formed. Assume the inflow to the reactor is Q_{in} (m^3/s) and that the concentration of A in the inflow is c_0 (mol/m^3). In the reactor, A reacts according to $A \to B$ and is thereby consumed with the rate kc, where k (s^{-1}) is the rate constant of the reaction and c (mol/m^3) is the concentration of A in the reactor. Assume there is no outflow from the reactor. Then the volume will be increasing. The concentration of A in the reactor at time $t = 0$ is $c(0) = 0$ and the volume at $t = 0$ is $V(0) = V_0$. Using the conservation law for the number of moles cV of A in the reactor and the volume V of the mix, we obtain

$$\frac{d(cV)}{dt} = Q_{in}\, c_0 - kcV, \quad c(0) = 0 \tag{9.23a}$$

$$\frac{dV}{dt} = Q_{in}, \quad V(0) = V_0 \tag{9.23b}$$

9.3.2 Some Examples of Distributed Models

Example 9.5. (The continuity equation in 1D). Assume we have a pipe with cross-section area A (m^2) through which some fluid flows (see Figure 9.5). The density of the fluid is $\rho(x, t)$ (kg/m^3) and the flux is $q(x, t)$ [kg/(s $\cdot$ m^2)]. We use the relation (9.21) to derive the continuity equation. Consider the small volume ΔV between x and $x + \Delta x$. At the left end, the flux is $q(x, t)$ and the density is $\rho(x, t)$. At the right end, the flux is $q\,(x + \Delta x, t)$ and the density is $\rho\,(x + \Delta x, t)$.

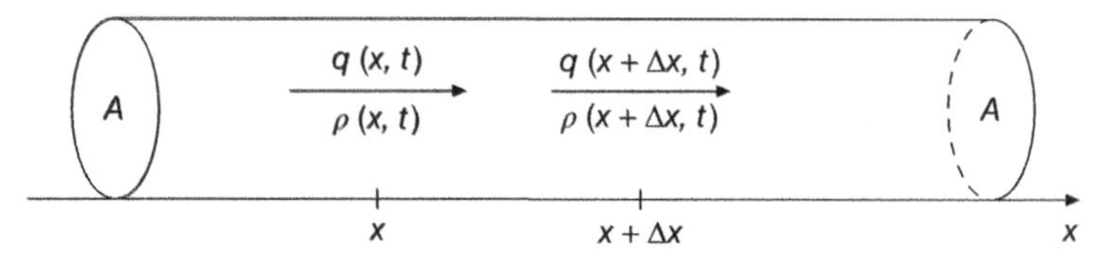

Figure 9.5 Illustration of the continuity equation

The accumulated amount of mass in ΔV at time t is $\int_x^{x+\Delta x} \rho(x,t)A\,dx$. The continuity equation (9.21) gives

$$\frac{d}{dt}\int_x^{x+\Delta x} \rho(x,t)\,A\,dx = q(x,t)A - q(x+\Delta x,t)A \tag{9.24}$$

Using the mean value theorem of integral calculus gives

$$\Delta x \frac{d\rho(x+\theta\Delta x,t)}{dt} = -(q(x+\Delta x,t) - q(x,t)), \quad 0 < \theta < 1$$

Dividing by Δx, and letting $\Delta x \to 0$ gives

$$\frac{\partial \rho}{\partial t} + \frac{\partial q}{\partial x} = 0 \tag{9.25}$$

which is the PDE of mass conservation called the *continuity equation* in 1D.

However, (9.25) is mathematically insufficient. There are two dependent variables, ρ and q, but only one equation. Hence, we need one more equation, e.g., a constitutive equation giving a second relation between ρ and q. The following alternatives are possible:

1. mass flow according to a general nonlinear flux function $q = f(\rho)$
2. mass flow with constant velocity v (m/s) through the pipe: $q = v\rho$
3. mass flow governed by mass diffusion: $q = -D\frac{d\rho}{dx}$
4. a combination of (2) and (3): $q = v\rho - D\frac{d\rho}{dx}$

Inserting (1) into (9.25) gives

$$\frac{\partial \rho}{\partial t} + \frac{\partial f(\rho)}{\partial x} = 0 \tag{9.26}$$

This equation is often called the scalar conservation law and is a nonlinear hyperbolic PDE (see Chapter 8).

Inserting (2) into (9.25) gives

$$\frac{\partial \rho}{\partial t} + v\frac{\partial \rho}{\partial x} = 0 \tag{9.27}$$

Hence, we obtain the advection equation (1.4) presented in Chapter 1. This is a linear hyperbolic PDE.

Inserting (3) into (9.25) gives

$$\frac{\partial \rho}{\partial t} = D\frac{\partial^2 \rho}{\partial x^2} \tag{9.28}$$

This equation we recognize as the *diffusion equation*, which is a parabolic PDE (see Chapter 6).

Inserting (4) into (9.25) gives

$$\frac{\partial \rho}{\partial t} + v\frac{\partial \rho}{\partial x} = D\frac{\partial^2 \rho}{\partial x^2} \tag{9.29}$$

This equation we recognize as the *convection-diffusion* equation, classified as a hyperbolic–parabolic PDE.

If there is also production of the substance in the pipe (through, e.g., chemical reaction) with the rate $r(c)$ [mol/(m$^3 \cdot$ s)] depending on the concentration c, we obtain a PDE known as the *convection–diffusion–reaction* equation

$$\frac{\partial c}{\partial t} + v\frac{\partial c}{\partial x} = D\frac{\partial^2 c}{\partial x^2} + r(c) \tag{9.30}$$

Example 9.6. (Energy conservation in a cooled cylindrical pipe).

Assume a hot fluid is flowing through a cylindrical pipe with radius R (m) in the z-direction according to Figure 9.6. The temperature $T(z, t)$ of the fluid at time t and at the space point z changes as the pipe is cooled on the outside. The temperature of the cooling medium is constant and has the value T_{out}. The thermal energy flux through the pipe is $q(z, t)$ [J/(m$^2 \cdot$ s)]. The density of the fluid is ρ (kg/m^3), the heat capacity is C [J/(K $\cdot$ kg)], and the convection heat transfer coefficient of the wall is k [J/(K $\cdot$ m$^2 \cdot$ s)]. The four parameters ρ, C, k, and T_{out} are constants, while the state variables temperature $T(z, t)$ and energy flux $q(z, t)$ of the fluid vary with time t and space z. Following the balance principle (9.24), we get

$$\begin{aligned}\frac{d}{dt}\int_z^{z+\Delta z} T(z,t)\rho C\pi R^2\, dz &= q(z,t)\pi R^2 - q(z+\Delta z,t)\pi R^2 \\ &\quad - k(T - T_{out})2\pi R\Delta z(1)\end{aligned} \tag{9.31}$$

Taking limit values with $\Delta z \to 0$, we obtain the PDE

$$\rho C\frac{\partial T}{\partial t} = -\frac{\partial q}{\partial z} - \frac{2k}{R}(T - T_{out}) \tag{9.32}$$

However, we have only one equation but two unknowns $T(z, t)$ and $q\,(z, t)$. The missing equation is obtained from a constitutive equation that relates the flux q to the

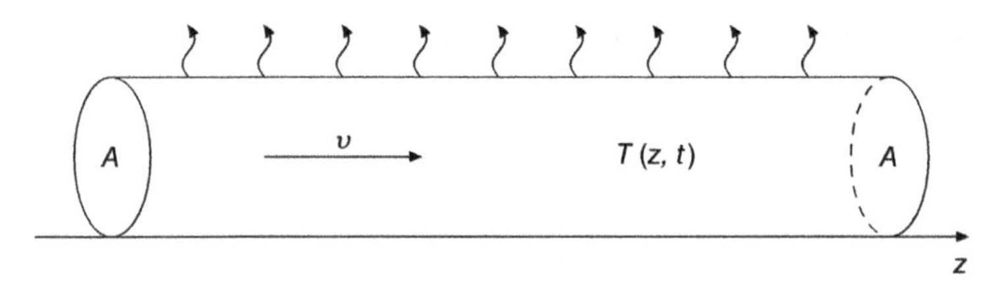

Figure 9.6 Hot fluid in a cylindrical pipe

temperature T. If we assume that the heat transport takes place through convection and Diffusion, we have

$$q = v\rho CT - \kappa \frac{dT}{dz}$$

Inserting this expression into the PDE, we finally obtain

$$\rho C \frac{\partial T}{\partial t} + v\rho C \frac{\partial T}{\partial z} = \kappa \frac{\partial^2 T}{\partial z^2} - \frac{2k}{R}(T - T_{out}) \tag{9.33}$$

which is the PDE presented in Example 6.1, where an IC and two BCs are added to this parabolic PDE.

If the problem is time independent, we obtain the BVP in Example 4.1.

If we reduce the model further by assuming that the diffusion transport is negligible compared to the convective transport, we obtain the first-order ODE

$$v\rho C \frac{dT}{dz} = -\frac{2k}{R}(T - T_{out}) \tag{9.34}$$

for which we need only one condition, an IC.

If we assume that the heat leaking out of the wall is negligible ($k = 0$), the problem is reduced to the simple ODE

$$v\rho C \frac{dT}{dz} = 0 \tag{9.35}$$

which with the IC $T(0) = T_0$ gives the solution $T = T_0$. Hence, by increasing the number of reduction assumptions, the model soon becomes uninteresting!

On the other hand, the model can be made more complex by taking into account transport in more than one space dimension. If we assume that heat is transported by diffusion also in the r-direction, we have to add a corresponding term to (9.33) and move the heat loss through the wall to a BC

$$\rho C \frac{\partial T}{\partial t} + v\rho C \frac{\partial T}{\partial z} = \kappa \frac{\partial^2 T}{\partial z^2} + \kappa \frac{1}{r}\frac{\partial}{\partial r}\left(r \frac{\partial T}{\partial r}\right) \tag{9.36}$$

This time-dependent 2D parabolic PDE is also presented in Example 6.2. Now the flow in a pipe problem has reached a level detail, which may be unnecessarily high for being an engineering problem. Hence, modeling is also about how to choose sufficient complexity in the model.

Exercise 9.3.1. Derive the time-dependent counterflow heat exchanger model in Example 8.2

9.4 SCALING OF DIFFERENTIAL EQUATIONS TO DIMENSIONLESS FORM

When the differential equation of a scientific or engineering process is formed through modeling, it is natural to think of all variables introduced in the model as physical or chemical quantities. Hence, the variables have some *dimension*, i.e., they are expressed in units such as kg, J/($m^2 \cdot s$), and mol/($m^3 \cdot s$). The parameters in the model, coming, e.g., from the constitutive equations, have dimensions as well.

Observe that the word dimension has many meanings: (i) the dimension of a vector is the number of elements in the vector, (ii) the dimension of space can be modeled in 1D, 2D, and 3D, (iii) the dimension of a physical variable or parameter is a combination of elementary units, s.a. m, s, kg, etc.

However, when the model is to be treated mathematically and/or numerically, it is often advantageous to scale the variables of the model to *dimensionless* form. This means that for the variable x with dimension (m), we introduce a dimensionless variable $\xi = x/L$, where L (m) is a characteristic length of the problem. Likewise, the time variable t (s) is scaled to dimensionless time through $\tau = t/t_0$, where t_0 is a characteristic time of the problem.

Through the scaling, the derivatives of the original model will be transformed to derivatives of the dimensionless variables. We show here how the first and second derivatives are changed.

$$\frac{dx}{dt} = \frac{d(L\xi)}{d(t_0\tau)} = \frac{L}{t_0}\frac{d\xi}{d\tau} \tag{9.37}$$

$$\frac{d^2x}{dt^2} = \frac{d}{dt}\left(\frac{dx}{dt}\right) = \frac{d}{d(t_0\tau)}\left(\frac{L}{t_0}\frac{d\xi}{d\tau}\right) = \frac{L}{t_0^2}\frac{d^2\xi}{d\tau^2} \tag{9.38}$$

We see that a consequence of scaling is that certain parameter combinations will appear in the scaled model. Usually these combinations can be arranged so that dimensionless combinations are formed.

Scaling of the original differential equation model can have several advantages:

- If the scaling is done correctly, i.e., with scaling factors characteristic of the problem, the new dimensionless variables for, e.g., length and time will be of order 1.
- The number of parameters in the scaled model will be smaller than in the original model.

- When comparing the size of the individual terms of the differential equation model, it is easier to judge which terms are important and which can be neglected in the scaled model. This may help to reduce the original model to a model that is mathematically simpler.

Example 9.7. The continuously stirred tank reactor model in Example 9.4. The following differential equation model is given:

$$\frac{d(cV)}{dt} = q_{in}\, c_0 - kcV, \quad c(0) = 0 \tag{9.39a}$$

$$\frac{dV}{dt} = q_{in}, \quad V(0) = V_0 \tag{9.39b}$$

The units of the variables and parameters are

c, c_0	V, V_0	t	q_{in}	k
(mol/m^3)	(m^3)	(s)	(m^3/s)	(s^{-1})

Introduce the scaled dimensionless variables x, y, and τ

$$x = c/c_0, \quad y = V/V_0, \quad \tau = t/T$$

where T is a characteristic time of the problem, yet to be determined. Express the original problem (9.23) in the scaled variables

$$\frac{c_0\ V_0}{T}\frac{d(xy)}{d\tau} = q_{in}\ c_0 - kc_0\ V_0 xy, \qquad x(0) = 0$$

$$\frac{V_0}{T}\frac{dy}{dt} = q_{in}, \qquad y(0) = 1$$

Multiply the first ODE by $T/c_0 V_0$ and the second ODE by T/V_0

$$\frac{d(xy)}{d\tau} = \frac{q_{in}T}{V_0} - kTxy, \qquad x(0) = 0$$

$$\frac{dy}{d\tau} = \frac{q_{in}T}{V_0}, \qquad V_0 y(0) = V_0$$

We see that there are two possibilities to make time dimensionless: either $T = V_0/q_{\text{in}}$ or $T = 1/k$. The first choice leads to the ODE system

$$\frac{d(xy)}{d\tau} = 1 - a_1 xy, \qquad x(0) = 0$$

$$\frac{dy}{d\tau} = 1, \qquad y(0) = 1$$

where a_1 is the dimensionless parameter $a_1 = V_0 k/q_{\text{in}}$.

The other possibility, $T = 1/k$, leads to the ODE system

$$\frac{d(xy)}{d\tau} = a_2 - xy, \qquad x(0) = 0$$

$$\frac{dy}{d\tau} = a_2, \qquad y(0) = 1$$

where $a_2 = V_0 k/q_{\text{in}}$. We see that in both Cases, the original number of parameters is reduced from four (c_0, V_0, q_{in}, k) to only one (a_1 or a_2).

Exercise 9.4.1. A 1D version of Navier–Stokes equations is

$$\rho\left(\frac{\partial v}{\partial t} + v\frac{\partial v}{\partial x}\right) = -\frac{\partial p}{\partial x} + \mu\frac{\partial^2 v}{\partial x^2} + \rho g \tag{9.40}$$

with appropriate ICs and BCs. In (9.40), ρ is the density, v the velocity, p the pressure, μ the dynamic viscosity, and g the gravitational constant. The variables and the parameters have the following units:

ρ	v	t	x	p	μ	g
(kg/m^3)	(m/s)	(s)	(m)	(N/m^2)	(N $\cdot$ s/m^2)	(m/s^2)

Introduce dimensionless variables: $v^* = v/V$, $x^* = x/L$, $p^* = (p - p_0)/\rho V^2$, $t^* = tV/L$, where V is a characteristic velocity, L a characteristic length, and p_0 is a reference pressure. All the parameters are assumed to be constant. The original differential equation will be expressed in the dimensionless variables and the dimensionless parameters Reynold's number $\text{Re} = LV\rho/\mu$ and Froude's number $\text{Fr} = V^2/gL$

Exercise 9.4.2. A model for stationary 1D heat transport presented in Example 4.1 is the following differential equation:

$$\rho C_p v\frac{dT}{dz} = \kappa\frac{d^2T}{dz^2} - \frac{2k}{R}(T - T_{out}) \quad T(0) = T_0, T(L) = T_{out} \tag{9.41}$$

The units of the variables and parameters are

T, T_0, T_{out}	z, L	v	κ	ρ	C_p
(K)	(m)	(m/s)	[J/(K $\cdot$ m $\cdot$ s)]	(kg/m^3)	[J/(K $\cdot$ kg)]

Introduce the scaling

$$u = T/T_{out}, \qquad \xi = z/L$$

and bring the differential equation on dimensionless form. Which dimensionless parameter combinations will appear in the scaled model?

Exercise 9.4.3. Given the vibration equation in mechanics

$$m\frac{d^2u}{dt^2} + c\frac{du}{dt} + ku = F\sin(\omega t) \tag{9.42}$$

where all parameters m, c, k, F, and ω are positive quantities. Which are the dimensions of the variables and the parameters? Let $\omega_0 = \sqrt{k/m}$, $c_0 = 2\sqrt{km}$ $\alpha = c/c_0$, and $\beta = \omega/\omega_0$. Introduce dimensionless time $\tau = \omega_0 t$ and dimensionless deviation $y = u/u_0$, where u_0 is a suitable combination of some parameters in (9.42).

Find this scaling factor u_0 and show that in the new variables y and τ, the differential equation (9.42) can be written (α is called the *damping factor*)

$$\frac{d^2y}{d\tau^2} + 2\alpha\frac{dy}{d\tau} + y = \sin(\beta\tau)$$

Exercise 9.4.4. The following chemical reactions occur in enzyme catalysis:

$$S + E \longleftrightarrow C \longrightarrow P + E$$

The kinetics of the system is modeled by the ODE system

$$\begin{aligned}
\dot{S} &= -k_1SE + k_{-1}C, & S(0) &= S_0\\
\dot{E} &= -k_1SE + k_{-1}C + k_2C, & E(0) &= E_0\\
\dot{C} &= k_1SE - k_{-1}C - k_2C, & C(0) &= 0\\
\dot{P} &= k_2C, & P(0) &= 0
\end{aligned}$$

where S, E, C, and P represent the *concentrations* (mol/m^3) of the four species taking part in the reactions. The parameters k_1 [m^3/(mol · s)] k_{-1} (1/s) and k_2 (1/s) are the *rate constants* of the reactions. Note first that there are linear relationships between the right hand side expressions, so we need differential equations for only two variables, say S and C, the other two variables depend linearly on these.

1. Find the ODE system for S and C (eliminate E and P).
2. Transform by scaling of S, C, and t the system in (1) into dimensionless form. Use the scaled variables $s = S/S_0$ and $c = C/E_0$. The time t should be scaled to dimensionless time τ as $\tau = t/T$, where T is chosen so that the scaled ODE system contains as few parameters as possible.

BIBLIOGRAPHY

1. E. Baltram, Mathematics for Dynamic Modeling, Academic Press, 1987
2. N. Fawkes, J. Mahony, An Introduction to Mathematical Modeling, Wiley, 1994
3. J.D. Logan, Applied Mathematics, Wiley, 1997

10

APPLIED PROJECTS ON DIFFERENTIAL EQUATIONS

The following projects have been used for a long time in a first advanced course on Numerical Solution of Differential Equations at KTH, Stockholm. Most of the projects have been developed and modified during several years by my colleague Gerd Eriksson, and I have her permission to present them in updated form in this book. Projects 4 and 5 are taken from one of the early manuals of COMSOL MULTIPHYSICS®, see also Appendix B.2. They have kindly given me permission to publish them in updated form.

Project 1. Signal propagation in a long electrical conductor

Given a 10 km long conductor with resistance R, inductance L, and capacitance C. From $x = 0$, signals of amplitude 1V are sent frequently during time intervals of different lengths. Denote this time-dependent voltage by $u_0(t)$. The voltage in the conductor is a function $u(x, t)$ of the position x on the conductor and time t and is modeled by the following hyperbolic partial differential equation (PDE)

$$\frac{\partial^2 u}{\partial t^2} + \frac{R}{L}\frac{\partial u}{\partial t} = \frac{1}{LC}\frac{\partial^2 u}{\partial x^2}, \quad 0 \leq x \leq X, \quad t > 0$$

which is the wave equation with damping. At $t = 0$, the initial conditions (ICs) are

$$u(x, 0) = 0, \qquad \frac{\partial u}{\partial t}(x, 0) = 0, \quad 0 < x \leq X$$

where $X = 10^4$.

Introduction to Computation and Modeling for Differential Equations, Second Edition. Lennart Edsberg.
© 2016 John Wiley & Sons, Inc. Published 2016 by John Wiley & Sons, Inc.

The boundary condition (BC) at $x = 0$ is the given signal $u_0(t)$, i.e.,

$$u(0,t) = u_0(t)$$

At $x = X$, the conductor is open, i.e., the signal is not reflected but disappears out. The BC fulfilling this condition is the advection equation, i.e.,

$$\frac{\partial u}{\partial t}(X,t) + \frac{1}{\sqrt{LC}}\frac{\partial u}{\partial x}(X,t) = 0$$

Discretize the conductor into 100 subintervals and use central difference approximations for the space and time derivatives in the damped wave equation. As for the approximation of the BC at $x = X$, the upwind method Forward-Time-Backward-Space (FTBS) is appropriate.

For the conductor, the following parameter values are given: $R = 0.004$ ohm, $L = 10^{-6}$ H, and $C = 0.25 \cdot 10^{-8}$ F. Simulate the signal propagation during a sufficiently long time, e.g., during 3 ms. First use the maximum allowed time step fulfilling the stability condition, then use a time step being 80% of the maximum time step. What is your comments on the result of the two time steps that have been used? Is the smoothing of the solution the effect of damping or of the method used?

Start your simulations by testing $u_0(t)$ with the MATLAB function given below. It corresponds to three short signals repeatedly sent with a period of 0.0004 s.

```
function usignal=uzero(t)
tau=rem(t,0.0004);
T0=0; T1=0.00005; T2= 0.00010; T3= 0.00013; T4= 0.00018; T5= 0.00025;
u0=1;
m=length(t);u1=zeros(m,1);
ind=find(t0<=tau & tau<=T1 | T2<=tau & tau <=T3 | T4<=tau & tau<=T5);
u1(ind)=u0*ones(size(ind));
usignal=u1;
```

You can of course try your own signal sequences.

Project 2. Flow in a cylindrical pipe

In a long straight pipe with circular cross section with radius R, a fluid is streaming and we want to find out how the flow velocity varies in the pipe. Let the velocity vector be $(u, v)^T$ where u is the velocity in the length direction of the pipe and v the velocity in the radial direction. The flow is assumed to be circular symmetric, i.e., in cylindrical coordinates $u(r,z)$ and $v(r,z)$ depend only on r and z, not φ.

At the inlet $z = 0$, the velocity of the fluid is $u_0 = 0.1$ m/s in the z direction, i.e., $u(r,0) = u_0$, $v(r,0) = 0$, $0 \leq r \leq R$. The radius $R = 0.05$ m. The fluid has density $\rho = 1000$ kg/m^3 and the viscosity is $\nu = 10^{-5}$ m^2/s. In fluid problems, the Reynolds number $Re = u_0 R/\nu$ is an important constant. When $Re >> 1$, which is the case here, it is possible to simplify the complicated Navier–Stokes PDEs that are used to model $u(r,z)$ and $v(r,z)$.

In this flow problem, Navier–Stokes equations are approximated by

$$u\frac{\partial u}{\partial z} = \nu\left(\frac{\partial^2 u}{\partial r^2} + \frac{1}{r}\frac{\partial u}{\partial r}\right) - v\frac{\partial u}{\partial r} - \frac{1}{\rho}\frac{dp}{dz} \tag{10.1}$$

$$\frac{\partial u}{\partial z} + \frac{1}{r}\frac{\partial(vr)}{\partial r} = 0 \tag{10.2}$$

The pressure p is only z dependent. At $z = 0$, $p = p_0$, where $p_0 = 10^4$ Pa. At $r = 0$, we have the following BC

$$\frac{\partial u}{\partial r}(0, z) = 0, \quad v(0, z) = 0$$

At the wall of the pipe, the two velocities are zero, i.e.,

$$u(R, z) = 0, \quad v(R, z) = 0$$

Far away in the pipe, several meters from the inlet, the velocity will have a stationary distribution with parabolic shape

$$u = 2u_0\left(1 - \left(\frac{r}{R}\right)^2\right), \quad v = 0 \tag{10.3}$$

The task is to compute how the velocities vary with r and z. It is also included to compute how far away from the inlet the stationary distribution (10.3) is attained with an acceptable tolerance.

The differential equations (10.1) and (10.2) must be treated specially at $r = 0$ where both are singular. Show that with l'Hospital's rule they turn into

$$u\frac{\partial u}{\partial z} = 2\nu\frac{\partial^2 u}{\partial r^2} - \frac{1}{\rho}\frac{dp}{dz}, \quad \frac{\partial u}{\partial z} + 2\frac{\partial v}{\partial r} = 0, \quad \text{at} \quad r = 0$$

For numerical treatment, the MoL with $n = 50$ subintervals in the r direction is used. Then there will be $2n$ unknown variables being functions of z. These are $u_1(z), u_2(z), \ldots, u_n(z), \sigma(z), v_2(z), v_3(z), \ldots, v_n(z)$. $u_1(z)$ and $v_1(z)$ are velocity functions at $r = 0$. We know that $v_1(z) = 0$, hence this component is not among the unknowns. The variable σ is defined as

$$\sigma(z) = \frac{1}{\rho}\frac{dp}{dz}$$

With difference approximations of all derivatives in the r direction, we will have n ODEs (ordinary differential equations) each of (10.1) and (10.2), hence $2n$ in total. To solve, these Euler's implicit method shall be used with stepsize h_z. The equations (10.1) and (10.2) will then be approximated by

$$u_i\frac{u_i - u_i^{old}}{h_z} - F_i(u_{i-1}, u_i, u_{i+1}, \sigma, v_i) = 0, \qquad \frac{u_i - u_i^{old}}{h_z} + G_i(v_{i-1}, v_i, v_{i+1}) = 0$$

where u_i and v_i denote the unknown velocity components at $z + h_z$ and u_i^{old} denotes the already computed velocity component at the previous value z. Find the expressions for F_i and G_i.

At the end of the day, there will be a nonlinear system of $2n$ algebraic equations to be solved at each step in the z direction. This is done with Newton's method. If the unknowns are written in the order $u_1, u_2, \ldots, u_n, \sigma, v_2, v_3, \ldots, v_n$, the jacobian J will have a block structure consisting of four $n \times n$ matrices

$$J = \begin{pmatrix} J_1 & J_2 \\ J_3 & J_4 \end{pmatrix}$$

where J_1 is tridiagonal, J_2 has nonzero elements only in the first column and in the diagonal, J_3 is diagonal, and J_4 is tridiagonal (perhaps with exception of the first row, depending on the difference approximation of $2(\partial v/\partial r)_{r=0}$).

Close to the inlet, there will be large variations in the velocities. The first steps should be $h_z = 0.001$ up to $z = 0.005$. Continue with $h_z = 0.005$ to $z = 0.04$. Further up in the pipe, the step h_z can be successively larger. Continue the calculations of u and v and the pressure p up to the value of z where the velocity has approximately has attained its stationary distribution. Present the result graphically!

Project 3. Soliton waves

From Wikipedia, we get the following information about soliton waves:

"In mathematics and physics, a soliton is a self-reinforcing solitary wave (a wave packet or pulse) that maintains its shape while it travels at constant speed. Solitons are caused by a cancellation of nonlinear and dispersive effects in the medium. (The term *dispersive effects* refers to a property of certain systems where the speed of the waves varies according to frequency.) Solitons arise as the solutions of a widespread class of weakly nonlinear dispersive partial differential equations describing physical systems."

The soliton phenomenon was first described in 1834 by John Scott Russell (1808–1882) who observed a solitary wave in the Union Canal in Scotland. He reproduced the phenomenon in a wave tank and named it the "Wave of Translation".

Solitons are modeled by the following nonlinear PDE formulated by Korteweg and de Vries in 1895

$$\frac{\partial u}{\partial t} = -6u\frac{\partial u}{\partial x} - \frac{\partial^3 u}{\partial x^3}$$

We study the x interval $-12 \leq x \leq 12$ with periodic BCs

$$u(12, t) = u(-12, t), \quad \frac{\partial^k u}{\partial x^k}(12, t) = \frac{\partial^k u}{\partial x^k}(-12, t), k = 1, 2, \ldots$$

For a soliton to appear, the IC must have a special form and amplitude. The following IC can be used

$$u(x, 0) = \frac{27}{\cosh(x) + \cosh(3x)}$$

Use the Method of Lines with 240 subintervals for the discretization of the x interval. The problem then turns into a system of ODEs

$$\frac{d\mathbf{u}}{dt} = \mathbf{F}(\mathbf{u})$$

Perform computer simulations from $t = 0$ to $t_{end} = 1.2$ with the Runge–Kutta classical method RK4 with the time step $h_t = 0.001$.

Show and explain what happens if you take a larger step $h_t = 0.002$.

Test how sensitive the soliton wave is when the amplitude of the IC is changed. How much enlargement/reduction in the amplitude is allowed for keeping the soliton effect?

An implicit method like the trapezoidal method is an alternative to RK4. What are the advantages and disadvantages when using this method?

Project 4. Wave scattering in a waveguide

A plane wave is sent into a waveguide at the left end. Due to reflexes from the walls, the plane wave is scattered. We want to investigate how the bend of the waveguide affects the wave for different frequencies at the end of the waveguide.

The waves are modeled by the 2D wave equation

$$\frac{\partial^2 U}{\partial t^2} = \frac{\partial^2 U}{\partial x^2} + \frac{\partial^2 U}{\partial y^2}$$

The time dependence is eliminated if the wave is monochromatic with a given wavelength λ. Let $\omega = 2\pi/\lambda$ and make the ansatz

$$U(x, y, t) = u(x, y)e^{i\omega t}$$

The wave equation is then reduced to Helmholtz' equation

$$\frac{\partial^2 u}{\partial x^2} + \frac{\partial^2 u}{\partial y^2} = -\omega^2 u$$

with appropriate BCs. On the reflective walls of metal, there are no waves, hence $u = 0$ at the walls. At the exit of the waveguide, we have an absorbing BC, which means that the wave disappears out from the guide. This is modeled by the advection equation

$$\frac{\partial U}{\partial t} + \frac{\partial U}{\partial y} = 0$$

if the y-axis is directed downwards. If the time dependence is eliminated, the BC at $y = y_{out}$ is

$$\frac{\partial u}{\partial y} + i\omega u = 0$$

At the inlet of the guide where $x = 0$, the wave $U(0, y, t)$ consists of two parts, one incoming plane wave $e^{i\omega(t-x)}$ and one representing reflections from the interior, $V = U(0, y, t) - e^{i\omega(t-x)}$. For the wave part V, there is the same kind of absorbing BC as at the exit, i.e.,

$$\frac{\partial V}{\partial t} - \frac{\partial V}{\partial x} = 0$$

Inserting the expression for V gives after some formula manipulation the BC at $x = 0$

$$\frac{\partial u}{\partial x} - i\omega u + 2i\omega = 0$$

We now have a 2D elliptic PDE with mixed Dirichlet and Neumann conditions.

The geometrical form of the waveguide is inserted into a square with the side 0.20 m. The width of the waveguide is 0.04 m . The contour of the waveguide is obtained from the following (x, y) coordinates, where the y-axis is directed downwards

```
x=[0,0.16,0.20,0.20,0.16,0.16,0.14,0]
y=[0,0,0.04,0.20,0.20,0.06,0.04,0.04]
```

It is interesting to investigate cases where the wavelength of the incoming wave has about the same size as the width of the waveguide. Experiment with wavelengths λ in the region 30 to 80 mm.

Use first the stepsize $h_x = h_y = 0.005$ then $h_x = h_y = 0.0025$. What advantages and disadvantages are there with another halving of the stepsize?

The problem involves complex-valued quantities, which implies that the numerical solution components $u_{i,j}$ have complex values. In the graphical presentation plot the real part of u, which can be interpreted as the wave propagation in the waveguide at a frozen time point.

Project 5. Metal block with heat source and thermometer

Given a homogeneous metal block with rectangular cross section 0.30×0.20 m^2. In the block, there is a stationary heat distribution modeled by Poisson's equation in 2D with appropriate BCs. Inside the block, there is a heat source with cross section 0.04×0.04 m^2 with its left side situated 0.04 m from the left side of the rectangle and symmetrically positioned in the y direction. The metal block has heat exchange with the environment through its left wall ($x = 0$) through a thin layer of glass. The remaining three walls are heat insulated.

The task here is to compute the temperature distribution in the metal block at stationary conditions. The temperature $u(x, y)$ is modeled by Poisson's equation

$$\beta \left(\frac{\partial^2 u}{\partial x^2} + \frac{\partial^2 u}{\partial y^2} \right) = -q(x, y)$$

where $\beta = 45$ $\left[\mathrm{W/(m \cdot K)}\right]$ is the thermal conductivity of the metal. This value is also valid in the quadratic region with the heat source. We assume that the heat source gives an evenly distributed heat flow of $q(x, y) = 20{,}000/0.04^2\ W/m^3$ inside the quadratic region. Outside of this region $q(x, y) = 0$. In grid points on the boundary of the quadratic region, half the q value is used.

The temperature of the environment is $u_{out} = 20°C$. The heat transfer coefficient of the glass layer is $K_c = 900$. Hence, the BC at $x = 0$ is

$$\beta \frac{\partial u}{\partial x}(0, y) = -K_c(u_{out} - u(0, y))$$

The isolated walls of the metal block gives the BCs $\partial u/\partial y = 0$ at the upper ($y = 0.20$) and the lower ($y = 0$) wall and $\partial u/\partial x = 0$ at the right wall ($x = 0.30$).

Discretize the problem with three different stepsizes, $h_x = h_y = 0.02, 0.01$, and 0.005. Plot contour curves of the temperature distribution and note specially the maximum temperature of the metal block (which is found inside the quadratic region). Also note the temperature at the position of the thermometer, which is the point $(0.16, 0.10)$.

Project 6. Deformation of a circular metal plate

A circular metal plate with thickness $t = 5$ m and the radius $R = 50$ mm is simply supported on a circular frame with the same radius. The plate is loaded transversally with an equally distributed pressure q (Pa). In this problem, there is no φ dependence in the deformation $u(r, \varphi)$, when described in polar coordinates. Hence, the dependent variable in the ODE modeling the deformation is r.

$$\frac{d^4u}{dr^4} + \frac{2}{r}\frac{d^3u}{dr^3} - \frac{\gamma^2}{r^2}\frac{d^2u}{dr^2} + \frac{\gamma^2}{r^3}\frac{du}{dr} = -\frac{q}{D_r}, \quad D_r = \frac{t^3}{12}\frac{E_r}{(1 - \nu_{\varphi r}^2/\gamma^2)}$$

The material constants are $E_\varphi = 40,000$ MPa, $E_r = 10,000$ MPa, $\gamma^2 = E_\varphi/E_r$, $\nu_{\varphi r} = 0.24$. The load pressure is constant $q = 0.15$.

The BCs of this fourth-order ODE is at $r = 0$ $u'(0) = 0$, $u'''(0) = 0$. At $r = R$, the deformation and the moment M_r are both zero. Since $M_r = -D_r(u'' + \nu_{\varphi r}u'/r)$, we have the BCs $u(R) = 0$ and $u''(R) + \nu_{\varphi r}u'(R)/R = 0$.

Discretize the problem in N subintervals and approximate all derivatives in the problem with difference quotients of second order. This will lead to a linear system of equations with a band matrix of band width five. Use the symmetry of the problem, i.e., that $u_{-1} = u_1$ and $u_{-2} = u_2$. Start with $N = 50$ giving the stepsize $h = R/N$ and continue with the stepsizes $h/2$ and $h/4$. Plot the result in a graph showing the three stepsize approximations of $u(r)$ in the same figure. To solve the linear system of equations use sparse technique available in, e.g., MATLAB(R).

The ODE is singular at $r = 0$. Use l'Hopital's rule to find the form of the ODE at $r = 0$.

In the problem, discretizations of $u'''(0)$ and $u^{IV}(r_i)$ are needed. Use the ansatz

$$u'''(0) = \frac{au(2h) + bu(h) + cu(0) + du(-h) + eu(2h)}{h^3}$$

and determine the coefficients a, b, c, d, e so that the difference quotient is of second order. Make a similar ansatz to approximate the fourth derivative.

Project 7. Cooling of a crystal ball

The cooling of crystal glass needs attention in the production process. Quick cooling is wanted to keep the cost lower but may cause the glass to break. A crystal ball with radius R shall be cooled down from 980°C to room temperature. The temperature lowering in the oven takes place in a controlled way and depends on the adjusted temperature parameter T that starts at $t = 0$ until $t = T$ according to

$$f(t) = 980e^{-3.9t/T}$$

For $t > T$, we have $f(t) = 980e^{-3.9} = 19.8$, hence normal room temperature.

The heat equation for a homogeneous sphere follows the PDE

$$\frac{\partial u}{\partial t} = \frac{D}{r}\frac{\partial^2(ru)}{\partial r^2}, \quad 0 < r < R, \quad t > 0$$

The thermal diffusivity for crystal glass is $D = 4.0 \cdot 10^{-7}$ m^2/s. At $r = 0$, we have the BC $\frac{\partial u}{\partial r} = 0$. Hence, at this Boundary, the heat equation takes the form

$$\frac{\partial u}{\partial t} = 3D\frac{\partial^2 u}{\partial r^2}$$

Prove this with the help of l'Hopital's rule. The BC at $r = R$ is $u(R, t) = f(t)$. When the cooling starts, the temperature of the ball is 980°C.

Of special interest is the temperature gradient in the r direction—the glass may break if $|\frac{\partial u}{\partial r}|$ is too large. In our case, we assume that the glass will break if the gradient on any occasion exceeds 6000°C, which will happen if the cooling is too quick. Make numerical experiments with the parameter T and find the smallest value of T in the cases $R = 6, 12$, and 18 cm.

Compute the temperature $u(r, t)$ in the glass ball from $t = 0$ until its surface has cooled down to room temperature.

Try discretization with 60 intervals in the r direction. Use the MoL and an appropriate ODE solver or use the implicit Euler method with small time steps in the beginning and then increase the time step.

Visualize the result graphically.

Project 8. Rotating fluid in a cylinder

In a cylinder, there is a viscous fluid. The radius of the cylinder is $R = 40$ mm. The cylinder with its content is rotating with an angular velocity ω, so that $u_\varphi = \omega r$, $0 \le r \le R$. The fluid has no velocity in the r or z direction, only the velocity component $u = u_\varphi$.

Suddenly at $t = 0$, the cylinder stops. The movement of the fluid for $t > 0$ is modeled by the PDE

$$\frac{\partial u}{\partial t} = \nu \left(\frac{1}{r}\frac{\partial}{\partial r}\left(r\frac{\partial u}{\partial r} \right) - \frac{u}{r^2} + \frac{\partial^2 u}{\partial z^2} \right), \quad 0 < r < R, \quad 0 < z < H$$

where H is the height of the cylinder and ν is the viscosity coefficient. The BCs are $u = 0$ at $r = 0$ and $r = R$. Also $u = 0$ at $z = 0$ and $z = H$. The parameter ν has the value 10^{-6} m^2/s (the value for water). The angular velocity has the value $\omega = 1$.

We want to compute and plot curves to see how the velocity $u(r, z, t)$ of the fluid changes with time.

1. Assume that the cylinder is very high. Then the velocity is independent of z. This model simplification makes the PDE depend on only t and r and hence 1D in space. Use, e.g., the Crank–Nicolson method to compute the velocity distribution $u(r, t)$ in the fluid. Plot curves with the velocity of the fluid as function of the radius at different time points.
2. Now assume the cylinder has the height $H = 8$ cm. The fact that the velocity is zero at $z = 0$ and $z = H$ will influence the velocity distribution in the cylinder. Use, e.g., the Crank–Nicolson method again for the numerical solution of this 2D problem. Plot the velocity distribution at different time points with the `contourf` command in MATLAB (R). Discuss the difference of the two results.

APPENDIX A

SOME NUMERICAL AND MATHEMATICAL TOOLS

A.1 NEWTON'S METHOD FOR SYSTEMS OF NONLINEAR ALGEBRAIC EQUATIONS

A.1.1 Quadratic Systems

Numerical solution of the nonlinear system of equations

$$\mathbf{f}(\mathbf{u}) = 0 \tag{A.1}$$

can be achieved with Newton's method.

Assume that $\mathbf{f} : \mathbb{R}^n \rightarrow \mathbb{R}^n$ and that $\mathbf{f}$ is twice continuously differentiable. A solution $\mathbf{u}^*$ is called a *root* of (A.1). Often there are several roots to a nonlinear system.

The jacobian $J(\mathbf{u})$ of $\mathbf{f}(\mathbf{u})$ is a *quadratic* matrix

$$J(\mathbf{u}) = \frac{\partial \mathbf{f}}{\partial \mathbf{u}} = \begin{pmatrix} \frac{\partial f_1}{\partial u_1} & \frac{\partial f_1}{\partial u_2} & \cdots & \frac{\partial f_1}{\partial u_n} \\ \frac{\partial f_2}{\partial u_1} & \frac{\partial f_2}{\partial u_2} & \cdots & \frac{\partial f_2}{\partial u_n} \\ \vdots & \vdots & \cdots & \vdots \\ \frac{\partial f_n}{\partial u_1} & \frac{\partial f_n}{\partial u_2} & \cdots & \frac{\partial f_n}{\partial u_n} \end{pmatrix} \tag{A.2}$$

Introduction to Computation and Modeling for Differential Equations, Second Edition. Lennart Edsberg.
© 2016 John Wiley & Sons, Inc. Published 2016 by John Wiley & Sons, Inc.

Newton's method is based on successive approximation using Taylor's expansion formula at a point $\mathbf{u}^{(i)}$

$$\mathbf{f}(\mathbf{u}) = \mathbf{f}(\mathbf{u}^{(i)}) + J(\mathbf{u}^{(i)})(\mathbf{u} - \mathbf{u}^{(i)}) + hot \tag{A.3}$$

where *hot* denotes higher order terms. The point $\mathbf{u}^{(i)}$ should be regarded as an approximation of $\mathbf{u}^*$ and the aim is to find a more accurate approximation.

If the term *hot* is neglected, (A.3) corresponds to a *linear* approximation of $\mathbf{f}(\mathbf{u})$ in the neighborhood of the point $\mathbf{u}^{(i)}$, i.e.,

$$\mathbf{f}(\mathbf{u}) \approx \mathbf{f}(\mathbf{u}^{(i)}) + J(\mathbf{u}^{(i)})(\mathbf{u} - \mathbf{u}^{(i)}) \tag{A.4}$$

If $\mathbf{f}(\mathbf{u})$ in (A.1) is replaced by the right hand side of (A.4), we obtain

$$\mathbf{f}(\mathbf{u}^{(i)}) + J(\mathbf{u}^{(i)})(\mathbf{u} - \mathbf{u}^{(i)}) = 0 \tag{A.5}$$

This is a *linear* system of equations, the solution $\mathbf{u}^{(i+1)}$ of which can be written as

$$\mathbf{u}^{(i+1)} = \mathbf{u}^{(i)} - \left(J(\mathbf{u}^{(i)})\right)^{-1}\mathbf{f}(\mathbf{u}^{(i)}) \tag{A.6}$$

The formula (A.6) together with a start value $\mathbf{u}^{(0)}$ defines an *iterative* sequence for finding an approximate solution to $\mathbf{u}^*$ and is called *Newton's method.* The iteration formula (A.6) is based on inversion of the jacobian. Since matrix inversion is more inefficient and more sensitive to rounding errors than Gaussian elimination, one iteration step is instead computed in two computational steps

$$\mathbf{u}^{(i+1)} = \mathbf{u}^{(i)} + \mathbf{h}^{(i)} \tag{A.7}$$

where $\mathbf{h}^{(i)}$, the correction term, is obtained by solving the linear system of equations

$$J(\mathbf{u}^{(i)})\mathbf{h}^{(i)} = -\mathbf{f}(\mathbf{u}^{(i)}) \tag{A.8}$$

by Gaussian elimination.

By using the Matlab operator \, the two steps (A.7) and (A.8) can be written in one step (Gaussian elimination is used by the operator \).

$$\mathbf{u}^{(i+1)} = \mathbf{u}^{(i)} - J(\mathbf{u}^{(i)})\backslash\mathbf{f}(\mathbf{u}^{(i)}) \tag{A.9}$$

The iteration is proceeded (i) until some *convergence* criterium like $\|\mathbf{h}^{(i)}\| < \epsilon$ is fulfilled or (ii) until some maximum number of iterations have been reached or (iii) until the iterates $\mathbf{u}^{(i)}$ have become too large in value. In the cases (ii) and (iii), the sequence is assumed to *diverge*.

One important reason for divergence is that the start value $\mathbf{u}^{(0)}$ is not close enough to the root we want to determine. Finding a start value is often not trivial.You can try to neglect the nonlinear terms and/or neglect terms with small coefficients to see whether a simpler system occurs. You may also try some "standard" values, e.g., $\mathbf{u}^{(0)} = (0, 0, \dots, 0)^T$ or $\mathbf{u}^{(0)} = (1, 1, \dots, 1)^T$ or try to utilize the applicational background to find a suitable start value.

Example A.1. Solve the following system of equations with Newton's method:

$$\mathbf{f}(\mathbf{u}) = \begin{pmatrix} +15u_1 & +u_2 & +u_3^2 & -30 \\ -u_1 & +30u_2 & +u_3 & -30 \\ -u_1^2 & +u_2 & +100u_3 & -20 \end{pmatrix} = 0$$

For this example, a start value can be obtained by neglecting nonlinear terms and terms with small coefficients, which gives

$$15u_1 = 30, \quad 30u_2 = 30, \quad 100u_3 = 20 \quad \Longrightarrow \quad u_1 = 2, \quad u_2 = 1, \quad u_3 = 0.2$$

We obtain

$$\mathbf{u}_0 = \begin{pmatrix} 2 \\ 1 \\ 0.2 \end{pmatrix}, \quad \Longrightarrow \quad \mathbf{f}_0 = \begin{pmatrix} 1.04 \\ -1.8 \\ -3 \end{pmatrix}, \quad \mathbf{J}_0 = \begin{pmatrix} 15 & 1 & 0.4 \\ -1 & 30 & 1 \\ -4 & 1 & 100 \end{pmatrix}$$

The linear system of equations to be solved in the first iteration is

$$\begin{pmatrix} 15 & 1 & 0.4 \\ -1 & 30 & 1 \\ -4 & 1 & 100 \end{pmatrix} \begin{pmatrix} h_1 \\ h_2 \\ h_3 \end{pmatrix} = - \begin{pmatrix} 1.04 \\ -1.8 \\ -3 \end{pmatrix}$$

Solving this system gives the first correction $\mathbf{h}_0$ and the first iterate $\mathbf{u}_1$

$$\mathbf{h}_0 = \begin{pmatrix} -0.0738 \\ 0.0567 \\ 0.0265 \end{pmatrix}, \quad \Longrightarrow \quad \mathbf{u}_1 = \mathbf{u}_0 + \mathbf{h}_0 = \begin{pmatrix} 1.9262 \\ 1.0567 \\ 0.2265 \end{pmatrix}$$

After two more iterations, the sequence has converged to four decimal accuracy $\mathbf{u}^* = (1.9251, 1.0717, 0.2263)^T$.

When the start value is close to the solution, the convergence is very fast. Newton's method has *quadratic convergence*, which means that the correction terms $||\mathbf{h}^{(i)}||$ behave as

$$||\mathbf{h}^{(i+1)}|| \approx C||\mathbf{h}^{(i)}||^2 \tag{A.10}$$

A.1.2 Overdetermined Systems

Newton's method can be modified to solve overdetermined systems of nonlinear equations

$$\mathbf{r}(\mathbf{p}) \approx 0 \tag{A.11}$$

where $\mathbf{r} \in \mathbb{R}^m$, $\mathbf{p} \in \mathbb{R}^n$, and $m > n$. Hence, there are more equations than unknowns so we have to specify in what sense the solution is wanted. Note that the notation has been changed. The reason for this change is that overdetermined systems often occur in *least squares* problems, where we want to minimize the Euclidean length of the *residual* vector $\mathbf{r}$ with respect to the components of a parameter vector $\mathbf{p}$.

The algorithm described here, *Gauss–Newton's method*, will converge (when it converges) to a solution in the *least squares* sense, i.e., to a point $\mathbf{p}^*$ giving a (local) minimum to the sum of squares of the residuals $\mathbf{r}(\mathbf{p})^T\mathbf{r}(\mathbf{p})$.

Systems of type (A.11) occur, e.g., in curve-fitting problems when a parameter vector $\mathbf{p}$ in a nonlinear algebraic model $g(\mathbf{p}, t)$ is to be fitted to measurements (t_k, g_k)

$$g(\mathbf{p}, t_k) \approx g_k, \quad k = 1, 2, \ldots, m \tag{A.12}$$

which can also be written in the form (A.11)

$$\mathbf{r}(\mathbf{p}) \equiv \mathbf{g}(\mathbf{p}) - \mathbf{g} \approx 0 \tag{A.13}$$

where $\mathbf{g}(\mathbf{p}) = (g(\mathbf{p}, t_1), g(\mathbf{p}, t_2), \ldots, g(\mathbf{p}, t_m))^T$. The linear approximation of $\mathbf{r}(\mathbf{p})$ at the point $\mathbf{p}^{(i)}$ is of the same form as (A.4), but the jacobian is now a *rectangular* $m \times n$ matrix. Replacing the left hand side of (A.11) by its linear approximation gives an overdetermined linear system of equations

$$\mathbf{r}(\mathbf{p}^{(i)}) + J(\mathbf{p}^{(i)})(\mathbf{p} - \mathbf{p}^{(i)}) \approx 0 \tag{A.14}$$

This system is solved in the least squares sense giving the solution $\mathbf{p}^{(i+1)}$

$$\mathbf{p}^{(i+1)} = \mathbf{p}^{(i)} - (J(\mathbf{p}^{(i)}))^+\mathbf{r}(\mathbf{p}^{(i)}) \tag{A.15}$$

where $J(\mathbf{p}))^+$ is the *pseudoinverse* of the jacobian

$$(J(\mathbf{p}))^+ = (J(\mathbf{p})^T J(\mathbf{p}))^{-1} J(\mathbf{p})^T \tag{A.16}$$

The iterative method (A.15) is called *Gauss–Newton's method* for solution of the nonlinear least squares problem.

The iteration formula (A.15) can be written without the pseudoinverse using the Matlab operator \ as in (A.9)

$$\mathbf{p}^{(i+1)} = \mathbf{p}^{(i)} - J(\mathbf{p}^{(i)}) \backslash \mathbf{r}(\mathbf{p}^{(i)}) \tag{A.17}$$

Like quadratic systems it is important to have a good start value $\mathbf{p}^{(0)}$ for convergence of Gauss–Newton's method. Note, however, that Gauss–Newton's method has only *linear convergence*.

A.2 SOME FACTS ABOUT LINEAR DIFFERENCE EQUATIONS

When a discretization method is used to solve a differential equation, the ordinary differential equation (ODE) or the partial differential equation (PDE) is approximated by a *difference* equation, i.e., a relation between successive solution points $u_0, u_1, \ldots, u_n, u_{n+1}, \ldots$ where n takes only integer values. As an example, we can take the second-order backward differentiation formula (BDF) method (see Chapter 3)

$$\nabla u_n + \frac{1}{2}\nabla^2 u_n = hf_n \tag{A.18}$$

If we use the definitions

$$\begin{aligned} \nabla u_n &= u_n - u_{n-1} \\ \nabla^2 u_n &= u_n - 2u_{n-1} + u_{n-2} \end{aligned} \tag{A.19}$$

we can also write the method as a relation between u_n values

$$\frac{3}{2}u_n - 2u_{n-1} + \frac{1}{2}u_{n-2} = hf_n \tag{A.20}$$

called a difference equation (or recurrence equation).

A general definition of a kth-order difference equation (Δ-eqn) is

$$F(n, u_n, u_{n-1}, u_{n-2}, \ldots, u_{n-k}) = 0 \tag{A.21}$$

The general solution of such an equation will depend on n and contain k arbitrary constants, i.e.,

$$u_n = f(n, C_1, C_2, \ldots, C_k) \tag{A.22}$$

where $C_1, C_2, \ldots, C_k$ are determined from the initial or boundary conditions (BCs) of the Δ-eqn.

An important special case of (A.21) is a linear Δ-eqn with constant coefficients. We show first the *homogeneous* form

$$u_n + a_1 u_{n-1} + a_2 u_{n-2} + \ldots + a_k u_{n-k} = 0 \tag{A.23}$$

Associate with this Δ-eqn, the *characteristic* equation

$$\mu^k + a_1 \mu^{k-1} + a_2 \mu^{k-2} + \ldots + a_k = 0 \tag{A.24}$$

The roots of this polynomial equation of degree k are denoted by

$$\mu_1, \mu_2, \ldots, \mu_k \tag{A.25}$$

If all the roots are different, the general solution of (A.23) is

$$u_n = C_1\mu_1^n + C_2\mu_2^n + \dots + C_k\mu_k^n \tag{A.26}$$

where $C_1, C_2, \dots, C_k$ are arbitrary constants (note that they enter linearly). Compare the technique of solving a linear Δ-eqn with constant coefficients with the technique for a linear ODE with constant coefficients. The fundamental solutions for the Δ-eqn is μ_i^n, for the ODE it is $e^{\mu_i x}$.

If there are multiple roots, the general solution expression must be modified. Say, e.g., that $\mu_1 = \mu_2$ (double root). Then the general solution is

$$u_n = (C_1 + C_2 n)\mu_1^n + C_3\mu_3^n + \dots + C_k\mu_k^n \tag{A.27}$$

The *inhomogeneous* form of (A.23) is

$$u_n + a_1 u_{n-1} + a_2 u_{n-2} + \dots + a_k u_{n-k} = b_n \tag{A.28}$$

Just as in the ODE case, we first have to find a particular solution p_n satisfying (A.28). The general solution is then (provided the characteristic roots are all different)

$$u_n = p_n + C_1\mu_1^n + C_2\mu_2^n + \dots + C_k\mu_k^n \tag{A.29}$$

From the general solution (A.26), we see that u_n is stable, i.e., the u_n values are bounded as $n \to \infty$, if

$$|\mu_i| \le 1 \tag{A.30}$$

If μ_i is a multiple root, the stability condition must be modified to

$$|\mu_i| < 1 \tag{A.31}$$

Example A.2. Given the difference equation

$$u_n + 3u_{n-1} + 2u_{n-2} = 0, \quad u_0 = 1, u_1 = 2$$

By inserting $n = 2, 3, 4, \dots$ we get $u_2 = -8$, $u_3 = 20$, $u_4 = -44$, ... What is u_n in general, i.e., a formula of type $u_n = f(n)$?

Solve first the characteristic equation

$$\mu^2 + 3\mu + 2 = 0$$

with the solution

$$\mu_1 = -1, \quad \mu_2 = -2$$

Hence, the general solution is

$$u_n = C_1(-1)^n + C_2(-2)^n$$

Insert the initial conditions

$$u_0 = C_1 + C_2 = 1$$
$$u_1 = -C_1 - 2C_2 = 2$$

which gives $C_1 = 4, C_2 = -3$ and the solution is

$$u_n = 4(-1)^n - 3(-2)^n$$

Example A.3. What is the stability area of the explicit midpoint method?

Apply the method to $u' = \lambda u$ and we get the Δ-eqn

$$u_{n+1} = u_{n-1} + 2h\lambda u_n$$

and the characteristic equation is

$$\mu^2 - 2\lambda h\mu - 1 = 0$$

where λ is a complex-valued parameter.

Write one root on polar form $\mu_1 = re^{i\phi}$. Since $\mu_1\mu_2 = -1$, we get $\mu_2 = -(1/r)e^{-i\phi}$.

For a stable solution, we must have $r = 1$. Since $\mu_1 + \mu_2 = 2\lambda h$, we get $e^{i\phi} - e^{-i\phi} = 2\lambda h$, i.e.,

$$\lambda h = i \sin\phi$$

Hence, the stability area of the explicit midpoint method is the interval $[-i, i]$ on the imaginary axis.

Example A.4. Compute the eigenvalues of the tridiagonal matrix

$$A = \text{tridiag}_n(-1, 2, -1) = \begin{pmatrix} 2 & -1 & 0 & \dots & 0 \\ -1 & 2 & -1 & \ddots & \vdots \\ 0 & \ddots & \ddots & \ddots & 0 \\ \vdots & \ddots & -1 & 2 & -1 \\ 0 & \dots & 0 & -1 & 2 \end{pmatrix}$$

The eigenvalue problem $A\mathbf{u} = \lambda\mathbf{u}$ for this problem is formulated as a Δ-eqn

$$-u_{k-1} + 2u_k - u_{k+1} = \lambda u_k, \quad k = 1, 2, \dots, n$$

Since there are no u_0- or u_{n+1} components in $\mathbf{u}$, we set these components to zero and obtain a boundary value problem for a Δ-eqn

$$u_{k+1} + (\lambda - 2)u_k + u_{k-1} = 0, \qquad u_0 = 0, \quad u_{n+1} = 0$$

The roots of the characteristic equation

$$\mu^2 + (\lambda - 2)\mu + 1 = 0$$

are denoted by μ_1 and μ_2. The roots fulfill the relations

$$\mu_1\mu_2 = 1, \quad \mu_1 + \mu_2 = 2 - \lambda$$

Hence, if $\mu_1 = re^{i\phi}$, then $\mu_2 = (1/r)e^{-i\phi}$ if we use the polar form. The general solution of the Δ-eqn (if $\mu_1 \neq \mu_2$) is

$$u_k = C_1\mu_1^k + C_2\mu_2^k$$

Insert the BCs

$$u_0 = C_1 + C_2 = 0 \rightarrow C_2 = -C_1$$
$$u_{n+1} = C_1\mu_1^{n+1} + C_2\mu_2^{n+1} = 0$$

The condition $C_2 = -C_1$ above is used and we get

$$C_1\mu_1^{n+1} = C_1\mu_2^{n+1}$$

We want a nontrivial solution, i.e., $C_1 \neq 0$, which gives

$$\left(\frac{\mu_1}{\mu_2}\right)^{n+1} = 1$$

or if we use the polar form of μ_1 and μ_2

$$r^2 e^{2i\phi(n+1)} = 1$$

which gives

$$r = 1, \quad 2i\phi(n+1) = 2\pi ik, \quad k = 1, 2, \ldots, n$$
$$\phi_k = \frac{\pi k}{n+1}, \quad k = 1, 2, \ldots, n$$

The values $k = 0$ and $k = n + 1$ must be excluded in the formula above, since we then obtain $\mu_1 = \mu_2 = 1$ or $\mu_1 = \mu_2 = -1$, respectively, i.e., double roots, in which case $u_k = (C_1 + C_2k) \rightarrow C_1 = C_2 = 0$, i.e., the trivial solution.

As a final result, we obtain

$$2 - \lambda = \mu_1 + \mu_2 = e^{i\phi} + e^{-i\phi} = 2\cos\phi$$

Hence, the eigenvalues are

$$\lambda_k = 2(1 - \cos \phi_k) = 4\sin^2(\phi_k/2), \quad k = 1, 2, \dots, n$$

The smallest eigenvalue is obtained from $k = 1$: $\lambda_1 \approx \pi^2/(n+1)^2$ and the largest for $k = n$: $\lambda_n \approx 4$.

A.3 DERIVATION OF DIFFERENCE APPROXIMATIONS

Difference approximation of derivative terms in ODEs and BCs are based on Taylor's formula

$$f(x+h) = f(x) + hf'(x) + \frac{h^2}{2}f''(x) + \frac{h^3}{6}f'''(x) + \frac{h^4}{24}f^{(4)}(x) + \mathcal{O}h^5 \qquad \text{(A.32)}$$

The Taylor expansion is suitable since a difference approximation should be valid *locally* in the neighborhood of a point x.

We have already seen the following examples of central difference approximations:

$$\frac{du}{dx}(x) = \frac{u(x+h) - u(x-h)}{2h} + \mathcal{O}h^2 \qquad \text{(A.33)}$$

$$\frac{d^2u}{dx^2}(x) = \frac{u(x+h) - 2u(x) + u(x-h)}{h^2} + \mathcal{O}h^2 \qquad \text{(A.34)}$$

How are similar formulas derived when approximations to higher order derivatives or unsymmetric formulas for the derivatives are needed? As a first example, assume we need a difference approximation to $f'(x)$ when the function values $f(x)$, $f(x+h)$, and $f(x+2h)$ are available. Make the following linear ansatz:

$$f'(x) = af(x) + bf(x+h) + cf(x+2h) + \mathcal{O}h^p \qquad \text{(A.35)}$$

where a, b, and c are to be determined so that the approximation order p is as high as possible. Use Taylor's formula on $f(x+h)$ and $f(x+2h)$

$$f'(x) = af(x) + b(f(x) + hf'(x) + \frac{h^2}{2}f''(x)) + c(f(x)$$

$$+ 2hf'(x) + \frac{(2h)^2}{2}f''(x)) + \mathcal{O}h^3$$

Since there are three unknown coefficients a, b, and c to determine, we need three equations. They are obtained by identifying coefficients in front of $f(x)$, $f'(x)$, $f''(x)$, … in the left and right hand side

$$a + b + c = 0$$

$$hb + 2hc = 1$$

$$\frac{h^2}{2}b + \frac{(2h)^2}{2}c = 0$$

The solution is

$$\begin{pmatrix} a \\ b \\ c \end{pmatrix} = \frac{1}{2h} \begin{pmatrix} -3 \\ 4 \\ -1 \end{pmatrix}$$

Hence, the unsymmetric difference approximation is

$$f'(x) = \frac{-3f(x) + 4f(x+h) - f(x+2h)}{2h} + e(x) \tag{A.36}$$

where $e(x)$ is the approximation error. What is the approximation order p? Look at the next term in the Taylor expansion and insert the solution a, b, and c

$$e(x) = \left(b\frac{h^3}{6} + c\frac{(2h)^3}{6} \right) f'''(x) = -\frac{h^2}{3} f'''(x) = \mathcal{O}h^2 \tag{A.37}$$

Hence, the approximation (A.36) is of second order.

With the same technique, the following difference approximations can be derived:

$$f'''(x) = \frac{f(x+2h) - 2f(x+h) + 2f(x-h) - f(x-2h)}{2h^3} + \mathcal{O}h^2$$

$$f^{(4)}(x) = \frac{f(x+2h) - 4f(x+h) + 6f(x) - 4f(x-h) + f(x-2h)}{h^4} + \mathcal{O}h^2$$

Deriving difference approximations based on Taylor's formula can be systemized with an elegant technique called *operator calculus*, see [1]. Define the following operators

$$Ef(x) = f(x+h) \tag{A.38}$$

$$\Delta f(x) = f(x+h) - f(x) \tag{A.39}$$

$$\nabla f(x) = f(x) - f(x-h) \tag{A.40}$$

$$Df(x) = f'(x) \tag{A.41}$$

Taylor's formula can be written

$$Ef(x) = \left(1 + hD + \frac{(hD)^2}{2} + \frac{(hD)^3}{6} + \dots \right) f(x) \tag{A.42}$$

i.e.,

$$E = \mathrm{e}^{hD} \tag{A.43}$$

Since $E = \Delta + 1$, we have

$$hD = \log(1 + \Delta) = \Delta - \frac{\Delta^2}{2} + \frac{\Delta^3}{3} + \dots \tag{A.44}$$

With this operator formalism, (A.36) can be derived by retaining only the first two terms in (A.44)

$$D \approx \frac{1}{h}\left(\Delta - \frac{\Delta^2}{2}\right) = \frac{1}{h}\left(E - 1 - \frac{1}{2}(E-1)^2\right)$$

$$f'(x) \approx \frac{1}{h}\left(f(x+h) - f(x) - \frac{1}{2}(f(x+2h) - 2f(x+h) + f(x))\right)$$

$$= \frac{-3f(x) + 4f(x+h) - f(x+2h)}{2h}$$

The approximation error is obtained from the first neglected term in (A.44)

$$e(x) = \frac{1}{h}\frac{\Delta^3}{3}f(x) = \frac{1}{3h}(e^{hD} - 1)^3 f(x) = \ \dots \ = -\frac{h^2}{3}f'''(x)$$

The algebra of operator calculus can be generalized to inversion. Since

$$E^n f(x) = f(x+nh) \rightarrow E^{-1}f(x) = f(x-h)$$

we get

$$\nabla = 1 - E^{-1} \tag{A.45}$$

With this relation, we can derive other approximations to $f'(x)$ from

$$hD = -\log(1-\nabla) = \nabla + \frac{\nabla^2}{2} + \frac{\nabla^3}{3} + \dots \tag{A.46}$$

Retaining only the first two terms gives

$$D \approx \frac{1}{h}\left(\nabla + \frac{\nabla^2}{2}\right) = \frac{3f(x) - 4f(x-h) + f(x-2h)}{2h}$$

BIBLIOGRAPHY

1. G. Dahlquist and Å. Björck, "Numerical Methods," Dover, 2003

A.4 THE INTERPRETATIONS OF GRAD, DIV, AND CURL

The gradient. Let $\Phi(\mathbf{r})$ be a scalar-valued field, where $\mathbf{r} = (x, y)^T$. From a fixed point $\mathbf{r}_0$, let $\mathbf{d}$ be a vector of unit length 1, i.e., $\|\mathbf{d}\|_2^2 = \mathbf{d}^T\mathbf{d} = 1$, defining a direction from $\mathbf{r}_0$. With Taylor expansion, we obtain

$$\Phi(\mathbf{r}_0 + \mathbf{d}) = \Phi(\mathbf{r}_0) + \nabla_{\mathbf{d}}\Phi(\mathbf{r}_0) + hot$$

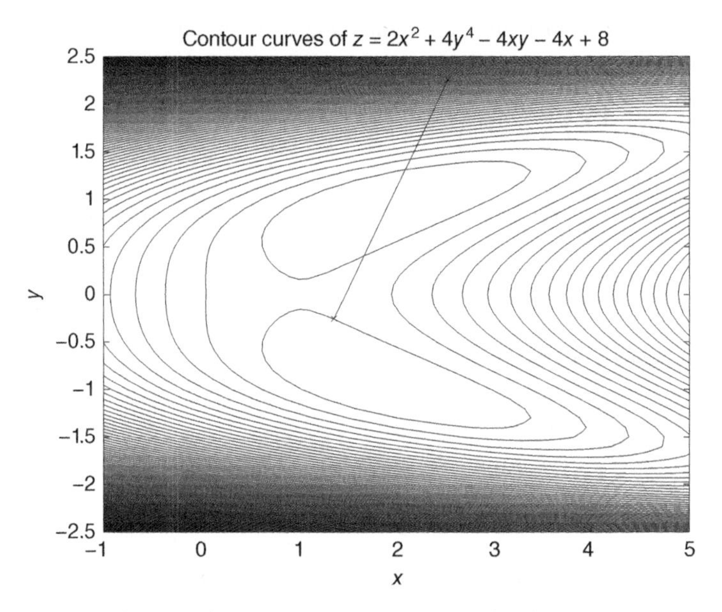

Figure A.1 Contour curves with a gradient inserted

where $\nabla_\mathbf{d}\Phi = \mathbf{d}^T\nabla\Phi$ is called the direction derivative of Φ in the direction $\mathbf{d}$. $\nabla_\mathbf{d}\Phi$ can be interpreted as the velocity with which Φ changes when moving away from $\mathbf{r}_0$ in the direction $\mathbf{d}$. The largest values of the velocity is obtained for $\mathbf{d} = \pm\nabla\Phi(\mathbf{r}_0)$. Hence, Φ increases fastest in the direction of the gradient and decreases fastest in the direction of the negative gradient, also called the steepest descent direction, see Section A.5.2 (Figure A.1).

A contour curve of Φ is a curve along which the function has a constant value. The gradient at a point is always orthogonal to the contour curve through that point. In a point where the direction $\mathbf{d}$ is orthogonal to $\nabla\Phi(\mathbf{r}_0)$, i.e., $\nabla_\mathbf{d}\Phi(\mathbf{r}_0) = 0$, Φ does not change (if *hot* is neglected). This direction is in fact the tangent of the contour curve at the point.

The divergence. Let $\partial\Omega$ be a positively oriented curve enclosing the region Ω in R^2. Let $P(x, y)$ and $Q(x, y)$ be two continuously differentiable functions in Ω. Then use Green's theorem

$$\oint_{\partial\Omega} Pdx + Qdy = \int\int_{\Omega}\left(\frac{\partial Q}{\partial x} - \frac{\partial P}{\partial y}\right) dxdy$$

How can this theorem be used? To be a little more precise, assume that the boundary curve is defined by the vector $\mathbf{r}(s)$, where the parameter s is the *arclength* of the curve measured from some initial point $\mathbf{r}_0 = \mathbf{r}(0)$. When s is increased, the curve moves counterclockwise (see Figure A.2).

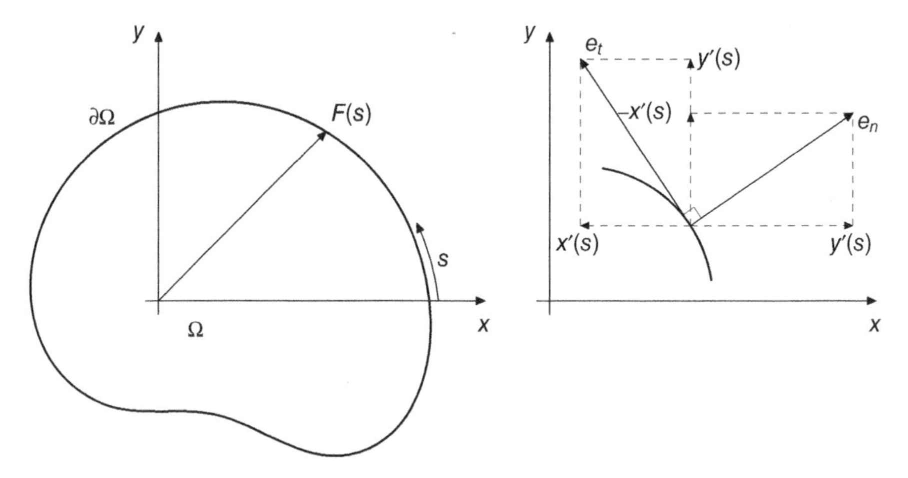

Figure A.2 Tangent and normal directions at a point on a curve

The unit vector in the tangent direction, $\mathbf{e}_t$, is

$$\mathbf{e}_t = \frac{d\mathbf{r}}{ds} = (x'(s), y'(s))^T$$

The unit vector in the normal direction pointing out of Ω, $\mathbf{e}_n$, is

$$\mathbf{e}_n = (y'(s), -x'(s))^T$$

We now show the meaning of the operator *div* applied to heat flow:

Assume there is a heat flux $\mathbf{F} = (P(x, y), Q(x, y))^T$, measured in $J/(m^2 \cdot s)$, in a thin plate of thickness h (m). The shape of the plate corresponds to the region Ω. The total flow of heat out of Ω is obtained if we sum the contributions of flow out from each little area section hds along $\partial\Omega$. The heat flow, measured in J/s, in the outward normal direction from such a section is $\mathbf{F} \cdot \mathbf{e}_n \cdot h \cdot ds$. The net heat flow, i.e., the difference between the outflow out of Ω and the inflow into Ω is (Figure A.3)

$$h \oint_{\partial\Omega} \mathbf{F} \cdot \mathbf{e}_n \, ds$$

Using Green's theorem on this integral, we arrive at Gauss' theorem:

$$\oint_{\partial\Omega} \mathbf{F} \cdot \mathbf{e}_n \, ds = \oint_{\partial\Omega} (Py'(s) - Qx'(s))ds$$
$$= \iint_{\Omega} \left(\frac{\partial P}{\partial x} + \frac{\partial Q}{\partial y} \right) dxdy = \iint_{\Omega} div \, \mathbf{F} \, dxdy$$

Hence, if $div \, \mathbf{F} = 0$ in the whole of Ω, the net flow of heat out of Ω is zero, in other words: the heat flow into Ω = the heat flow out of Ω. What is the interpretation if

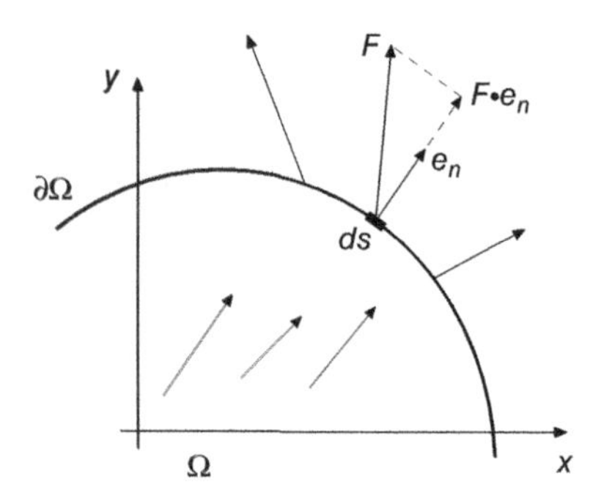

Figure A.3 Flux vector at a point on the boundary projected to the normal direction

div $\mathbf{F} \neq 0$? Apply Gauss' theorem to a small area A of Ω, so small that div $\mathbf{F}$ can be considered constant in A. The net flow out of the small volume hA is

$$h \oint_{\partial A} \mathbf{F} \cdot \mathbf{e}_n \, ds \approx hA \; div \; \mathbf{F} \Rightarrow$$

$$div \; \mathbf{F} \approx \frac{1}{hA} h \oint_{\partial A} \mathbf{F} \cdot \mathbf{e}_n \, ds$$

Hence, div $\mathbf{F}$ can be interpreted as the amount of heat energy produced (div $\mathbf{F} > 0$) or consumed (div $\mathbf{F} < 0$) per unit time and unit volume $[\mathrm{J}/(\mathrm{m}^3 \cdot \mathrm{s})]$ by the vector field $\mathbf{F}$ in Ω. Hence, div $\mathbf{F}$ is the *source strength* of the field $\mathbf{F}$.

The curl (rotation). In a similar way, *curl* can be given an interpretation if we apply Green's theorem on a velocity field $\mathbf{v}(x, y) = (P(x, y), Q(x, y))^T$ defined on a small circle C_r with radius r in Ω, i.e.,

$$\oint_{\partial C_r} \mathbf{v} \cdot \mathbf{e}_t \, ds = \oint_{\partial C_r} (Px'(s) + Qy'(s))ds = \oint_{\partial C_r} Pdx + Qdy = \iint_{C_r} curl \; \mathbf{v} \; dxdy$$

If the circle is so small that *curl* $\mathbf{v}$ is approximately constant in C_r, we get

$$\oint_{\partial C_r} \mathbf{v} \cdot \mathbf{e}_t \, ds \approx \pi r^2 \; curl \; \mathbf{v}$$

The mean value of v_t, the tangential component of $\mathbf{v}$ along ∂C_r, is (Figure A.4)

$$v_{t_{mean}} = \frac{1}{2\pi r} \oint_{\partial C_r} v_t \, ds$$

Since $v_t = r\omega$, where ω is the angular velocity, $\dot{\varphi}$, the mean value of the angular velocity along ∂C_r is

$$\omega_{mean} \approx \frac{1}{2\pi r} \oint_{\partial C_r} \frac{v_t}{r} ds = \frac{1}{2\pi r^2} \oint_{\partial C_r} \mathbf{v} \cdot \mathbf{e}_t \, ds \approx \frac{1}{2} curl \; \mathbf{v}$$

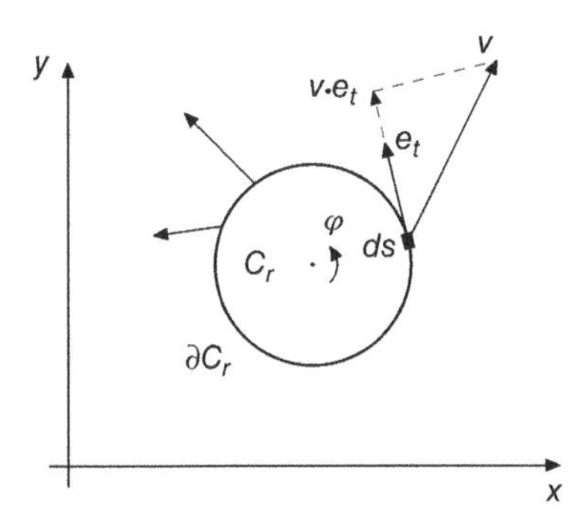

Figure A.4 Velocity vector at a point on the boundary projected to the tangential direction

Hence, *curl* **v** is interpreted as the *vorticity strength* of the velocity field **v**.

A.5 NUMERICAL SOLUTION OF ALGEBRAIC SYSTEMS OF EQUATIONS

Numerical solution of systems of equations occurs frequently as part of the numerical treatment of differential equations. The system can be linear

$$A\mathbf{u} = \mathbf{b}, \quad \text{where} \quad A \quad \text{is } n \times n$$

or nonlinear

$$\mathbf{f}(\mathbf{u}) = 0, \quad \text{where the jacobian} \quad J(\mathbf{u}) = \frac{\partial \mathbf{f}}{\partial \mathbf{u}} \quad \text{is } n \times n$$

For a nonlinear system, an iterative method must be used. If Newton's method (see Section A.1) is chosen, a linear system of equations has to be solved in each iteration.

The parameter n is the number of unknowns = the number of equations. When A has no special structure, it is called a *full* matrix. For a full matrix, all n^2 elements must be stored.

However, when differential equations are solved numerically, *sparse* systems of equations occur frequently. The system is sparse if A (or J) has a small number $n_z \ll n^2$ of nonzero elements. Often in a sparse system, there are a few unknowns in each equations, i.e., $n_z = \mathcal{O}(n)$. Hence, all elements of a sparse matrix must not be stored but essentially only those which are nonzero.

For efficiency reasons, it is important to utilize the sparsity of the system, both when it is comes to how the storage is organized and how the solution algorithm is designed.

A.5.1 Direct Methods

A.5.1.1 Some Facts about Gaussian Elimination The standard method for solving $A\mathbf{u} = \mathbf{b}$ is Gaussian elimination with row pivoting. The algorithm gives both the solution $\mathbf{u}$ and the LU factorization of A. If no pivoting is needed, we obtain

$A = LU$, where L is lower triangular with ones in the main diagonal and U is upper triangular. The computational work to compute the solution and the factorization amounts to about $n^3/3$ flops.

If row pivoting is necessary, the LU factorization is valid for a row permuted A, i.e., $PA = LU$, where P is a permutation matrix.

If A is symmetric and positive definite (SPD), $A = A^T$ and $\lambda_i(A) > 0, i = 1, 2, \ldots, n$, the factorization can be written $A = LL^T$, also called the Cholesky factorization of A, and is achieved in $n^3/6$ flops.

A.5.1.2 Direct Methods for Sparse Linear Systems For sparse matrices, there are modified versions of Gaussian elimination that give the solution more efficiently than Gaussian elimination for a full matrix.

A.5.1.2.1 Methods for Band Structured Matrices. A common sparse structure is the *band* structure with bandwidth p, e.g.,

$$A = \begin{pmatrix} a_{11} & a_{12} & 0 & 0 & 0 & 0 \\ a_{21} & a_{22} & a_{23} & 0 & 0 & 0 \\ a_{31} & a_{32} & a_{33} & a_{34} & 0 & 0 \\ 0 & a_{42} & a_{43} & a_{44} & a_{45} & 0 \\ 0 & 0 & a_{53} & a_{54} & a_{55} & a_{56} \\ 0 & 0 & 0 & a_{64} & a_{65} & a_{66} \end{pmatrix}, \qquad p = 4$$

The number of bands is called the bandwidth of A. A bandmatrix with bandwidth p can be stored as p columns in an $n \times p$ matrix B in the following way:

$$B = \begin{pmatrix} 0 & 0 & a_{11} & a_{12} \\ 0 & a_{21} & a_{22} & a_{23} \\ a_{31} & a_{32} & a_{33} & a_{34} \\ a_{42} & a_{43} & a_{44} & a_{45} \\ a_{53} & a_{54} & a_{55} & a_{56} \\ a_{64} & a_{65} & a_{66} & 0 \end{pmatrix}$$

Hence, np elements are needed to store the elements of A instead of n^2. The LU factorization is computed in np^2 flops and stored in the same matrix by overwriting the elements of B

$$(L, U) = \begin{pmatrix} 0 & 0 & r_{11} & r_{12} \\ 0 & l_{21} & r_{22} & r_{23} \\ l_{31} & l_{32} & r_{33} & r_{34} \\ l_{42} & l_{43} & r_{44} & r_{45} \\ l_{53} & l_{54} & r_{55} & r_{56} \\ l_{64} & l_{65} & r_{66} & 0 \end{pmatrix}$$

Since the elements in the main diagonal of L are ones, there is no need to store that diagonal.

If the matrix A is symmetric, only the main diagonal and the diagonals below need to be stored in B. The Cholesky factorization can be done on the same matrix space as B and the result, the band-stored L-matrix, overwrites the original B-matrix. Observe that the Cholesky factor L has a main diagonal with elements not equal to one. Hence, this diagonal must also be stored. The number of flops needed is $np^2/2$. Following is an example:

$$B = \begin{pmatrix} 0 & 0 & a_{11} \\ 0 & a_{21} & a_{22} \\ a_{31} & a_{32} & a_{33} \\ a_{42} & a_{43} & a_{44} \\ a_{53} & a_{54} & a_{55} \\ a_{64} & a_{65} & a_{66} \end{pmatrix}, \qquad L = \begin{pmatrix} 0 & 0 & l_{11} \\ 0 & l_{21} & l_{22} \\ l_{31} & l_{32} & l_{33} \\ l_{42} & l_{43} & l_{44} \\ l_{53} & l_{54} & l_{55} \\ l_{64} & l_{65} & l_{66} \end{pmatrix}$$

Band matrices occur, e.g., when the finite difference method (FDM) or finite element method (FEM) is used to discretize a PDE problem defined on a rectangular region, e.g., Poisson's equation defined on a square (see Chapter 7).

A special but important type of band matrix is the tridiagonal $n \times n$ matrix A

$$A = \text{tridiag}_n(a, d, c)$$

where a and c are vectors of dimension $n - 1$ (the sub- and superdiagonals) and d a vector of dimension n (the main diagonal). Tridiagonal matrices occur when FDM or FEM is used to solve boundary value problem (BVP) (see Chapter 4).

Obviously much computational work can be saved if a sparse matrix A can be transformed into a bandmatrix with small bandwidth. The transformations must be done in an economic way in order not to waste flops on this part of the algorithm. Examples of simple "cheap" transformations are permutations of rows and columns. Let P and Q be permutation matrices. The effect of multiplying A from the left with P is changing the order of the rows of A, and multiplying A from the right with Q will change the order of the columns of A, e.g.,

$$PA = \begin{pmatrix} - & \mathbf{a}_2 & - \\ - & \mathbf{a}_4 & - \\ - & \mathbf{a}_1 & - \\ - & \mathbf{a}_3 & - \end{pmatrix}, \qquad AQ = \begin{pmatrix} | & | & | & | \\ \mathbf{a}_3 & \mathbf{a}_1 & \mathbf{a}_4 & \mathbf{a}_2 \\ | & | & | & | \end{pmatrix}$$

Do the following series of transformations of the system $A\mathbf{u} = \mathbf{b}$:

$$PA\mathbf{u} = P\mathbf{b}, \qquad \mathbf{u} = Q\mathbf{v}, \qquad PAQ\mathbf{v} = P\mathbf{b}$$

The ordering of rows and columns in A, resulting in a matrix PAQ that has a smaller bandwidth than A, can be done with different algorithms. Examples of such algorithms are the minimum degree permutation or the reverse Cuthill MacKee permutation (see the Bibliography for more information).

A.5.1.2.2 Methods for Profile Structured Matrices. An alternative way of storage, applicable for an SPD matrix A, is the *profile* storage of the main diagonal and the lower part of A. The profile is the border for which all elements to the left of the border are zero-elements. Profile structured matrices occur, e.g., when FEM is applied to an elliptic PDE problem defined on an irregular region.

It can be shown that the profile of the Cholesky factorized L-matrix has the same profile as the original matrix, i.e., the profile is invariant under Cholesky factorization

$$A = \begin{pmatrix} x & & & & & & & \\ x & x & & & & & & \\ & x & x & & & & & \\ x & x & x & x & & & & \\ & & x & x & x & & & \\ & & & & x & x & & \\ & & & x & x & x & x & \\ & & & & & & x & x \end{pmatrix}, \quad L = \begin{pmatrix} x & & & & & & & \\ x & x & & & & & & \\ & x & x & & & & & \\ x & x & x & x & & & & \\ & & x & x & x & & & \\ & & & & x & x & & \\ & & & x & x & x & x & \\ & & & & & & x & x \end{pmatrix}$$

It is not efficient to store A as a full matrix. Instead, only the nonzeros of A are stored in a vector $\mathbf{a}$ together with a pointer vector $\mathbf{p}$ the elements of which show the locations of the diagonal elements of A. See the example below:

$$A = \begin{pmatrix} 25 & & & & \\ 3 & 21 & & & \\ 0 & 2 & 23 & & \\ 0 & 4 & 0 & 22 & \\ 0 & 0 & 1 & 0 & 20 \end{pmatrix}$$

$$\mathbf{a} = (25, 3, 21, 2, 23, 4, 0, 22, 1, 0, 20)$$

$$\mathbf{p} = (1, 3, 5, 8, 11)$$

A.5.1.2.3 Methods for General Sparse Matrices. A general way of storing a sparse matrix A is to store only its nonzero elements in a vector together with two index vectors containing the row and column indices of the nonzero elements.

$$A = \begin{pmatrix} a_{11} & 0 & a_{13} & 0 & a_{15} \\ a_{21} & 0 & 0 & a_{24} & 0 \\ 0 & a_{32} & 0 & 0 & a_{35} \\ 0 & a_{42} & 0 & a_{44} & 0 \\ a_{51} & 0 & 0 & a_{54} & 0 \end{pmatrix}$$

is stored in the three vectors

$$a = (a_{11}, a_{13}, a_{15}, a_{21}, a_{24}, a_{32}, a_{35}, a_{42}, a_{44}, a_{51}, a_{54})$$

$$ia = (1, 1, 1, 2, 2, 3, 3, 4, 4, 5, 5)$$

$$ja = (1, 3, 5, 1, 4, 2, 5, 2, 4, 1, 4)$$

Observe that storing this matrix A as a band matrix is not efficient, since the full matrix then must be stored.

The solution and factorization of a general sparse system $A\mathbf{u} = \mathbf{b}$ is more complicated, since the sparseness structure in A will not be preserved in L and U after the Gaussian elimination process. Usually L and U will be much less sparse than A. For more information, see the Bibliography.

A.5.2 Iterative Methods for Linear Systems of Equations

An alternative method for solving $A\mathbf{u} = \mathbf{b}$, where A is $n \times n$ matrix, is to use an *iterative* method. For such a method to be more efficient than a direct method the matrix A should fulfill the following requirements:

- it is "cheap" to form $A\mathbf{x}$ (number of flops $\ll n^2$)
- the iterative method is "fast" (number of iterations $\ll n$)

The first property origins from the *structure* of A, while the second property is *algebraic*; the convergence rate depends on the eigenvalues of A. The iterative methods can be divided into *stationary* and *search* methods.

A.5.2.1 Stationary Iterative Methods The general idea with an iterative method is based on *splitting* of the matrix A

$$A = M - N \tag{A.47}$$

The iteration method is then defined as

$$M\mathbf{u}_{k+1} = N\mathbf{u}_k + \mathbf{b} \tag{A.48}$$

where $\mathbf{u}_0$ is a given start vector. The iterations are continued until $\|\mathbf{u}_{k+1} - \mathbf{u}_k\|$ is small enough, where $\|.\|$ is some norm, e.g., the maximum norm or the Euclidean norm. The method is called *stationary* if M and N are *constant* matrices. The iteration formula can also be written

$$\mathbf{u}_{k+1} = G\mathbf{u}_k + M^{-1}\mathbf{b} \tag{A.49}$$

where $G = M^{-1}N$ is the *iteration matrix*. As usual, however, inversion should not be used in numerical methods for solving systems of equations. Let $\mathbf{u}^*$ be the exact solution of $A\mathbf{u} = \mathbf{b}$. The *error* in $\mathbf{u}_k$ is defined as

$$\mathbf{e}_k = \mathbf{u}^* - \mathbf{u}_k \tag{A.50}$$

and satisfies the iteration formula

$$\mathbf{e}_{k+1} = G\mathbf{e}_k \tag{A.51}$$

The *residual* at $\mathbf{u}_k$ is defined as

$$\mathbf{r}_k = \mathbf{b} - A\mathbf{u}_k \tag{A.52}$$

and the relation between the error and the residual is

$$\mathbf{r}_k = A\mathbf{e}_k \tag{A.53}$$

The iteration method converges if $\mathbf{e}_k \to 0$ as $k \to \infty$. The condition for convergence is

$$\rho(G) < 1, \qquad \text{where} \qquad \rho(G) = \max_i |\lambda_i(G)| \tag{A.54}$$

Convergence is fast if $\rho(G) \ll 1$, slow if $\rho(G) \approx 1 - \epsilon$, where $0 < \epsilon \ll 1$, and divergent if $\rho(G) > 1$. $\rho(G)$ is called the spectral radius of G.

If the convergence is fast, i.e., the number of iterations needed to achieve a given accuracy is small, an iterative method is more efficient than a direct method. Observe that in a method based on splitting, as above, the number of flops in each iteration is essentially n_z if the matrix-vector multiplication is performed in a sparse way and M is a matrix that is cheap to "invert."

The splitting can be done in different ways. Classical methods are

- Jacobi's method, $A = D - B$, where D is the diagonal of A and B is the outer diagonal part of A. Hence, the iteration scheme is

$$D\mathbf{u}_{k+1} = B\mathbf{u}_k + \mathbf{b} \tag{A.55}$$

- Gauss–Seidel's method, $A = D - L - U$, where L is strictly lower triangular and U is strictly upper triangular. The iterative scheme is

$$(D - L)\mathbf{u}_{k+1} = U\mathbf{u}_k + \mathbf{b} \tag{A.56}$$

The classical methods above are usually not efficient on problems from ODEs and PDEs. Extensions of these methods to SOR (successive overelaxation) and SSOR (symmetric successive overelaxation SOR) have better convergence properties but are not competitive with modern methods. However, as preconditioners (see later) both Jacobi and SSOR are very useful. For further description of these methods, see the Bibliography.

A.5.2.2 Search Methods for SPD Matrices In this part of the appendix, we restrict the description to systems of equations $A\mathbf{u} = \mathbf{b}$, where A is an SPD matrix. For such matrices, the solution of the linear system of equations is equivalent to the minimization problem

$$\min_{\mathbf{u}} F(\mathbf{u}) \tag{A.57}$$

where the object function $F(\mathbf{u})$ is

$$F(\mathbf{u}) = \frac{1}{2}\mathbf{u}^T A\mathbf{u} - \mathbf{u}^T\mathbf{b} \tag{A.58}$$

In the context of numerical methods for minimization, it is appropriate to design *search methods* to find the minimum point $\mathbf{u}^*$. Such methods are defined by

$$\mathbf{u}_{k+1} = \mathbf{u}_k + \alpha_k \mathbf{d}_k \tag{A.59}$$

where $\mathbf{d}_k$ is the *search direction* and α is the *step length* taken in this direction. The step length is usually chosen so that $F(\mathbf{u})$ is minimized along the search direction $\mathbf{d}_k$. This minimization is performed with respect to $\alpha \geq 0$ and it is easy to show that the optimal value of α is

$$\alpha_k = \frac{\mathbf{r}_k^T \mathbf{d}_k}{\mathbf{d}_k^T A \mathbf{d}_k} \tag{A.60}$$

What differs between methods is the choice of the search direction. Two well-known methods are

- steepest descent (SD) method, where $\mathbf{d}_k = \mathbf{r}_k$, $\mathbf{u}_0 = 0$
- conjugate gradient (CG) method, where $\mathbf{d}_k$ is updated according to

$$\mathbf{d}_{k+1} = \mathbf{r}_{k+1} + \beta_k \mathbf{d}_k, \quad \text{where} \quad \beta_k = \frac{\mathbf{r}_{k+1}^T \mathbf{r}_{k+1}}{\mathbf{r}_k^T \mathbf{r}_k}, \quad \mathbf{u}_0 = 0, \quad \mathbf{d}_0 = \mathbf{r}_0 \tag{A.61}$$

It can be shown that the residuals $\mathbf{r}_0, \mathbf{r}_1, \dots, \mathbf{r}_k$ generated by CG are orthogonal and that both the residuals and the search directions $\mathbf{d}_0, \mathbf{d}_1, \dots, \mathbf{d}_k$ span the so-called Krylov space

$$\mathbf{K}_{k+1} = [\mathbf{b}, A\mathbf{b}, A^2\mathbf{b}, \dots, A^k\mathbf{b}] \tag{A.62}$$

What is most remarkable about the CG method is that, in the absence of rounding errors, it gives the exact solution after less than or equal to n iterations. In a way, therefore, CG could be classified as a direct method. However, often the convergence is so fast that only few iterations are needed to achieve sufficient accuracy.

Both SD and CG are *descent* methods, i.e., $F(\mathbf{u}_{k+1}) < F(\mathbf{u}_k)$ but CG converges faster to $\mathbf{u}^*$ than SD.

It can be shown that the convergence rate for search methods depends on the condition number $\kappa(A)$, defined as

$$\kappa(A) = \frac{\max_i \lambda_i(A)}{\min_i \lambda_i(A)} \tag{A.63}$$

For a large condition number the convergence is slow. The smallest condition number that can be achieved for all matrices A is equal to 1, and in that case the exact solution is obtained after 1 iteration. It can be shown that convergence to a specified accuracy can be expected in $\mathcal{O}(\kappa(A))$ iterations for SD and in $\mathcal{O}\left(\sqrt{\kappa(A)}\right)$ iterations for CG.

If the condition number is large, it can be reduced by *preconditioning*. Introduce the nonsingular $n \times n$ matrix E and the variable substitution $\mathbf{v} = E\mathbf{u}$. The function $F(\mathbf{u})$ in (A.58) is then changed to

$$\tilde{F}(\mathbf{v}) = F(\mathbf{u}) = F(E^{-1}\mathbf{v}) = \frac{1}{2}\mathbf{v}^T\tilde{A}\mathbf{v} - \mathbf{v}^T\tilde{\mathbf{b}}$$

where

$$\tilde{A} = E^{-T}AE^{-1}, \quad \tilde{\mathbf{b}} = E^{-T}\mathbf{b} \tag{A.64}$$

How should E be chosen so that $\kappa(\tilde{A}) \ll \kappa(A)$? In case $E = I$, $\tilde{A} = A$ and nothing has been gained. If $E = L^T$, where L is the Cholesky factor of A, $\tilde{A} = L^{-1}AL^{-T} = L^{-1}LL^TL^{-T} = I$, and $\tilde{A}$ is perfectly conditioned. However, using Cholesky factorization of A to get a preconditioner would be equivalent to using a direct method, which we wanted to avoid. In between $E = I$ and $E = L^T$, however, there are many good working preconditioners C, where $C = E^TE$ is called the *preconditioning matrix*.

A simple choice of the matrix C is $C = \text{diag}(A)$, which is similar to Jacobi's method and $e_{ii} = \sqrt{a_{ii}}$.

Another often-used preconditioner is the *incomplete Cholesky* factorization, $\tilde{L}\tilde{L}^T$, where $\tilde{L}$ is a modified Cholesky factor, allowed to have nonzeros only in positions where A has nonzeros. Hence, $E = \tilde{L}^T$ and $C = \tilde{L}\tilde{L}^T$.

To make the preconditioning methods efficient, it is not possible to form the $\tilde{A}$-matrix in (A.64) explicitly before the search method is started. This would involve inverting E and, as usual, the inverse should never be formed in numerical operations, but, e.g., $\tilde{\mathbf{b}}$ should be computed by solving $E^T\tilde{\mathbf{b}} = \mathbf{b}$.

Below it is shown how SD with preconditioning is designed efficiently. The SD iterations for the $\mathbf{v}_k$ sequence (A.64) is

$$\mathbf{v}_{k+1} = \mathbf{v}_k + \alpha_k(\tilde{\mathbf{b}} - \tilde{A}\mathbf{v}_k)$$

Express this iteration as a sequence in the $\mathbf{u}_k$ values

$$E\mathbf{u}_{k+1} = E\mathbf{u}_k + \alpha_k(\tilde{\mathbf{b}} - \tilde{A}E\mathbf{u}_k)$$

Multiply by E^{-1}

$$\mathbf{u}_{k+1} = \mathbf{u}_k + \alpha_k E^{-1}(E^{-T}\mathbf{b} - E^{-T}AE^{-1}E\mathbf{u}_k)$$

which can be written

$$\mathbf{u}_{k+1} = \mathbf{u}_k + \alpha_k C^{-1}(\mathbf{b} - A\mathbf{u}_k)$$

where $C = E^TE$ is the *preconditioner*.

In the discretization of PDEs on a fine grid, the systems of equations are huge. However, if the same problem is solved on a course grid, we have a small system to solve. The idea of switching between coarser and finer grids has developed to a class of methods called *multigrid* methods, where the number of flops needed to solve the system is $\mathcal{O}(n)$.

A.5.2.3 Search Methods for General Matrices When the matrix A is sparse but unsymmetric and/or semidefinite, there is no suitable corresponding minimization problem. If the system of equations is multiplied by A^T, we obtain a SPD system

$$A^T A\mathbf{u} = A^T \mathbf{b}$$

This system, however, has an even worse condition number, since

$$\kappa(A^T A) = \kappa(A)^2$$

and is therefore not always sufficiently effective. This method is known as the *CGN*, CG to normal equations.

An alternative widely used method is GMRES, the generalized minimum residual method. Like CG it generates a sequence of search directions $\mathbf{d}_k$ but requires all previous search vectors $\mathbf{d}_1, \mathbf{d}_2, \dots, \mathbf{d}_k$ to form $\mathbf{d}_{k+1}$ in contrast to CG, where $\mathbf{d}_{k+1}$ is formed from $\mathbf{d}_{k-1}$ and $\mathbf{d}_k$ only.

For more information about these methods, see the Bibliography.

A.6 SOME RESULTS FOR FOURIER TRANSFORMS

Fourier analysis is the common term for Fourier transforms and Fourier series, named after the French mathematician Joseph Fourier, active in the beginning of the 19th century.

Assume that $u(x)$ is defined on the entire real axis. The *continuous Fourier transform* $\hat{u}(\omega)$ of $u(x)$ is defined as the integral

$$\hat{u}(\omega) = \frac{1}{\sqrt{2\pi}} \int_{-\infty}^{+\infty} u(x) e^{-i\omega x} dx$$

The inverse transform is

$$u(x) = \frac{1}{\sqrt{2\pi}} \int_{-\infty}^{+\infty} \hat{u}(\omega) e^{i\omega x} d\omega$$

Here $\omega = 2\pi f$, where ω is the angular frequency and f the frequency. The second formula can be interpreted as a superposition of waves $e^{i\omega x}$ with the corresponding amplitudes $\hat{u}(\omega)$.

If the function $u(x)$ is known only at discrete equally spaced points $x_j = jh, j = \dots, -2, -1, 0, 1, 2, \dots$ the *discrete Fourier transform* is defined as

$$\hat{u}(\omega) = \frac{1}{\sqrt{2\pi}} \sum_{j=-\infty}^{j=+\infty} u(x_j) e^{-i\omega x_j} h$$

with the inverse

$$u(x_j) = \frac{1}{\sqrt{2\pi}} \int_{-\pi/h}^{+\pi/h} \hat{u}(\omega) e^{i\omega x_j} d\omega$$

The transforms exist if the original function $u(x)$ is square integrable, i.e., for the continuous and discrete case, respectively,

$$\int_{-\infty}^{+\infty} |u(x)|^2 dx < \infty$$

$$\sum_{-\infty}^{+\infty} |u(x_j)|^2 < \infty$$

The following relations between the original function and its transform are called *Parseval's equalities*:

$$\int_{-\infty}^{+\infty} |u(x)|^2 dx = \int_{-\infty}^{+\infty} |\hat{u}(\omega)|^2 d\omega$$

$$\sum_{-\infty}^{+\infty} |u(x_j)|^2 h = \int_{-\pi/h}^{+\pi/h} |\hat{u}(\omega)|^2 d\omega$$

When the Fourier transform is applied to the heat equation

$$u_t = \kappa u_{xx}, \quad u(x,0) = u_0(x), \quad -\infty < x < +\infty$$

we obtain the following ODE for the continuous Fourier transform $\hat{u}(\omega, t)$ of the solution $u(x, t)$:

$$\frac{d\hat{u}}{dt} = -\kappa\omega^2 \hat{u}, \quad \hat{u}(\omega, 0) = \hat{u}_0(\omega)$$

with the solution

$$\hat{u}(\omega, t) = \mathrm{e}^{-\kappa\omega^2 t} \hat{u}_0(\omega)$$

Applying the inverse transform

$$u(x,t) = \frac{1}{\sqrt{2\pi}} \int_{-\infty}^{+\infty} e^{-\kappa\omega^2 t} \hat{u}_0(\omega) e^{i\omega t} d\omega$$

we see that higher frequencies are quickly damped out as time increases.

When the FTCS method is applied to the heat equation, a sequence of values $u_{j,k}$ values are obtained, where $u_{j,k} \approx u(x_j, t_k)$. Applying the inverse transform gives

$$u_{j,k} = \frac{1}{\sqrt{2\pi}} \int_{-\pi/h}^{\pi/h} \hat{u}(\omega, t_k) e^{i\omega x_j} d\omega$$

This can be interpreted as a superposition of waves $e^{i\omega x_j}$ with amplitude $\hat{u}(\omega, t_k)$. Hence, the following ansatz for the discretized solution is motivated:

$$u_{j,k} = q_k e^{i\omega x_j}$$

Inserting this ansatz into the FTCS method gives

$$q_{k+1} e^{i\omega x_j} = q_k e^{i\omega x_j} + q_k \frac{h_t}{h_x^2}(e^{i\omega(x_j+h_x)} - 2e^{i\omega x_j} + e^{i\omega(x_j-h_x)})$$

$$q_{k+1} = q_k \left(1 - \frac{2h_t}{h_x^2}(1 + \cos(\omega h_x))\right)$$

The second factor in this relation can be seen as an amplification factor. The q_k sequence is bounded this factor fulfills the inequality

$$-1 \leq 1 - \frac{2h_t}{h_x^2}(1 + \cos(\omega h_x)) \leq 1$$

Since $-1 \leq \cos(\omega h_x) \leq 1$ this inequality leads to the stability condition

$$\frac{h_t}{h_x^2} \leq \frac{1}{2}$$

This stability analysis is called *von Neumann analysis* and gives the same result for the heat equation as the *eigenvalue analysis* in Chapter 6. This is not true, however, for hyperbolic PDEs.

BIBLIOGRAPHY

1. L. Trefethen, D. Bau, “Numerical Linear Algebra,” SIAM, 1997
2. G. Dahlquist, Å. Björck, “Numerical Methods,” Chapter 5, Dover, 2003

APPENDIX B

SOFTWARE FOR SCIENTIFIC COMPUTING

During the last decades, a lot of software has been developed for modeling and simulation of processes in science and engineering. The software can roughly be divided into the following three categories:

1. General programming languages suitable for construction of programs with optimal performance behavior on various computer architectures, e.g., vector or parallel architecture. Examples are Fortran90, C, and C++. Many numerical subroutine libraries written in these languages can be found as public domain code on the net, e.g., Netlib. There are also several textbooks containing subroutines written in Fortran, Pascal, C, C++, Python, etc.
2. Programming languages specially designed for general numerical calculations and symbolic mathematics. Examples of such tools are MATLAB®, Simulink, Maple, and Mathematica. Within this category can also be placed COMSOL MULTIPHYSICS®, which is a gui-based (graphical user interface) environment for simulation of processes modeled by partial differential equations (PDEs) from very many scientific and engineering applications.
3. Toolboxes for various applications often implemented for one specific area, e.g., fluid dynamics and designed in a way of communicating indata and outdata that is familiar for the user in the specific application area.

As was mentioned in the Preface, this textbook is based on a course where the labs and exercises are solved with MATLAB and COMSOL MULTIPHYSICS. Therefore, these software products will be given special attention.

Introduction to Computation and Modeling for Differential Equations, Second Edition. Lennart Edsberg.
© 2016 John Wiley & Sons, Inc. Published 2016 by John Wiley & Sons, Inc.

B.1 MATLAB

MATLAB is a programming language designed for scientific and engineering computation. The original MATrix LABoratory was developed by Prof. Cleve Moler some 30 years ago for interactive computer labs in linear algebra. In the middle of the 1980s, MATLAB became a commercial product manufactured by the US company MathWorks, founded by Cleve Moler and since then a continued success story, now having more than 1000 employees. Products from the MATLAB and Simulink family are spread all over the world, widely used by universities, industries, and high-tech companies.

One of the strengths of MATLAB is the fact that there is a large subroutine library with all kinds of high-quality numerical functions that can be used in your own programs. Another strength is the easy-to-use graphical routines for visualization and animation in 2D and 3D.

What further makes MATLAB a suitable programming language for scientific computing is the built-in vector/matrix facilities which make it easy to write compact and efficient programs without for-loops. Indeed, MATLAB is very convenient for making numerical experiments involving, e.g.,

- variation of stepsize/grids or tolerance parameters
- change of methods having different order of accuracy
- questions regarding numerical stability
- different algorithms, e.g., direct or iterative methods for solving linear systems of equations
- robustness of a method

MATLAB is good for solving moderately sized problems. For large problems requiring fast number crunching, it is more efficient to use a traditional programming language such as Fortran90 or C.

We will not give a detailed description of MATLAB in this book, but we show here some MATLAB codes, which can be used as example programs in the construction of similar programs. For a detailed description of MATLAB we refer the reader to some textbooks and manuals accessible on the Internet (see the Bibliography).

The templates presented here show solutions of some model problems taken from the chapters of this book.

B.1.1 Chapter 3: IVPs

In the MATLAB library, there are several functions solving initial value problems (IVPs). To mention a few, there are `ode23`, `ode45`, and `ode113` for nonstiff problems and `ode23s` and `ode15s` for stiff problems.

B.1.1.1 Runge–Kuttas Fourth-Order Method, Constant Stepsize The following program solves Van der Pol's equation:

```
%Program rkstep.m for Runge-Kutta's 4th order method
%Model problem is Van der Pol's equation, stored
%in fvdp.m
global epsilon          %parameter used in both script
                        %and fvdp
epsilon=1;              %parameter in VdPs equation
t0=0;tend=10;           %time interval for the solution
u0=[1 0]';              %initial value
N=100;                  %number of steps
h=(tend-t0)/N;          %stepsize
result=[u0'];           %collecting solution values
time=[t0];              %collecting time points
u=u0;t=t0;              %initialization of RK's method
for k=1:N
    k1=fvdp(t,u);
    k2=fvdp(t+h/2,u+h*k1/2);
    k3=fvdp(t+h/2,u+h*k2/2);
    k4=fvdp(t+h,u+h*k3);
    u=u+h*(k1+2*k2+2*k3+k4)/6;
    t=t+h;
    result=[result;u'];
    time=[time;t];
end
plot(time,result(:,1),'o',time,result(:,2),'+')
title('Van der Pols equation')
xlabel('time t')
ylabel('u (o) and du/dt (+)')
grid
```

The function file `fvdp.m` containing the right hand side of Van der Pol's equation:

```
%right hand side of Van der Pol's equation
function rhs=fvdp(t,u)
global epsilon          %parameter in VdPs equation
rhs=[u(2);
    -epsilon*(u(1)*u(1)-1)*u(2)-u(1)];
```

B.1.1.2 MATLAB's ODE Function ode45 The same problem solved with one of MATLAB's ordinary differential equation (ODE) functions using adaptive stepsize control:

```
%Program showing the use of Matlabs ode45 function
%Model problem is Van der Pol's equation, stored in
%fvdp.m
global epsilon
epsilon=1;          %parameter in VdPs equation
t0=0;tend=10;        \%time interval for the solution
u0=[1 0]';           \%initial value
[time,result]=ode45(@fvdp,[t0,tend],u0);
plot(time,result(:,1),'o',time,result(:,2),'+')
title('Van der Pols equation')
xlabel('time t')
ylabel('u (o) and du/dt (+)')
grid
```

B.1.1.3 Change of Tolerance Parameters When the default tolerance in a MATLAB ODE function is not enough, the relative and absolute tolerance can be changed from the preset values 10^{-3} and 10^{-6}, respectively,

```
%Example 1.2 in chapter 1. The vibration equation
%demo of tolerance parameter settings
global m c k f0 w
m=1;c=10;k=1e3;f0=1e-4;w=40;
t0=0;tend=5;
u0=[0 0]';
[time,result]=ode45(@fvib,[t0,tend],u0);
                            %not enough accuracy
options=odeset('RelTol',1e-6,'AbsTol',1e-9);
                            %set tolerances
[t,res]=ode45(@fvib,[t0,tend],u0,options);
                            %accurate solution
plot(time,result(:,1),'.',t,res(:,1))
title('Accurate and bad solution of the vibration
                                     equuation')
xlabel('time t')
ylabel('vibration y(t)')
```

The corresponding right hand side function:

```
%Right hand side of vibration equation in chapter 1
function rhs=fvib(t,u)
global m c k f0 w
rhs=[u(2);
     -(c/m)*u(2)-(k/m)*u(1)+(f0/m)*sin(w*t)];
```

B.1.2 Chapter 4: BVPs

Solution of the model problem in Example 4.8:

```
%Solution of the model problem BVP with FDM,
%constant stepsize h
%-u"=f(x), u(0)=u(1)=0, f(x)=sin(pi*x)
a=0;b=1;                 %interval bounds
N=63;                    %N+1=number of steps
h=(b-a)/(N+1);           %stepsize
xp=a:h:b;                %DG, discretize interval to grid
x=a+h:h:b-h;             %inner points generated
f=h*h*[sin(pi*x)]';      %right hand side generated
A=zeros(N,N);            %start generation of the full
                         %matrix A
for i=1:N-1
    A(i,i)=2;
    A(i,i+1)=-1;
    A(i+1,i)=-1;
end
A(N,N)=2;                %full matrix A generated
u=A\f;                   %solution of linear system of
                         %equations
```

```
up=[0,u',0];            %add the BCs to the solution
                        %vector
plot(xp,up)
title('Solution of -u"=sin(\pi *x), u(0)=u(1)=0')
xlabel('x')
ylabel('u')
grid
```

Example of a sparse version of the A matrix generation:

```
e=ones(N,1);
a=spdiags([-e 2*e -e],-1:1,N,N);
                              %sparse storage of A
u=a\f;                        %sparse solution of Au=f
```

B.1.3 Chapter 6: Parabolic PDEs

Solution of the parabolic model problem with the MoL:

```
%Solution of u_t=u_xx, u(x,0)=sin(pi*x),
%u(0,t)=u(1,t)=0
%using the MoL. Discretized system stored in fpar.m
global N hx       %discretization parameters
N=10              %number of inner x-points
hx=1/(N+1);       %stepsize in x-direction
x=hx:hx:1-hx;     %inner points
u0=[sin(pi*x)]'; %initial values
t0=0;tend=0.25;   %time interval
[time,result]=ode23s(@fpar,[t0,tend],u0);
                                  %stiff ode-solver
resultp=[zeros(size(time)) result zeros(size(time))]
                                  %add BCs
xp=0:hx:1;        %add boundary points
mesh(xp,time,resultp)
title('Solution of u_t=u_xx, u(x,0)=sin(\pi *x), ...
        u(0,t)=u(1,t)=0 using MoL')
xlabel('space variable x')
ylabel('time variable t')
zlabel('solution variable u(x,t)')
```

The MoL discretized PDE is stored in the MATLAB function file `fpar.m`:

```
%Right hand side of the MoL discretized model
%problem
function rhs=fpar(t,u);
global N hx       %discretization parameters
rhs=zeros(N,1);
rhs(1)=(-2*u(1)+u(2))/(hx*hx);
rhs(2:N-1)=(u(1:N-2)-2*u(2:N-1)+u(3:N))/(hx*hx);
rhs(N)=(u(N-1)-2*u(N))/(hx*hx);
```

B.1.4 Chapter 7: Elliptic PDEs

Solution of the elliptic model problem with the finite difference method (FDM):

```
%Numerical solution of the elliptic PDE Laplace u=1
%on the quadrangle Omega={(x,y),0<=x<=1,0<=y<=1}
%BC is u=0 on the boundary of Omega
N=100;                      %number of inner x and
                            %y-points
hx=1/(N+1); hy=1/(N+1);     %stepsize in x- and
                            %y-direction
x=0:hx:1; y=0:hy:1;         %gridpoints
d=4*ones(N*N,1);            %main diagonal of A
d_1=-ones(N*N,1);
d1=-ones(N*N,1);
for i=N:N:N*N-1             %sub- and superdiagonals
    d_1(i)=0;
    d1(i+1)=0;
end
d_2=-ones(N*N,1);           %subdiagonal from unit matrix
d2=-ones(N*N,1);            %superdiagonal
a=spdiags([d_2 d_1 d d1 d2],[-N,-1:1,N],N*N,N*N);
                            %sparse matrix A generated
f=ones(N*N,1);              %right hand side
uinner=a\f;                 %solution at inner points
Uinner=zeros(N,N);
for i=1:N                   %inner solution as a matrix
    Uinner(i,1:N)=uinner((i-1)*N+[1:N]);
end
U=zeros(N+2,N+2);            %add BCs to inner solution
for i=1:N
    U(i+1,2:N+1)=Uinner(i,1:N);
end
mesh(x,y,U)                  %3D plot of the solution
title('Solution of \Laplace u = 1 on the unit
                                     square') ...
xlabel('x')
ylabel('y')
zlabel('u(x,y)')
```

B.1.5 Chapter 8: Hyperbolic PDEs

Solution of the hyperbolic PDE model problem:

```
%Solution of the hyperbolic model problem
%u_t+u_x=0, u(x,0)=0, 0<x<=1, u(0,t)=1, t>=0
%Upwind method is used
clear,clf,hold off %clear variables and graph
N=100;              %number of gridpoints
hx=1/N;             %stepsize in x-direction
x=hx:hx:1;          %x-grid, inner points
xp=[0 x];           %x-grid plus boundary point
ht=0.01;            %timestep, try also 0.008 and 0.011
```

```
sigma=ht/hx;       %Courant number
u0=zeros(N,1);     %IV at inner points
up=[1;u0];         %IV+BV
plot(xp,up)
title('Initial state for the advection equation')
xlabel('x')
ylabel('u(x,t)')
u(1:N,1)=u0;       %initialization of upwind
for k=1:100        %take 100 timesteps with upwind
    u(1,k+1)=(1-sigma)*u(1,k)+sigma*1;%*1 is *BC
    u(2:N,k+1)=(1-sigma)*u(2:N,k)+sigma*u(1:N-1,k);
    up=[1;u(:,k+1)];
    plot(xp,up) %plot the front for each timestep
    hold on      %keep the previous plots
    pause(0.1)
end
```

B.2 COMSOL MULTIPHYSICS

While MATLAB is good for numerical computation of moderately sized problems, COMSOL MULTIPHYSICS is designed for model experimentation of problems from various applications. Its development started some twenty years ago by a Swedish provider company called COMSOL®. COMSOL was founded by Dr. H.C. Svante Littmarck and Farhad Saeidi and is another example of a success story within scientific computing. COMSOL has now around 420 employees and branch offices in several countries, including the United States.

COMSOL MULTIPHYSICS is an interactive program for simulation of time-dependent or stationary scientific/engineering PDE models in 1D, 2D, or 3D. The user and the program communicates with the help of a graphical user interface (gui), which provides advanced tools for geometric modeling. The models are based on three forms of PDEs, the coefficient form (for linear or quasi-linear PDEs), the general form (for nonlinear PDEs)

$$e\frac{\partial^2 u}{\partial t^2} + d\frac{\partial u}{\partial t} + \nabla \cdot (-c\nabla u - \alpha u + \gamma) + \beta \cdot \nabla u + au = f$$

and

$$d\frac{\partial u}{\partial t} + \nabla \cdot \Gamma = F$$

and the weak form (see section 7.4.2). Simulations can be done in one single application, e.g., heat transfer, and also in coupled models where several applications are mixed in an almost arbitrary way, e.g., heat conduction and electrical current strength. There are several physics interfaces that can be studied alone or in combination with others, e.g.,

- equation-base modeling
- chemical engineering

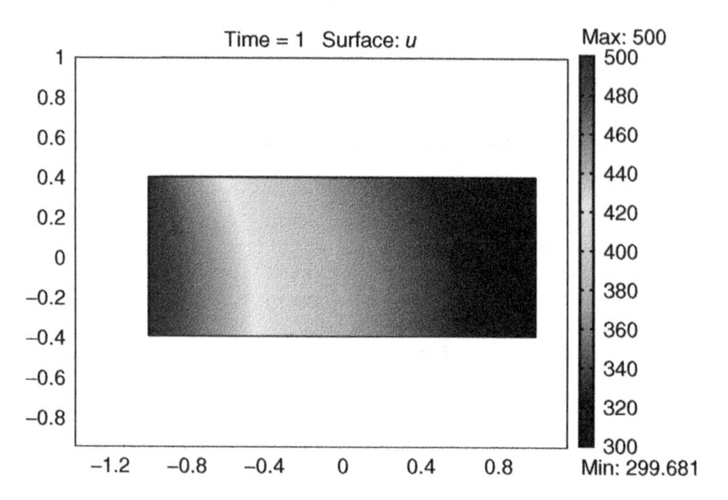

Figure B.1 Solution of heat conduction problem with COMSOL Multiphysics

- electromagnetics
- heat transfer
- structural mechanics
- acoustics

The solution process of a problem proceeds in short in the following way:

1. choose application mode and space dimensionality 1D, 2D, or 3D
2. generate the geometry
3. enter global parameter values and initial values
4. give the parameters of the PDE
5. give the boundary conditions
6. generate the grid
7. compute and solve
8. inspect the solution with the postprocessing facilities

After having completed the steps (1) – (8), you can go back to any step earlier to modify the model see Figure B.1 for a heat conduction simulation.

The underlying numerical method is based on the finite element method. For time-dependent problems, finite element method (FEM) is combined with a stiff ODE solver. The discretization is done with a grid generator that automatically produces a grid whose appearance depends on the geometry of the problem. By option, the grid can also be generated adaptively depending on the solution.

For more examples, see Ref. 1 and the webpage of this textbook http://www.csc.kth.se/~edsberg.

BIBLIOGRAPHY AND RESOURCES

1. COMSOL MULTIPHYSICS®, http://www.comsol.com
2. Introduction to Matlab, http//www.maths.dundee.ac.uk/ ftp/na-reports/MatlabNotes.pdf
3. Matlab, http://www.mathworks.com
4. D. J. Higham and N. J. Higham, Matlab Guide, 2000, SIAM
5. Netlib, http://www.netlib.org/
6. Numerical Recipes, W. Press, B. Flannery, S. Teukolsky, W. Vetterling, Cambridge University Press, 1986

APPENDIX C

COMPUTER EXERCISES TO SUPPORT THE CHAPTERS

C.1 COMPUTER LAB 1 SUPPORTING CHAPTER 2

C.1.1 ODE Systems of LCC Type and Stability

In this exercise, the following problems are treated:

- linear ordinary differential equation (ODE) systems with constant coefficients (LCC systems)
- stability of trajectories and critical points

LCC systems can be solved analytically (see Chapter 2). Hence, no numerical method is needed, so the problem class is suitable for a refresher in MATLAB programming.

We start by illustrating a MATLAB program for solution of an LCC problem. Given the following second-order differential equation:

$$\frac{d^2y}{dt^2} + 2\frac{dy}{dt} + 5y = 0, \quad y(0) = 0, \; \frac{dy}{dt}(0) = 1$$

compute the solution of this ODE problem on a suitable t interval. First rewrite this scalar problem to the standard form, i.e., as a system of first-order ODEs. This ODE system is of LCC type

$$\frac{d\mathbf{u}}{dt} = A\mathbf{u}, \quad \mathbf{u}(0) = \mathbf{u}_0$$

which is solved analytically with the `expm`-function. Input data to the MATLAB program consists of the matrix A, the initial vector $\mathbf{u}(0)$, and grid point data h and N for the points $t_i = ih, i = 0, 1, 2, \dots, N$, where the solution shall be computed and plotted. The output from the program is the graph of the solution at these grid

Introduction to Computation and Modeling for Differential Equations, Second Edition. Lennart Edsberg.
© 2016 John Wiley & Sons, Inc. Published 2016 by John Wiley & Sons, Inc.

points. In the MATLAB program below, the solution $\mathbf{u}(t)$ is stored in a resulting matrix `result` of size $(N+1)\times 2$. Each row in `result` corresponds to the solution vector at a grid point t_i. The first row is the initial vector $\mathbf{u}(0)$

```
A=[0 1;-5 -2];
u0=[0 1]';t0=0;
h=0.1;N=60;
result=[u0'];time=[t0];
for k=1:N
  t=k*h;
  u=expm(A*t)*u0;
  result=[result; u'];
  time=[time; t];
end
plot(time,result)
```

This MATLAB program plots two curves. The first column in the `result`-matrix corresponds to the solution $y(t)$, while the second column corresponds to the derivative of the solution $z(t) = \frac{dy}{dt}(t)$. The initial values of y and z are different in this case so it is easy to see which curve corresponds to which variable. Otherwise you can use different plot symbols for different curves.

Observe that the time step h must be chosen with some care if the plotted curve is to be smooth. When $h = 0.1$ is used as stepsize, the resulting curve is smooth, but with $h = 1$, however, the line segments building up the curves are clearly seen and the curve gets rough.

With too small time steps, on the other hand, say $h = 0.001$, the computation of the `result`-matrix will take an unreasonably long time. Hence, it is advisable to work out the grid point spacing interactively. For other problems, it may also be better to use `loglog`, `semilogx`, or `semilogy` instead of `plot` to display the important qualitative features of solution curves.

C.1.1.1 Solution of ODE Systems with Constant Coefficients

C.1.1.1.1 Electric Circuit. Given the following simple electric circuit with a voltage source of size E, a resistance R, an inductance L, and a capacitance C. The components are coupled in series (see Figure C.1). Assume that the circuit is first at rest. The switch is activated at $t = 0$ and a current i starts to go through the circuit. Introduce the variable q defined by the relation $\dot{q} = i$, and the following differential equation can be set up:

$$L\ddot{q} + R\dot{q} + \frac{1}{C}q = E, \quad q(0) = 0, \quad \dot{q}(0) = 0$$

Rewrite this scalar ODE as a system of first-order ODEs and then write a MATLAB program to solve the problem for the following parameter values: $E = 10, L = C = 0.1$, $R = 0.01, 0.1, 1, 10$. Observe that this ODE system is *inhomogeneous*.

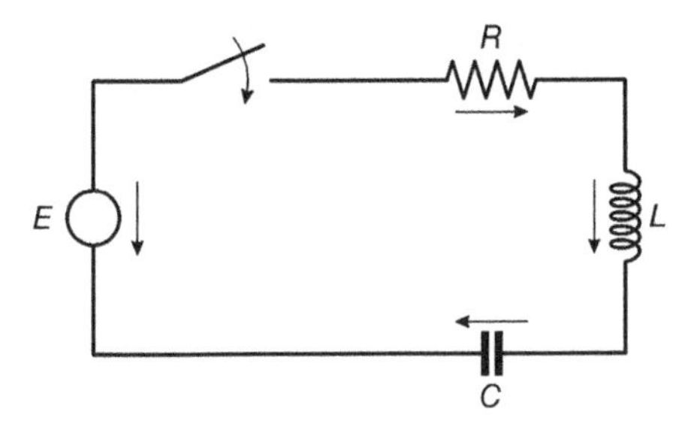

Figure C.1 Simple electric circuit

Plot the solutions $i(t)$ on suitable time intervals and with suitable time steps. Use the `subplot`-command to obtain the four graphs in the same figure.

C.1.1.2 Stability of ODE Systems and Equilibrium Points

C.1.1.2.1 Stability of the Solutions of an ODE System of LCC Type. In many technical applications where a differential equation is used to model a dynamic process, it is important to see how the solution curves change as a parameter in the model is changed. For a control system, it is of great interest to investigate how the stability properties change as a parameter is changed continuously.

Given the following third-order differential equation:

$$\frac{d^3y}{dt^3} + 3\frac{d^2y}{dt^2} + 2\frac{dy}{dt} + Ky = 0, \quad y(0) = 1, \quad \frac{dy}{dt}(0) = 1, \quad \frac{d^2y}{dt^2}(0) = 1$$

Investigate how the solution $y(t)$ behaves for values $K \geq 0$.

Rewrite this third-order ODE as a system of first-order ODEs. We then obtain a system

$$\frac{d\mathbf{u}}{dt} = A(K)\mathbf{u}, \quad \mathbf{u}(0) = (1, 1, 1)^T$$

Write a MATLAB program showing the solutions $y(t)$ for $K = 0, 1, 4, 8$. Use `subplot` to plot the four graphs in the same figure. Estimate from these plots an approximate value of K for which the solution becomes unstable.

The stability properties can also be shown in a so-called root locus, often used in control theory to visualize the stability properties for a regulator modeled by an ODE of LCC type. A root locus is a graph showing the paths of the eigenvalues in the complex plane as a parameter varies. Assume that all eigenvalues start in the left half plane for say $K = 0$, where we have a stable system. When K then increases, the eigenvalues will follow continuous paths and for a certain value of K a few eigenvalues eventually enter the right half plane and the system becomes unstable.

Write a MATLAB program that plots the root locus when $0 \leq K \leq 10$. Compute with two decimal accuracy the smallest value of K that gives an unstable system.

C.1.1.2.2 Stability of the Critical Points of a Nonlinear ODE System. Given the following system of nonlinear ordinary differential equations:

$$\frac{du_1}{dt} = 5u_1 + 4u_2 - u_1\, u_3$$
$$\frac{du_2}{dt} = u_1 + 4u_2 - u_2\, u_3$$
$$\frac{du_3}{dt} = u_1^2 + u_2^2 - 89$$

Write a MATLAB program where all critical points of the ODE system are computed. Also compute which of these critical points are stable.

Hint: use Newton's method to compute the critical points with at least five significant digits. There are four critical points and they are situated in the neighborhood of the following points in R^3: $(8, -5, 2)$, $(-8, 5, 2)$, $(9, 3, 7)$, and $(-9, -3, 7)$.

C.2 COMPUTER LAB 2 SUPPORTING CHAPTER 3

C.2.1 Numerical Solution of Initial Value Problems

In this lab, initial value problems (IVPs) are solved numerically and the following items are studied:

- accuracy and stability
- constant stepsize and adaptive (variable) stepsize
- stiff and nonstiff problems
- parameter study of the solutions of a system of ODEs

C.2.1.1 Accuracy of a Runge–Kutta Method In Chapter 3, different Runge–Kutta methods are presented. Make a numerical experiment to find the order of accuracy of the following RK method:

$$u_k = u_{k-1} + \frac{h}{6}(K_1 + K_2 + 4K_3), \quad t_k = t_{k-1} + h, \quad k = 1, 2, \ldots, N$$
$$K_1 = f(t_{k-1}, u_{k-1})$$
$$K_2 = f(t_{k-1} + h, u_{k-1} + hK_1)$$
$$K_3 = f(t_{k-1} + h/2, u_{k-1} + hK_1/4 + hK_2/4)$$

Implement the method on Van der Pol's differential equation

$$\frac{d^2y}{dt^2} + (y^2 - 1)\frac{dy}{dt} + y = 0, \quad y(0) = 1, \frac{dy}{dt}(0) = 0, \quad \epsilon = 1, \quad t \in [0, 1]$$

Run the problem with constant stepsizes using $N = 10$, 20, 40, 80, 160, and 320 steps in the t interval [0, 1]. Estimate the error at $t = 1$ by computation of

$e_N = y_N - y(1), N = 10, 20, 40, 80, 160$. Since $y(1)$ is not known exactly, use the approximation $y(1) \approx y_{Nmax}$, where $Nmax = 320$. Make a *loglog*-plot of $|e_N|$ as a function of h and estimate the order of accuracy from the graph.

Hint 1: Treat the problem as a system on *vector form*, both when you rewrite the second-order differential equation to a system of two first-order ODEs and when you program the method.

Hint 2: Be careful to take the correct number of steps to reach $t = 1$. If you get the answer $order = 1$, there is some mistake in your MATLAB code!

C.2.1.2 Stability Investigation of a Runge–Kutta Method The stability of a numerical method for IVPs is important when we want to solve *stiff* problems. The following ODE system modeling the kinetics of a set of three reactions, known as Robertson's problem, is studied here:

$$A \rightarrow B \quad (k_1) \qquad B + C \rightarrow A + C \quad (k_2) \qquad 2B \rightarrow B + C \quad (k_3)$$

In the reactions above k_1, k_2, and k_3 denote the *rate constants* of the three reactions. The following set of ODEs describe the evolution of (scaled) concentrations of A, B, and C as a function of time t:

$$\frac{dx_1}{dt} = -k_1\ x_1 + k_2\ x_2\ x_3, \qquad x_1(0) = 1$$

$$\frac{dx_2}{dt} = k_1\ x_1 - k_2\ x_2\ x_3 - k_3\ x_2^2, \qquad x_2(0) = 0$$

$$\frac{dx_3}{dt} = k_3\ x_2^2, \qquad x_3(0) = 0$$

The rate constants have the values: $k_1 = 0.04$, $k_2 = 10^4$, $k_3 = 3 \cdot 10^7$.

C.2.1.2.1 Constant Stepsize Experiment. If Robertson's problem is solved with an explicit method, the stepsize has to be very small to avoid numerical instability. Use the Runge–Kutta method given on Robertson's problem when the t interval is [0, 1]. Run the problem with constant stepsizes corresponding to $N = 125, 250, 500, 1000, 2000$ steps and find the smallest number of steps (from the 5 given) needed to obtain a stable solution. Plot the solution trajectory in a *loglog*-diagram for the solution computed with the smallest step.

C.2.1.2.2 Adaptive Stepsize Experiment Using MATLAB Functions. There are several IVP solvers in MATLAB. Use the command `»help funfun` to see which are available. To get more information about one of them, say `ode23`, give the command `»help ode23`. In order to control, e.g., accuracy parameters you also need to read about the function `odeset`. When the problem is stiff, you need a stiff IVP solver, e.g., `ode23s`.

There are several ways to find demo examples in Matlab. If you give the command `»demo` and follow the path MATLAB, Numerics, Differential Equations, you find a number of examples to look at, e.g., `vdpode` and `rigidode`. You can run the demo programs and see the graphical output and you can also see the MATLAB code.

Make the following numerical experiments on Robertson's problem:

- Use the nonstiff IVP solver `ode23` on the t interval [0, 1] for different relative tolerances: $RelTol = 10^{-3}, 10^{-4}, 10^{-5}, 10^{-6}$ and record the number of steps taken by `ode23`. Make a graph of the stepsize h as function of t for one of the tolerances.
- Run the stiff IVP solver `ode23s` on the t interval [0, 1000] for the same relative tolerances as above and record the number of steps taken by `ode23s`. Make a graph of the stepsize h as function of t for one of the tolerances.

C.2.1.3 Parameter Study of Solutions of an ODE System Make a parameter study for the following problems taken from applications. Choose a method (order of accuracy must be at least two) yourself. Present the result graphically in a suitable way. Think about the following possibilities and choose what you think is best:

- one or several graphs (using `subplot`) in the figure window?
- linear or logarithmic scales?
- in the graphs: `title`, `x-label`, `y-label`

C.2.1.3.1 Problem 1: Particle Flow Past a Cylinder. A long cylinder with radius $R = 2$ is placed in an incompressible fluid streaming in the direction of the positive x-axis. The axis of the cylinder is perpendicular to the direction of the flow. The position $(x(t), y(t))$ of a flow particle at time t is determined by the start position $(x(0), y(0))$ and the ODE system:

$$\frac{dx}{dt} = 1 - \frac{R^2(x^2 - y^2)}{(x^2 + y^2)^2}, \quad \frac{dy}{dt} = -\frac{2xyR^2}{(x^2 + y^2)^2}$$

At $t = 0$, there are four flow particles at $x = -4$ with the y-positions 0.2, 0.6, 1.0, and 1.6. Compute and make a graph of the flow curves of the particles in the t interval [0, 10]. Use `axis equal` in the graph!

C.2.1.3.2 Problem 2: Motion of a Particle. A particle is thrown from the position (0, 1.5) with an elevation angle α and the velocity $v_0 = 20$. The trajectory of the particle depends on α, the air resistance coefficient k, and the ODE system

$$\frac{d^2x}{dt^2} = -k\frac{dx}{dt}\sqrt{\left(\frac{dx}{dt}\right)^2 + \left(\frac{dy}{dt}\right)^2}, \quad x(0) = 0, \quad \frac{dx}{dt}(0) = 20\cos(\alpha)$$

$$\frac{d^2y}{dt^2} = -9.81 - k\left|\frac{dy}{dt}\right|\sqrt{\left(\frac{dx}{dt}\right)^2 + \left(\frac{dy}{dt}\right)^2}, \quad y(0) = 1.5, \quad \frac{dy}{dt}(0) = 20\sin(\alpha)$$

For two different values of k, say $k = 0.020$ and 0.065, the solution trajectories for $\alpha = 30°, 45°$, and $60°$. For the graphical presentation, observe that the model is valid only until the particle touches the ground, i.e., it is valid only while $y \geq 0$. The graph should show the motion in a xy-coordinate system with t as a parameter.

C.3 COMPUTER LAB 3 SUPPORTING CHAPTER 4

C.3.1 Numerical Solution of a Boundary Value Problem

Consider a long pipe of length L with small cylindrical cross section (Figure C.2). In the pipe, there is a fluid heated by an electric coil. The heat is spreading along the pipe and the temperature $T(z)$ at steady state is determined by the diffusion–convection ODE

$$-\frac{d}{dz}\left(\kappa\frac{dT}{dz}\right) + v\rho C\frac{dT}{dz} = Q(z) \qquad (*)$$

where all parameters are constant: κ is the heat conduction coefficient, v the fluid velocity in the z direction through the pipe, ρ the fluid density, and C the heat capacity of the fluid. The driving function $Q(z)$, modeling the electric coil, is defined as

$$Q(z) = \begin{cases} 0 & \text{if } 0 \leq z \leq a \\ Q_0 \cdot \sin\left(\frac{z-a}{z-b}\pi\right) & \text{if } a \leq z \leq b \\ 0 & \text{if } b \leq z \leq L \end{cases} \tag{C.1}$$

At $z = 0$, the fluid has the inlet temperature T_0

$$T(0) = T_0$$

At $z = L$, heat is leaking out to the exterior, having temperature T_{out}. This assumption gives the following boundary condition (BC):

$$-\kappa\frac{dT}{dz}(L) = k(T(L) - T_{out})$$

where k is a constant heat convection coefficient.

Discretize this boundary value problem (BVP) with the finite difference (FD) method using constant stepsize and write a Matlab program that solves the problem.

Use the following values of the parameters in the problem: $L = 10$, $a = 1$, $b = 3$, $Q_0 = 50$, $\kappa = 0.5$, $k = 10$, $\rho = 1$, $C = 1$, $T_{out} = 300$, $T_0 = 400$, and $v = 0, 0.1, 0.5, 1, 10$. The case $v = 0$ corresponds to no convection, only diffusion.

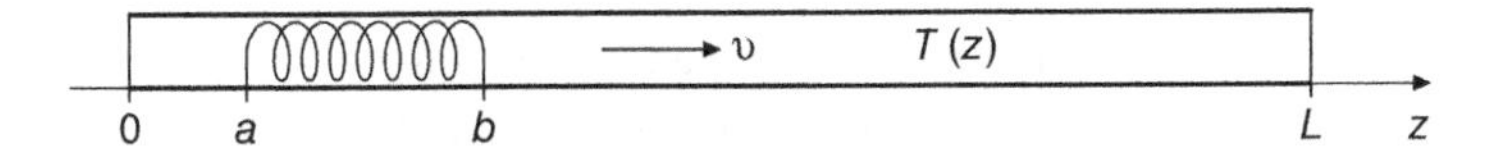

Figure C.2 Pipe with fluid heated by an electric coil

Discretize the z interval [0, L] with constant stepsize and use a node-numbering where $z_0 = 0$ and $z_N = L$.

Discretize the ODE and the BCs. A sparse system of linear equations is obtained.

Plot of the solution $T(z)$ for $v = 0$, $N = 10, 20, 40, 80$ in the same graph. Note the convergence of the curves in the graph.

$N = 40$ gives a solution that is accurate enough for our purposes. Use this discretization to solve the problem for $v = 0.1, 0.5, 1, 10$ in the same graph. When $v = 10$, spurious oscillations occur! Make another plot when $v = 10$, showing the solution for $N = 10, 20, 40$ in the same graph. The oscillations become more pronounced when h is increasing. We have a spurious oscillation problem! How do we get rid of these oscillations?

C.4 COMPUTER LAB 4 SUPPORTING CHAPTER 6

C.4.1 Partial Differential Equation of Parabolic Type

A metallic rod of length L (m) is initially of temperature $T = 0$ (C). At time $t = 0$, a heat pulse of temperature $T = T_0$ and duration t_P (s) hits the left end (at $x = 0$) of the rod. At the right end (at $x = L$), the rod is isolated. After some time, the rod will therefore be warmer in the right end and then cool off again. The following partial differential equation can be set up for the heat diffusion process through the rod:

$$\rho C_p \frac{\partial T}{\partial t} = k \frac{\partial^2 T}{\partial x^2}, \quad t > 0, \quad 0 < X < L,$$

The BCs are

$$T(0,t) = \begin{cases} T_0 & \text{if } 0 \leq t \leq t_P \\ 0 & \text{if } t > t_P \end{cases}, \qquad \frac{\partial T}{\partial x}(L,t) = 0 \tag{C.2}$$

and the initial condition (IC) is

$$T(x,0) = \begin{cases} T_0 & \text{if} \quad x = 0 \\ 0 & \text{if} \quad 0 < x \leq L \end{cases} \tag{C.3}$$

In the partial differential equation (PDE), ρ is the density [kg/m^3], C_p the heat capacity [$J/(kg{\cdot}C)$], and k the thermal conductivity [$J/(m{\cdot}s{\cdot}C)$] of the rod.

The purpose of this lab is to

- scale the problem to dimensionless form
- discretize the scaled problem with the Method of Lines (MoL)
- investigate stability properties of Euler's explicit method
- comparison between an explicit and an implicit adaptive method
- show how an implicit method can be made more efficient
- visualize the result in a two- and three-dimensional plot

1. Show that with the new variables u, ξ, and τ defined by

$$T = T_0 u, \qquad x = L\xi, \qquad t = t_p \tau$$

the problem can be transformed (scaled) into the following dimensionless form:

$$\frac{\partial u}{\partial \tau} = a\frac{\partial^2 u}{\partial \xi^2}, \quad \tau > 0, \quad 0 < \xi < 1$$

with BCs

$$u(0,\tau) = \begin{cases} 1 \text{ if } 0 \leq \tau \leq 1 \\ 0 \text{ if } \tau > 1 \end{cases}, \qquad \frac{\partial u}{\partial \xi}(1,\tau) = 0 \tag{C.4}$$

and IC

$$u(\xi,0) = \begin{cases} 1 \text{ if } \xi = 0 \\ 0 \text{ if } 0 < \xi \leq 1 \end{cases} \tag{C.5}$$

Show that the only remaining parameter a is dimensionless. From now on assume that a has the numerical value $a = 1$.

2. Discretize in space (the ξ-variable) using constant stepsize h and central differences to obtain an ODE system

$$\frac{d\mathbf{u}}{d\tau} = A\mathbf{u} + \mathbf{b}(\tau), \qquad \mathbf{u}(0) = \mathbf{u}_0,$$

where $\mathbf{u}_0$ is the zero vector. Show the dimensions and structures of A, $\mathbf{b}(\tau)$, and $\mathbf{u}_0$. Show also the discretized grid of the ξ-axis you have used and how the grid points are numbered.

3. *Numerical part:* Discretize the ODE system in (2) with Euler's explicit method. Use constant time step Δt. To make your calculations efficient, you should write your code so that the vector $A\mathbf{u} + \mathbf{b}$ is formed directly, not from component form. Store the whole approximate solution (including ICs and BCs) in a large matrix U, as

$$U = \begin{pmatrix} 0 & x & x & x & \dots & x \\ 0 & x & x & x & \dots & x \\ 0 & x & x & x & \dots & x \\ . & . & . & . & \dots & x \\ . & . & . & . & \dots & x \\ . & . & . & . & \dots & x \\ 0 & x & x & x & \dots & x \\ 1 & 1 & 1 & 1 & \dots & 0 \end{pmatrix}$$

Graphical part: Use, e.g., `surf` to draw a 3D plot of the solution. Experiment with different values of the discretization step h and Δt and study stability. Make one graph showing a stable solution and one graph with an unstable solution. Present the values of h, Δt, and $\Delta t/h^2$ in the two cases.

4. In this part of the lab, you shall compare the explicit method in `ode23` and the implicit method in `ode23s`, suitable for stiff problems, in the Matlab library. The two functions shall be run under similar conditions (same problem, same tolerance) and for three values of the stepsize h corresponding to $N = 10$, 20, 40 grid points on the ξ-axis.
   The comparison shall comprise
   a) the number of time steps needed to reach $\tau = 2$
   b) the cpu time needed to do each computation
   c) the maximal time step that each method could take
   Collect your statistics in three tables:

	Time Steps		*CPU – Time*		$\Delta t_{\max}$	
N	*ode23*	*ode23s*	*ode23*	*ode23s*	*ode23*	*ode23s*
10						
20						
40						

5. The conclusion from (4) is that the number of time steps is considerably smaller when using a stiff method. However, the default implementation is not efficient for stiff problems coming from parabolic PDEs. The main reason is that all the linear systems of equations $Ax = b$ that are to be solved in each time step need a lot of number crunching since they are based on using the backslash operator, i.e., Gaussian elimination on a full matrix is performed each time $Ax = b$ is solved. However, the system matrix of these equations is tridiagonal, which is not considered in the default implementation of `ode23s`. With the Matlab function `odeset`, options can be set by the user in order to make the computation more efficient. Do `help odeset` and study the options for "Jacobian," and "Jpattern." Also do `help odefile` and look at the example. With the help of these options, the number of flops can be reduced considerably. Experiment with these three options and collect statistics regarding the cpu time as was done in (4).
6. Visualize the result of a successful computation from (3) in graphs. One graph shall show two-dimensional plots of $u(\tau, \xi)\, 0 \leq \xi \leq 1$ at four time points $\tau \approx 0.5$, 1, 1.5, 2. Another graph shall show a three-dimensional plot of u as function of τ and ξ.
   Conclude your results by answering the following questions. If we want efficient numerical calculations for parabolic problems:
   1. What type of ODE method should be used, stiff or nonstiff?
   2. What structure does the jacobian of the ODE system have, banded or sparse in another way?
   3. What type of linear equation solver should be used in the implicit ODE method, solver for full matrices or for sparse matrices?

C.5 COMPUTER LAB 5 SUPPORTING CHAPTER 7

C.5.1 Numerical Solution of Elliptic PDE Problems

This lab is an exercise in solving an elliptic PDE, the applicational background of which is stationary heat conduction in 2D.

The physical background is as follows: Heat is conducted through a rectangular metal block, being the region $\Omega = [0 \leq x \leq 4, 0 \leq y \leq 2]$ when placed in a xy-coordinate system (see Figure C.3). Depending on the BCs, the temperature distribution $T(x, y)$ in the block will be different.

The following elliptic problem is formulated:

$$T = 0, \quad (x, y) \in \Omega$$

$$T(0, y) = 300, \quad T(4, y) = 600, \quad 0 \leq y \leq 2$$

$$\frac{\partial T}{\partial y}(x, 0) = 0, \frac{\partial T}{\partial y}(x, 2) = 0, \quad 0 \leq x \leq 4$$

1. Solve the elliptic partial differential equation problem. Use the finite difference method as described in Chapter 7. Discretize the rectangular domain into a quadratic mesh with the following two step sizes: $h = 0.2$, i.e., $N = 19$, $M = 11$, and $h = 0.1$, i.e., $N = 39$, $M = 21$ (see Figure C.4). Solve the problem for these two stepsizes and visualize the solution $T(x, y)$ with colors, using the MATLAB function `imagesc`. What is the T value at (2, 1) for the two stepsizes? Derive the analytic solution. Why is the numerical solution almost exact. *Hint*: From the graph of $T(x, y)$ make a suitable ansatz of the analytic solution and prove that it satisfies the PDE and the BCs.
2. Solve the same problem with COMSOL MULTIPHYSICS®. General recommendation: if the session turns out to be too messy, start the session again. Check your numerical result from (1): what is the T value at the point (2, 1)? How many nodes and triangles have been generated in the mesh? Make a refinement of the grid and find again T(2, 1). How many nodes and triangles are there now? Make the whole session again: draw geometry, impose boundary values, set the PDE values, initialize the mesh, solve the problem, and plot a 2D graph of the solution.

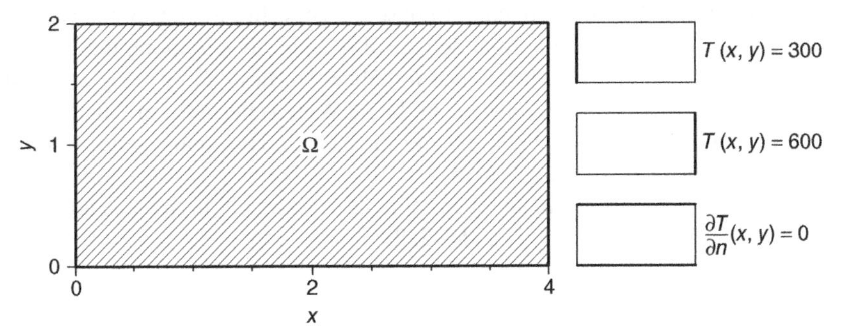

Figure C.3 Region and boundaries for a rectangular heat problem

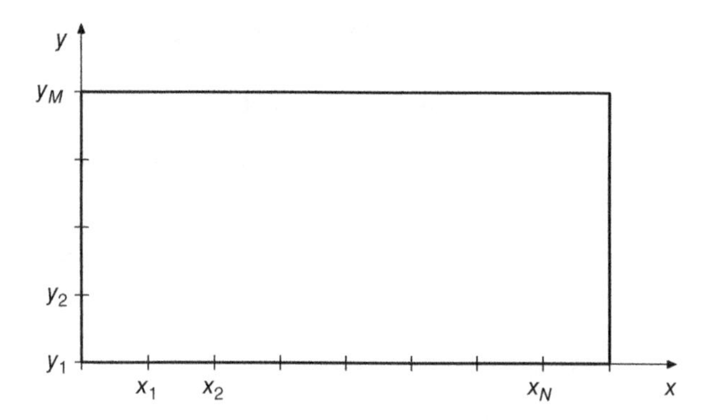

Figure C.4 Discretization of the rectangular region

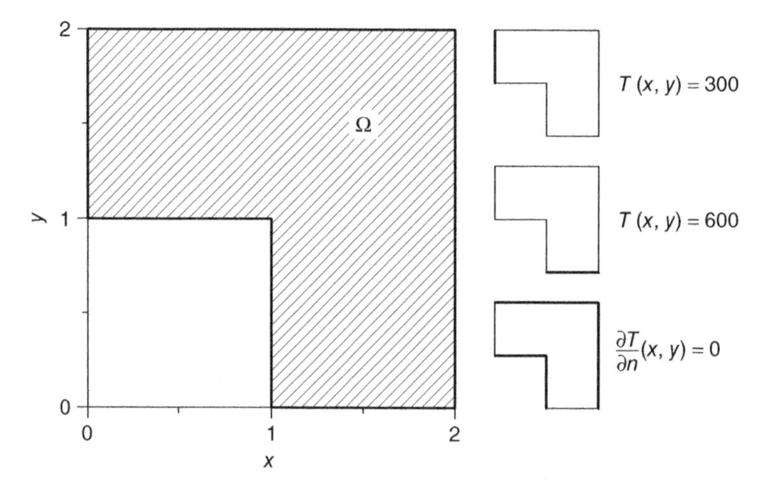

Figure C.5 Region and boundary conditions for the L-shaped area

3. Go back to COMSOL MULTIPHYSICS and generate the following L-shaped area with the Geometry builder. Impose the BCs. In Figure C.5, the walls are heat insulated, i.e., the normal derivative is equal to zero along these boundaries. Set the PDE parameters and initialize the mesh. How many nodes and triangles are generated? Solve the problem. What is the value of T in the points (1, 1) and (2, 2)? Refine the mesh and solve again. How many nodes and triangles? What is $T(1, 1)$ and $T(2, 2)$?
   Initialize the mesh again and then refine the mesh only in the area around the point (2, 2). What is $T(1, 1)$ and $T(2, 2)$?

C.6 COMPUTER LAB 6 SUPPORTING CHAPTER 8

C.6.1 Numerical Experiments with the Hyperbolic Model PDE Problem

Given the model problem for a hyperbolic PDE

$$\frac{\partial u}{\partial t} + a\frac{\partial u}{\partial x} = 0, \quad a > 0, \quad 0 < x < 2, \quad t > 0$$

with IC $u(x, 0) = 0, \quad 0 < x \leq 2$. As for the BC at $x = 0$, assume a square wave is entering from the left, i.e.,

$$u(0,t) = \begin{cases} 1, & -(n+1)\frac{T}{2} < t \leq -n\frac{T}{2}, \quad n = 0, 2, 4, \dots \\ -1, & -(n+1)\frac{T}{2} < t \leq -n\frac{T}{2}, \quad n = 1, 3, 5, \dots \end{cases}$$

where T is the period time.

Make a numerical experiment showing how the (i) upwind, (ii) Lax–Friedrich, and (iii) Lax–Wendroff methods behave on this PDE problem. Discretize the x interval into N equidistant subintervals and define grid points $x_i = ih_x, i = 0, 1, 2, \dots, N$, where $h_x = 2/N$. The Courant number σ is $\sigma = ah_t/h_x$.

1. Write programs and run the three methods on the time interval $(0, 2T)$. Let $N = 100$, $T = 1$, and $a = 1$. Present the results in graphs with $u(x, 2T)$ as a function of x. Show graphs for the σ values: $\sigma = 0.8$, 1, and 1.1. In all, there are nine graphs to be presented. Are your experimental results in agreement with the theoretical stability results? Which methods will smooth the solution and which will introduce spurious oscillations?
2. In this part, the exchanger in Example 8.1 is studied. A fluid of temperature $T(x, t)$ is flowing with constant speed v in a pipe. Outside the pipe, there is a cooling medium that keeps a constant low temperature T_{cool}. The temperature of the fluid in the pipe is initially cool, i.e., $T = T_{cool}$ but within a short time period hot fluid enters the pipe. The task is to study how the temperature $T(x, t)$ of the fluid in the pipe depends on x and t.

The following PDE model is given:

$$\frac{\partial T}{\partial t} + v\frac{\partial T}{\partial x} + a(T - T_{cool}) = 0, \quad 0 < x < L, \quad t > 0$$

with the IC

$$T(x, 0) = T_{cool}$$

and the BC

$$T(0,t) = \begin{cases} T_{cool} + (T_{hot} - T_{cool})\, sin(\pi t) & 0 \leq t \leq 0.5 \\ T_{hot} & 0.5 \leq t \leq 4 \\ T_{hot} + T_{cool} sin(\pi(t-4)) & t > 4 \end{cases}$$

The length of the heat exchanger is $L = 3$. The heat exchange parameter $a = 0.1$, the velocity of the fluid $v = 1$, the cooling temperature $T_{cool} = 5$, and the hot temperature $T_{hot} = 100$.

a) Use the upwind method to simulate the temperature in the pipe. Choose suitable stepsizes h_x and h_t and present the result in a 3D graph.
b) Work out a difference formula of Lax–Wendroff type that solves the PDE problem in (2). Present the result as a difference formula of the type

$$u_{i,k+1} = c_1 u_{i-1,k} + c_2 u_{i,k} + c_3 u_{i+1,k} + c_4$$

where the coefficients c_1, c_2, c_3 and c_4 are to be determined by you.
c) Program the Lax–Wendroff formula in (b), run it for a suitable choice of h_x and h_t and present the result in a 3D plot. Make a comparison with the graph from (a).

INDEX

Introduction to Computation and Modeling for Differential Equations, Second Edition. Lennart Edsberg.
© 2016 John Wiley & Sons, Inc. Published 2016 by John Wiley & Sons, Inc.

Printed and bound by CPI Group (UK) Ltd, Croydon, CR0 4YY

16/04/2025

14658521-0001